Applied mechanics O2

P. D. Collins
Principal
Worcester Technical College

and

A. Jackson
Senior Lecturer in Mechanical and Aeronautical Engineering
The Hatfield Polytechnic

Longman
London and New York

LONGMAN GROUP LIMITED
London

Associated companies, branches and representatives throughout the world

First published 1972
Second impression 1975

ISBN 0 582 42511·5 cased
ISBN 0 582 42512·3 paper

*Set in IBM Press Roman
and printed in Great Britian by
Lowe & Brydone (Printers) Ltd.,
Thetford, Norfolk*

Preface

This text deals with the topics which must be covered in the Applied Mechanics Syllabus at the O2 stage of the Ordinary National Certificate in Engineering, Naval Architecture or Mining. It should also be useful to students taking an Ordinary National Diploma course in Engineering or Part II of the Mechanical Engineering Technicians course.

Throughout the text units have been presented in squared brackets to separate them from the numerical part of the calculation. This should enable the student to become as familiar with units as with the actual numerical calculation.

Many worked examples are included in the text, the solutions being given in full and it is hoped that the layout of these worked examples will assist students to achieve a better presentation of their work.

We wish to express our thanks to Dr. J. M. Howe for reading and offering helpful advice on the complete text and to Mr. R. J. Frost for taking various photographs.

P. D. COLLINS
A. JACKSON

Acknowledgements

We wish to thank the following for permission to reproduce questions from their past examination papers:
East Midland Educational Union (E.M.E.U.)
Northern Counties Technical Examination Council (N.C.T.E.C.)
Union of Educational Institutions (U.E.I.)
Union of Lancashire and Cheshire Institutes (U.L.C.I.)
Yorkshire Council for Further Education (Y.C.F.E.)
Welsh Joint Education Committee (W.J.E.C.)
The photographs were taken by courtesy of the Hatfield Polytechnic.

Symbols and abbreviations

SYMBOLS

Term	*Symbol*
length	l
change in length	δ
distance	s
radius	r, R
diameter	d, D
area	A
volume	V
time	t
velocity	v, V
angular velocity	ω
acceleration	a
angular acceleration	α
mass	m
density	ρ
force	F
shear force	Q
direct stress	σ
shear stress	τ
direct strain	ϵ
shear strain	ϕ
modulus of elasticity	E
modulus of rigidity	G
pressure	p
work	W
strain energy	U
bending moment	M
torque	T
power	P
coefficient of friction	μ
coefficient of restitution	e
second moment of area, moment of inertia	I
polar second moment of area	J
section modulus	Z
polar modulus of section	Z_p
radius of gyration	k
periodic time	T
frequency	f
coefficient of linear expansion	α
temperature	T
coefficient of contraction	C_c
coefficient of velocity	C_v
coefficient of discharge	C_d

ABBREVIATIONS

Term or quantity	*Abbreviation*
metre	m
square metre	m^2
cubic metre	m^3
litre	l
second (time)	s
minute (time)	min
hour	h
degree, minute, second (angle)	° ′ ″
radian	rad
radian per second	rad/s
revolution per minute	rev/min
gramme	g
newton	N
joule	J
watt	W

Contents

	Preface	iii
	Acknowledgements	iii
	Symbols and abbreviations	iv
1.	**Statics**	1
1.1	Equilibrium condition	1
1.2	Force polygon	1
1.3	Funicular or link polygon	2
1.4	Graphical determination of beam reactions	3
1.5	Simple framed structures	9
1.6	Structural equilibrium	11
2.	**Shear force and bending moment**	37
2.1	Definitions of shear force and bending moment	37
2.2	Uniformly distributed loading	44
2.3	Contraflexure or inflexion	53
3.	**Stress and strain**	66
3.1	Transmission of forces	66
3.2	Direct stress	66
3.3	Direct strain	68
3.4	Elasticity and Hooke's law	68
3.5	The tensile test	69
3.6	Results from a tensile test	71
3.7	Proof stress	74
3.8	Compound bars	77
3.9	Temperature stresses	81
3.10	Stresses in thin cylindrical shells	86

3.11 Stresses in thin rotating rings 89
3.12 Strain energy 91
3.13 Shear stress 93
3.14 Shear strain 94
3.15 Modulus of rigidity 95
3.16 Complementary shear stress 95
3.17 Riveted joints 96

4. Bending of beams 110
4.1 Pure bending 110
4.2 Bending stresses 111
4.3 Position of neutral axis 113
4.4 Moment of resistance 115
4.5 Second moments of area 116
4.6 Effect of web of an I-section 126

5. Torsion of circular shafts 140
5.1 Torsional stress 140
5.2 Moment of resistance 142
5.3 Polar second moment of area 144
5.4 Transmission of power 144
5.5 Comparison of solid and hollow shafts 147
5.6 Compound shafts 153

6. Materials testing 161
6.1 Purpose of material testing 161
6.2 The tensile test 161
6.3 Extensometers 164
6.4 The compression test 165
6.5 The torsion test 166
6.6 Hardness tests 168
6.7 Impact testing 172

7. Kinematics 174
7.1 Kinematics 174
7.2 Revision of terms 174
7.3 Derivation of equations of motion 174
7.4 Relationship between linear and angular motion 188
7.5 Centripetal acceleration 189

8. Kinetics – translatory motion 196
8.1 Kinetics 196
8.2 Momentum 196
8.3 Newton's laws of motion 196
8.4 The kinetic equation of motion 197
8.5 The law of universal gravitation 198
8.6 Mass and weight 199
8.7 Linear translation 200
8.8 Translation in a curved path 203
8.9 Vehicle on a curved horizontal track 205
8.10 Vehicle on an inclined curved track 210
8.11 The conical pendulum 215
8.12 Energy 217
8.13 Kinetic energy of translation 217
8.14 Linear impulse 218
8.15 Conservation of linear momentum 219
8.16 Impact 219
8.17 Loss of kinetic energy due to impact 221
8.18 Oblique impact 222
8.19 Impact of a fluid jet 225
8.20 Motion under a varying force 228

9. Kinetics – angular motion 236
9.1 Rotation of a body about a fixed axis 236
9.2 Radius of gyration 237
9.3 Centre of percussion 237
9.4 Kinetic energy of rotation 242
9.5 Kinetic energy of a body possessing translation and rotation 246
9.6 Angular impulse 250
9.7 Angular momentum 250

10. Simple harmonic motion 258
10.1 Periodic motion 258
10.2 Simple harmonic motion 258
10.3 Mass on a vertical spring 262
10.4 Static deflection of a spring 263
10.5 The simple pendulum 271
10.6 The compound pendulum 273

11. Fluids at rest 280
11.1 Fluids 280
11.2 Liquids and solids 280
11.3 Density and relative density 280
11.4 Intensity of pressure and thrust 281
11.5 Pressure at a point in a liquid 281
11.6 Variation of pressure with depth 282
11.7 Measurement of pressure 283
11.8 Atmospheric pressure and equivalent head 286
11.9 Thrust on an immersed surface 289
11.10 Centre of pressure 290

12. Liquids in motion 300
12.1 Introduction 300
12.2 Rate of flow 301
12.3 Equation of continuity 301
12.4 Energy of a liquid in motion 302
12.5 Bernoulli's equation 304
12.6 Flow measurement 308

Answers 324

Index 331

1

Statics

1.1 Equilibrium condition

If a body is in equilibrium under the action of a system of co-planar forces which are either concurrent (pass through one point) or are parallel then the force and funicular polygons must both close.

However, in the more general case of a body which is acted upon by a system of co-planar forces that are not concurrent, the condition of zero resultant force (i.e., the force polygon closes) is not sufficient on its own to specify that the body is in equilibrium. In instances where the body is not in equilibrium the funicular polygon does not close and equilibrium can then only be achieved by the application of a couple.

1.2 Force polygon

When a body is acted upon by a system of co-planar forces the 'polygon of forces' may be used to determine the magnitude and direction of the resultant force acting on the body.

Consider a body which is acted upon by forces F_1, F_2, and F_3 as shown in Fig. 1.1(*a*). This figure, which shows the physical arrangement and positions of the forces, is called the 'space diagram'. The force polygon (*b*) is drawn according to Bow's notation. This consists of specifying the forces on the space diagram by labelling the intervening spaces A, B, C, etc., and representing the force vectors by ***ab***, ***bc***, ***cd***, etc. Thus, force F_1 is denoted by AB on the space diagram and by vector ***ab*** on the force polygon.

From (*b*) we obtain the resultant of F_1, F_2, and F_3 to be vector ***ad*** in both magnitude and direction.

If, in addition to F_1, F_2, and F_3, the body was acted upon by a force equal in magnitude but opposite in direction to the resultant then the

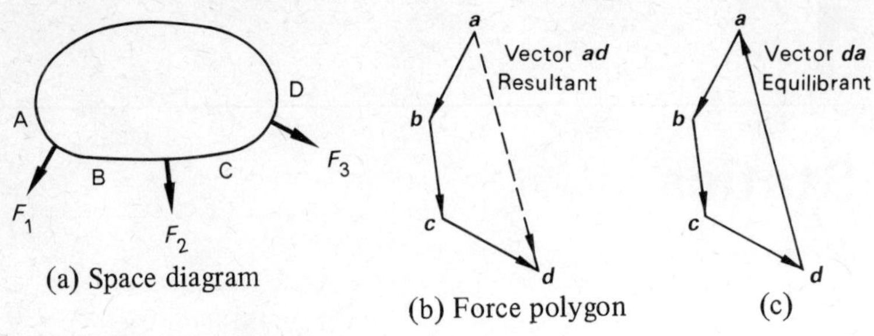

Fig. 1.1 Forces acting on a body

force polygon would close, as shown in (*c*). In this case the arrowheads follow in the same direction round the diagram and there is no resultant force on the body. This force, given by vector ***da*** in magnitude and direction, is known as the equilibrant.

1.3 Funicular or link polygon

Although the magnitude and direction of the resultant (or equilibrant) of a system of forces is obtained from the force polygon, this diagram does not indicate the position on the space diagram at which such a force must be applied. This will now be determined by means of the link polygon.

The procedure is as follows:
Choose any point ***O*** at a convenient distance from the force polygon and draw lines ***Oa***, ***Ob***, etc., as indicated in Fig. 1.2(*b*). Point ***O*** is known as the pole point.

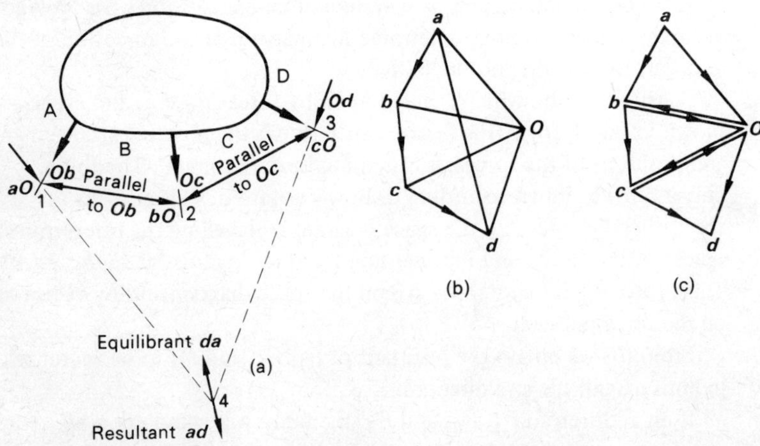

Fig. 1.2 The Funicular or Link Polygon

Referring to triangle ***Oab*** we can consider ***aO*** and ***Ob*** to be the components of ***ab***, i.e., forces ***aO*** and ***Ob*** acting together will produce the same effect as force ***ab*** acting alone. Similarly, for ***bc*** and ***cd***, the component triangles being shown in (*c*). It will be noticed that the components ***bO*** and ***Ob*** cancel each other, as do ***cO*** and ***Oc***, and that the three forces acting on the body can be replaced by components ***aO*** and ***Od***. Furthermore, the resultant of ***aO*** and ***Od*** is ***ad***.

Then, at any position 1 on the line of action of force AB component lines are drawn parallel to ***aO*** and ***Ob***. This is repeated for other forces, as shown on (*a*), it being observed that ***Ob*** and ***bO*** are coincident across space B, etc. The lines of action of ***aO*** and ***Od*** are then extended to intersect at 4. This is the point through which the resultant of ***aO*** and ***Od*** must pass, also being parallel to ***ad*** on the force polygon (*b*). Thus, the position of the resultant is now known.

The closed figure 1–2–3–4 is known as the funicular polygon. Closure of the funicular polygon indicates that equilibrium of the body can be achieved by the application of the single force ***da*** (the equilibrant). It follows that if a body is in equilibrium the force polygon and funicular polygon will both close. Although closure of the force polygon indicates that there is no resultant force on the body it does not indicate that the body is in equilibrium since there may be a resultant moment or couple acting on the body (see example 1.3).

1.4 Graphical determination of beam reactions

Although the support reactions of a beam can be obtained by the application of moment equilibrium, it is also possible to obtain the reactions by means of the funicular polygon. This is illustrated for a simply supported beam in Fig. 1.3 – the weight of the beam being neglected.

A force diagram, consisting of the straight line ***abcd*** only, is drawn at (*b*) and a convenient pole point ***O*** is chosen and joined to ***a***, ***b***, ***c***, and ***d***. The funicular polygon is commenced at any suitable position (1) on the line of the left-hand reaction; 1–2 is drawn across space A parallel to ***aO*** on the force polygon; 2–3 is drawn across space B parallel to ***bO***, and the procedure continued until point 5 is reached on the line of the right-hand reaction. Line 1–5 is now drawn across space E and ***eO*** is drawn parallel to it on the force polygon. From section 1.3 the force represented by ***aO*** is equivalent to the vertical force ***ae*** and a force ***eO*** along the closing line 1–5 of the funicular polygon. Similarly, ***Od*** is equivalent to ***ed*** and ***Oe***. It will now be seen that the components ***eO***

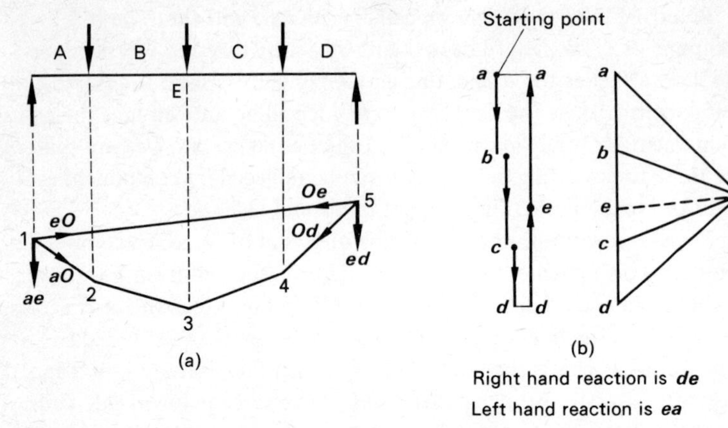

Fig. 1.3 Graphical determination of beam reactions by funicular polygon

and ***Oe*** cancel out and the three forces AB, BC and CD can therefore be replaced by forces ***ae*** and ***ed***. Thus, the supporting forces to balance the beam must then be equal and opposite, i.e., the reactions are ***de*** at the right-hand side and ***ea*** at the left-hand side.

Example 1.1

Determine the reactions R_1 and R_2 for the beam given in Fig. 1.4 by means of the funicular polygon.

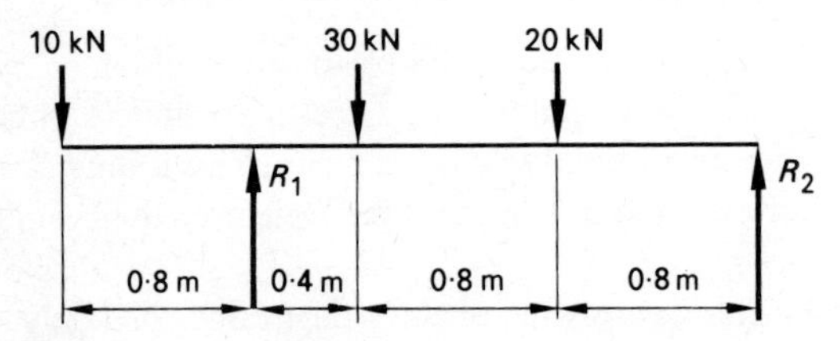

Fig. 1.4 Example 1.1

Solution

The space diagram and force polygon ***abcd***, are drawn to scale in Fig. 1.5. A suitable pole point ***O*** is selected and joined to ***a***, ***b***, ***c***, and ***d***. The funicular polygon is commenced at point 1 and line 1–2 drawn across space B parallel to ***Ob***; 2–3 across space C parallel to ***Oc***, 3–4 across space D parallel to ***Od*** and line 1–5 is drawn across space A parallel to ***Oa***.

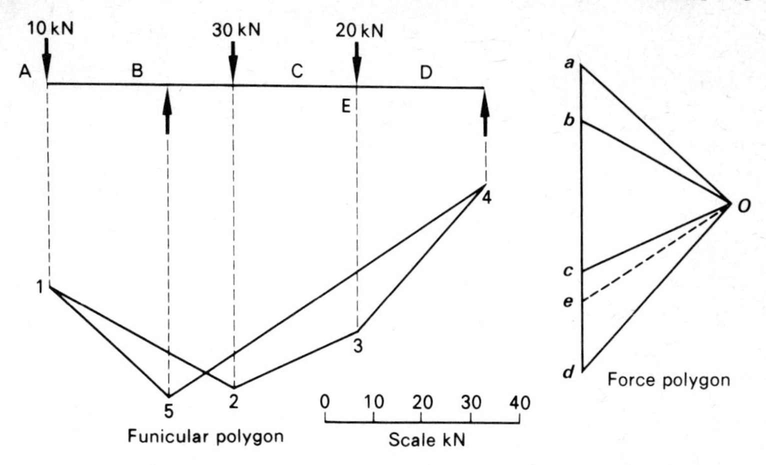

Fig. 1.5 Force and funicular polygons for Example 1.1

The line 4–5 is the closing line for the funicular polygon and ***Oe*** is drawn on the force polygon parallel to this closing line. This gives reactions ***de*** and ***ea***.

From the force polygon:

Reaction R_1 = ***ea*** = 46 kN

Reaction R_2 = ***de*** = 14 kN

Example 1.2

The beam XY shown in Fig. 1.6 is hinged at X and maintained in a horizontal position by a simple support at Y. Use the funicular polygon to determine for the applied loading shown:

(*a*) the magnitude and direction of the reaction at hinge X;

(*b*) the magnitude, direction, and position of application of the resultant force on the beam.

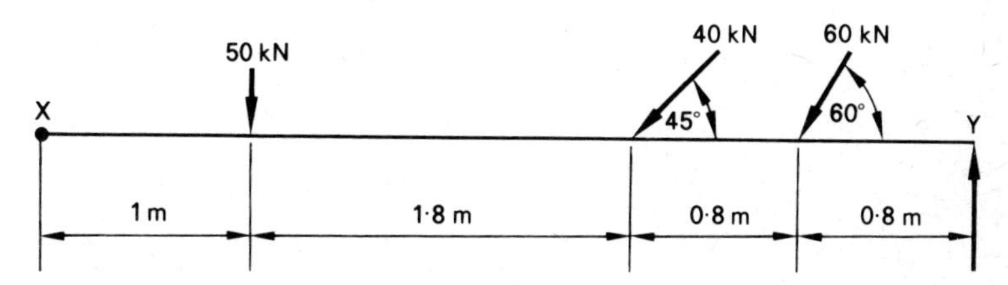

Fig. 1.6 Example 1.2

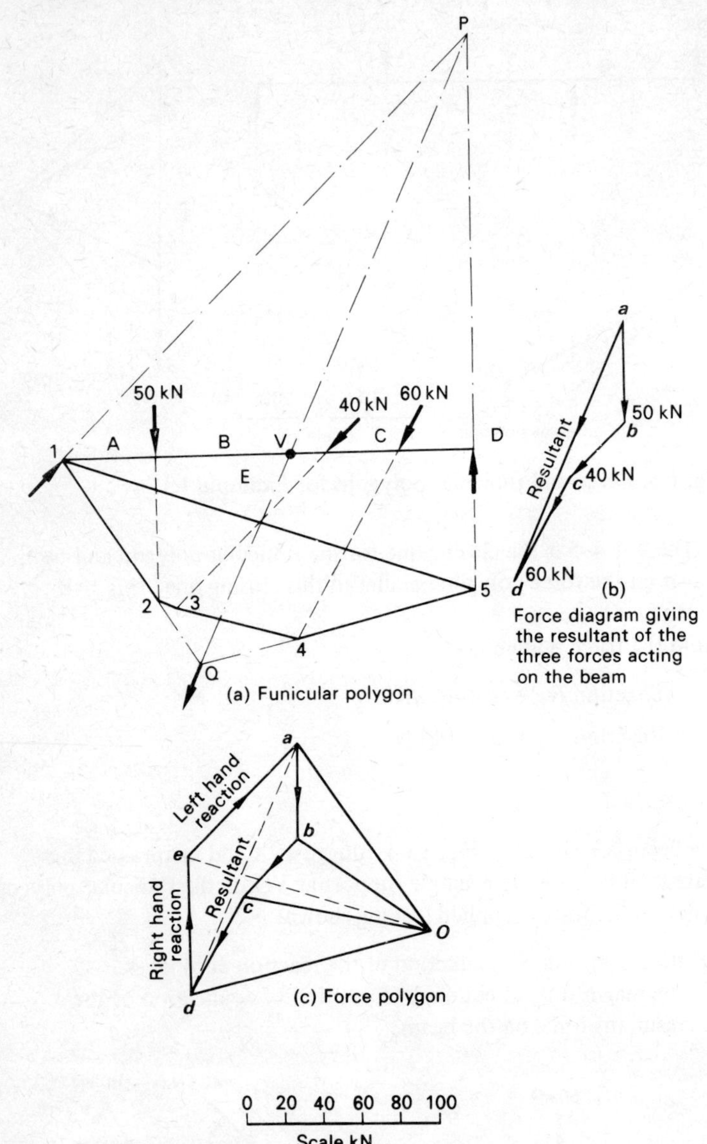

Fig. 1.7 Force and funicular polygons for Example 1.2

Solution

The inclination of the 40 kN and 60 kN forces does not affect the method which has been given and this is indicated by the force diagram drawn in Fig. 1.7(*b*). On this diagram ***ab*** is drawn parallel to the force of 50 kN, ***bc*** parallel to the force of 40 kN between the spaces B and C and ***cd*** parallel to the force of 60 kN. The forces are, of course, drawn to a suitable scale. The resultant of these three forces is given by vector ***ad***, in both magnitude and direction.

The only known condition regarding the hinge reaction at X is that it must actually pass through the hinge. For this reason, the funicular polygon is commenced at the position (1) and the remaining construction follows on as previously indicated until point (5) is reached. The closing line 5–1 over space E is transferred to the force polygon through ***O*** to meet a vertical through ***d*** (parallel to the right-hand reaction at Y) at ***e***. The vector ***ea*** must, likewise, be parallel to the reaction at X and also gives the magnitude of this reaction. This direction is now transferred to the space diagram.

The point of application of the resultant force on the beam is now obtained by extending lines 1–2 and 5–4 to intersect at Q. The resultant force must act through Q, its direction and magnitude being represented by vector ***ad***. Hence, the point on the beam through which the load acts is given by V.

Alternatively, as the beam is in equilibrium it follows that the end reactions and the resultant force on the beam will form a closed triangle of forces. Thus, the resultant force on the beam must act through P – the intersection of the lines of action of the two end reactions. This will again give point V.

From the force and funicular polygons the following results are obtained:

Reaction at X:	83 kN at 43·3° to horizontal
Resultant force on beam:	143 kN at 65° to the horizontal applied at 2·48 m from the left-hand end

Example 1.3

Use the funicular polygon to determine the magnitude of the reaction at the wall for the cantilever shown in Fig. 1.8.

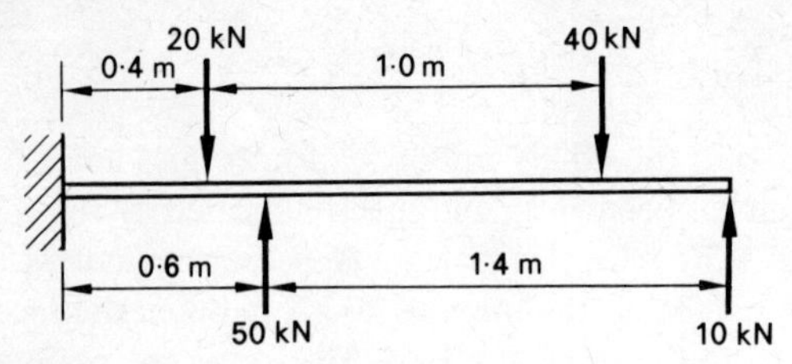

Fig. 1.8 Example 1.3

Solution

It will be seen that there is no resultant vertical force on the beam and therefore the force polygon is a closed figure. This is drawn to scale as ***abcde*** in Fig. 1.9(*b*). A suitable pole position ***O*** is selected as shown.

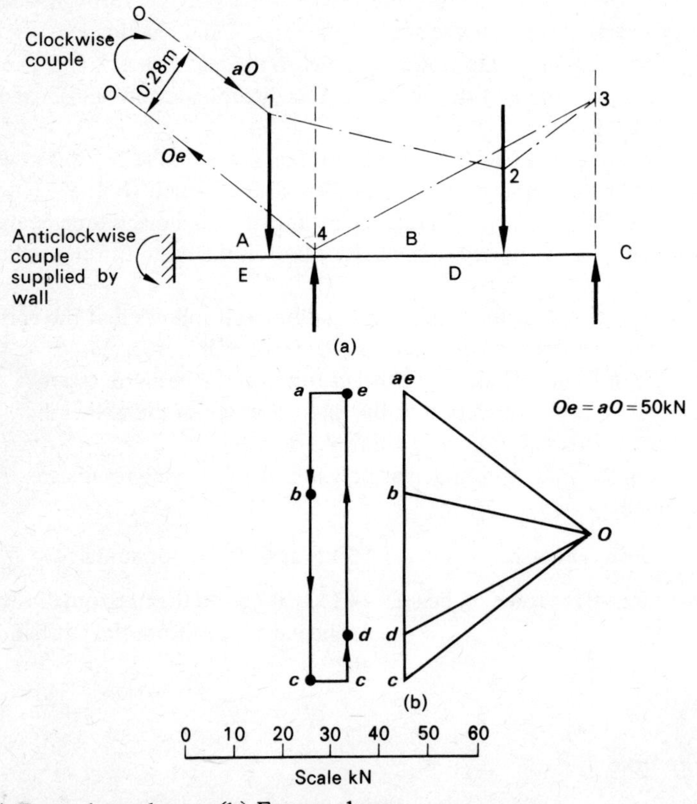

(a) Funicular polygon (b) Force polygon
Fig. 1.9 Force and funicular polygons for Example 1.3

The funicular polygon is commenced at any position (1) on the line of action of force AB and line 0–1 drawn parallel to ***aO*** across space A, 1–2 parallel to ***bO*** across space B, 2–3 parallel to ***cO*** across space C, 3–4 parallel to ***dO*** across space D and 4–0 parallel to ***Oe*** across space E.

It will be evident that the funicular polygon does not close. Note that ***aO*** and ***Oe*** are parallel by virtue of the fact that they are identical lines on the force polygon. Thus, these two forces constitute a couple of magnitude 50 [kN] x 0·28 [m] = 14 000 [Nm].

For equilibrium of the cantilever the wall must supply an equal, but opposing couple, as illustrated. This result can be quickly verified by applying the *principle of moments.*

1.5 Simple framed structures

The design of any structure, such as a space frame or a plane roof truss, which is made up of members joined at their ends, involves the determination of the forces in each member under a given system of loading. These forces can be determined for 'simple structures' providing two assumptions are made which will first be discussed in some detail.

1. 'All the connections between individual members in a simple structure will be considered as pin-jointed'.

When a member is pinned at the end, it is free to rotate, and therefore there can be no moment at the joint. Hence the significance of the term pin-joint – in a pin-jointed frame there is no bending of the members. Figure 1.10(*a*) illustrates a typical riveted connection between members which, for ease of calculation, is frequently replaced by a pin-joint as shown at (*b*).

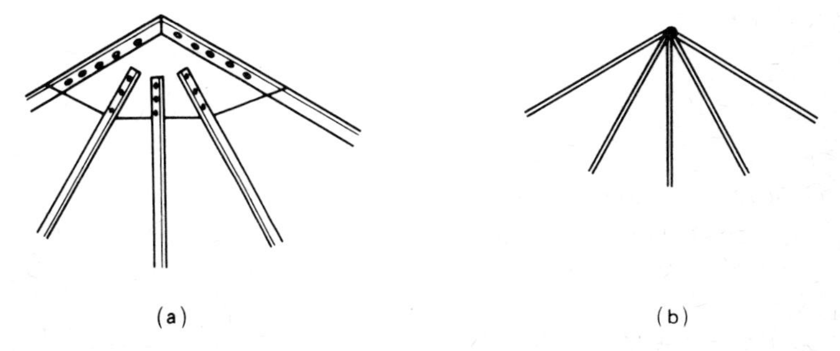

Fig. 1.10 A riveted joint (a) replaced by a pin joint (b)

The term 'simple structure' applies to those frames which are made up of individual triangular elements as shown in Fig. 1.11.

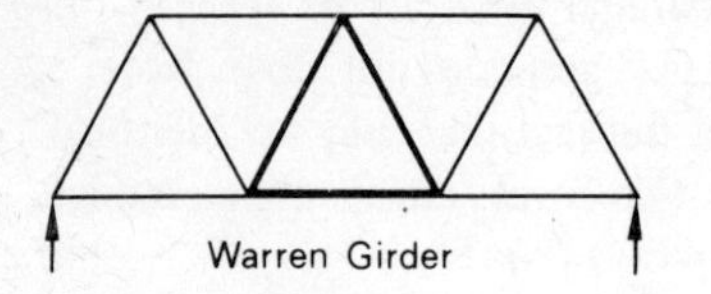

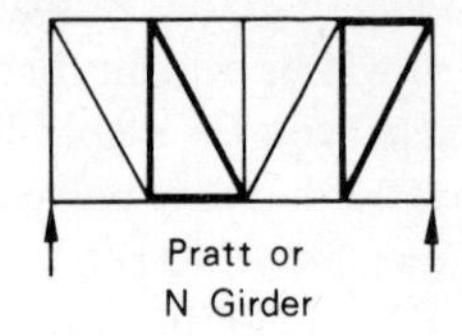

Fig. 1.11 Simple triangulated structures

Since each individual triangle – pin-jointed at its corners – is capable of carrying load it follows that a framework which is built up in this way will also be capable of carrying load, i.e., it will not collapse unless the load carrying capabilities of its members are exceeded.

Consider the simple framework shown in Fig. 1.12. There are just sufficient members to give triangulation and therefore the frame is

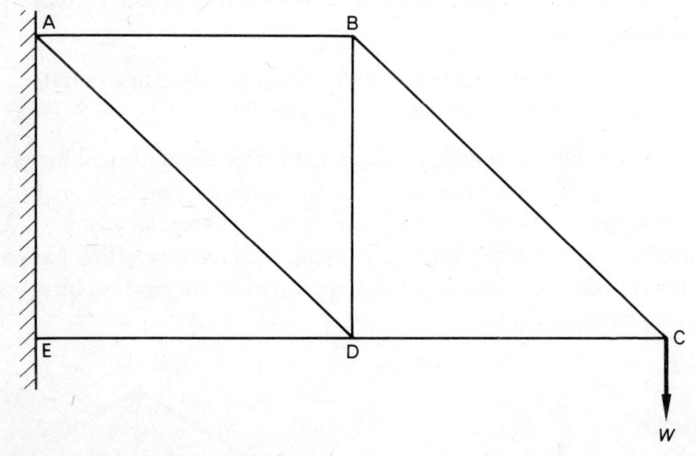

Fig. 1.12 A simple framework

capable of carrying load. Removal of member AD would cause the frame to collapse, i.e., it would act as a mechanism under load. This is not the case in Fig. 1.13, where removal of either AD or BE would leave a triangulated structure capable of carrying load. Any structure which contains members such as AD or BE, that can be removed without affecting the load-carrying capability of the structure, is said to be

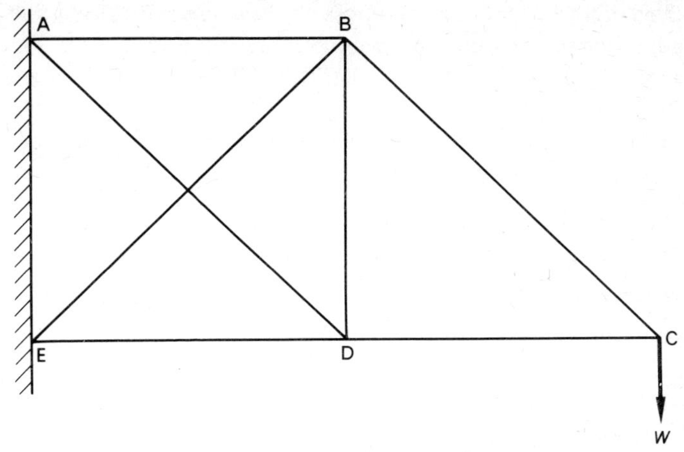

Fig. 1.13 A framework with a redundant member

a 'redundant structure'. The analysis of redundant structures is beyond the scope of this text.

2. 'The external loads are applied only at the pin-joints. This means that individual members of the structure are subject only to tensile or compressive loadings'.

1.6 Structural equilibrium

If a structure is in a state of equilibrium as a whole then each individual joint will be in equilibrium under the action of the forces which act on the joint. Furthermore, the external forces applied to any member must be balanced by internal forces created within that member if equilibrium is to be maintained.

When the external forces acting on a member tend to increase its length then internal forces are set up as shown in Fig. 1.14(*a*). The member is then in TENSION and is known as a TIE.

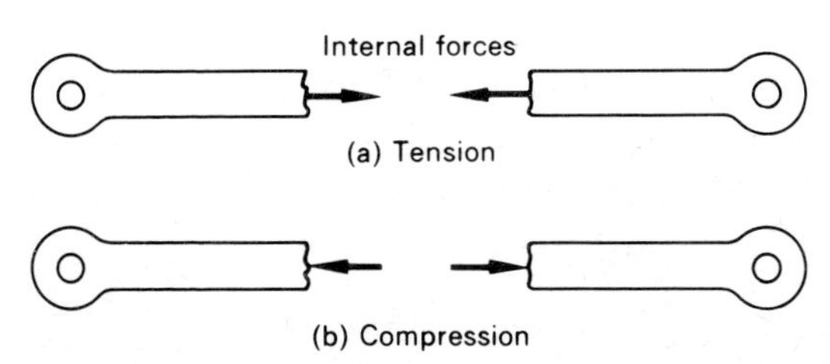

Fig. 1.14 Internal forces in a member

Similarly, when the external forces have a tendency to reduce the length of the member then internal forces are set up as shown in Fig. 1.14(*b*). The member is then in COMPRESSION and is known as a STRUT.

The member forces obtained from a force vector diagram are the internal forces in the member. Thus,

A member is a TIE when the internal forces pull on a joint.

A member is a STRUT when the internal forces push on a joint.

There are three methods that can be used to determine the forces in the members of a simple pin-jointed structure:

(*i*) Resolution at the joints.
(*ii*) Force vector diagram.
(*iii*) Method of sections.

Each of these methods will now be illustrated with reference to the framework shown in Fig. 1.15.

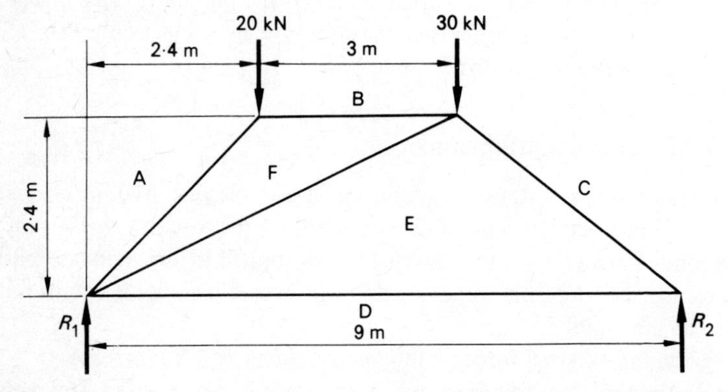

Fig. 1.15

In this problem a convenient first step for all three methods is to determine the reactions R_1 and R_2. These may be obtained by taking moments about one support or by drawing a funicular polygon.

Thus, by moments about R_1:

$$(20 \times 2{\cdot}4) + (30 \times 5{\cdot}4) = 9R_2$$

$$R_2 = 23{\cdot}33 \text{ kN}$$

and $$R_1 = 26{\cdot}67 \text{ kN}$$

(*i*) Resolution at the joints

This consists of determining the forces at each joint, in turn, that are necessary to produce equilibrium at that joint. The necessary equilibrium condition is that the resultant force at the joint is zero and it is usual to resolve the forces into two directions at right angles.

The joint equilibrium condition then becomes:

$$\Sigma \text{ Vertical forces} = \text{zero}$$

$$\Sigma \text{ Horizontal forces} = \text{zero}$$

When determining the forces at a particular joint the initial assumption is made that the unknown forces are tensile.

The resolution must be commenced at a joint where there are not more than two unknown forces. Thus, referring to the space diagram, Fig. 1.15, which is lettered according to Bow's notation, we can commence at joint CDE.

JOINT CDE (FIG. 1.16):
Member EC must have a vertically downward component to balance R_2. The force in DE must then balance the horizontal component of F_{EC}.

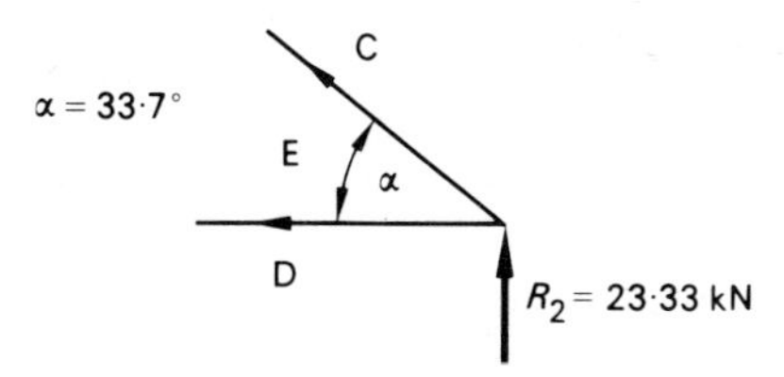

Fig. 1.16 Joint CDE

Resolving vertically:

$$F_{EC} \sin \alpha + 23{\cdot}33 = 0$$

$$F_{EC} = -42 \text{ kN} - \text{Strut (the negative sign indicating that EC is a strut)}$$

Resolving horizontally:

$$F_{EC} \cos \alpha + F_{ED} = 0$$

$$-42 \cos \alpha + F_{ED} = 0$$

$$F_{ED} = +35 \text{ kN} - \text{Tie}$$

JOINT BCEF (FIG. 1.17):

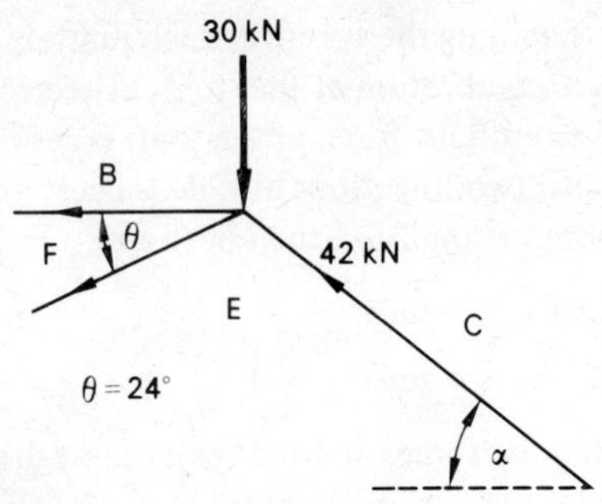

Fig. 1.17 Joint BCEF

Resolving vertically:

$$30 - 42 \sin \alpha + F_{FE} \sin \theta = 0$$

$$30 - 23{\cdot}33 + F_{FE} \sin 24^\circ = 0$$

$$F_{FE} = -16{\cdot}4 \text{ kN} - \text{Strut}$$

Resolving horizontally:

$$42 \cos \alpha + F_{FE} \cos \theta + F_{BF} = 0$$

$$+35 - 16{\cdot}4 \cos 24^\circ + F_{BF} = 0$$

$$F_{BF} = -20 \text{ kN} - \text{Strut}$$

JOINT ABF (FIG. 1.18):

The only unknown force at this joint is F_{AF}.

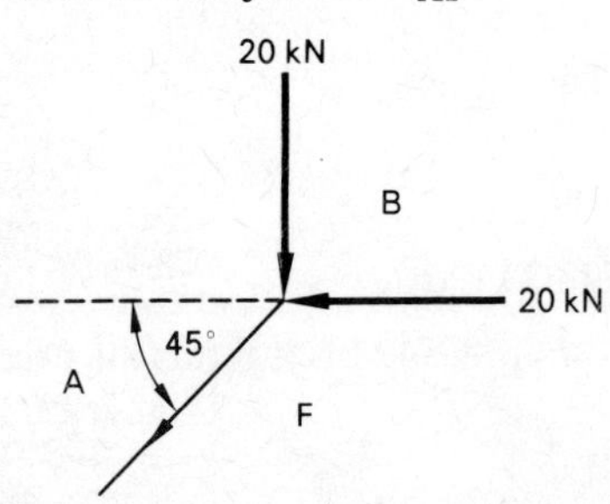

Fig. 1.18 Joint ABF

Resolving horizontally:

$$F_{AF} \cos 45^\circ + 20 = 0$$

$$F_{AF} = -28{\cdot}3 \text{ kN} - \text{Strut}$$

(*ii*) Force vector diagram

The force vector diagram for the complete framework is simply a combination into a single diagram of all the separate force diagrams for the individual joints.

The space diagram is drawn to scale and lettered according to Bow's notation and in Fig. 1.19 the separate force diagrams and complete force polygon are shown.

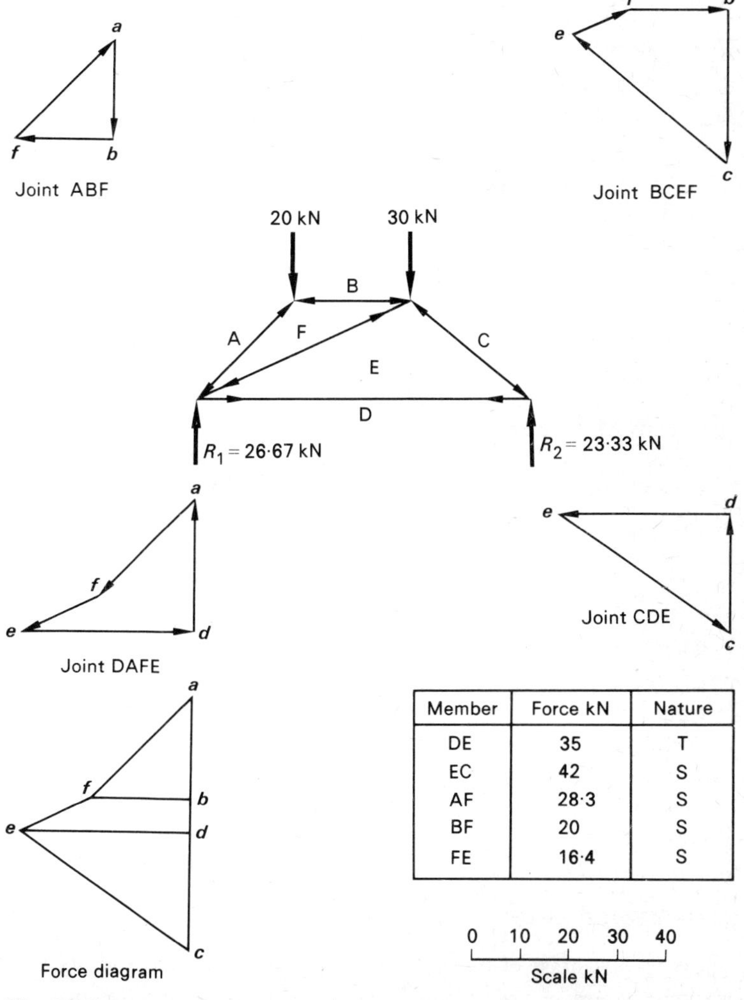

Member	Force kN	Nature
DE	35	T
EC	42	S
AF	28·3	S
BF	20	S
FE	16·4	S

Fig. 1.19

With parallel loads and reactions as in this case, the polygon of external forces for the frame is the straight line ***abc*** in which ***cd*** and ***da*** represent the support reactions.

The diagram is commenced with the right-hand reaction joint. From ***c*** on the force polygon draw ***ce*** parallel to CE on the space diagram to intersect a line drawn from ***d*** parallel to ED at point ***e***. To determine if the members meeting at this joint are struts or ties the members are considered in a clockwise direction round the joint – i.e., in the order C–D–E. The force which member DE exerts on the right-hand joint is represented by vector ***de*** on the force polygon. The direction of ***de*** is from right to left on the force diagram, see Fig. 1.20. Therefore, the direction of ***de*** is inwards away from the joint, indicating that DE is in tension.

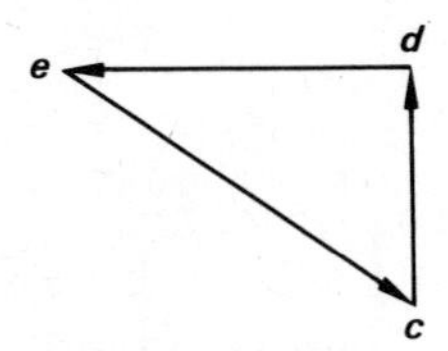

Fig. 1.20 Direction of forces at joint CDE

Now consider the next joint DAFE at the left-hand reaction. From ***e*** on the force polygon draw ***ef*** parallel to EF to intersect a line from ***a*** parallel to AF at the point ***f***.

The polygon for this joint is now ***dafe*** in that order. The direction of vector ***af*** is downwards to the left, and, therefore, the arrowhead on member AF acts downwards towards the joint indicating a strut.

It should be observed that arrowheads are omitted from the complete force diagram. The reason for this can be seen from the two separate polygons *cde* and *dafe* in Fig. 1.19.

The member ED features in both joints, and therefore the direction of the vector in the force diagram depends on which of the two joints is being considered – e.g., when considering the right-hand reaction joint the bottom member becomes DE, and when considering the left-hand one it is ED.

This method is completed by tabulating the magnitude and the nature of the force in each of the members.

(*iii*) Method of sections

The main advantage of this method is that the force in almost any member can usually be obtained by cutting the frame through that member and applying the principle of moments.

For example, suppose that we require the forces in the three members cut by the imaginary plane XX in Fig. 1.21.

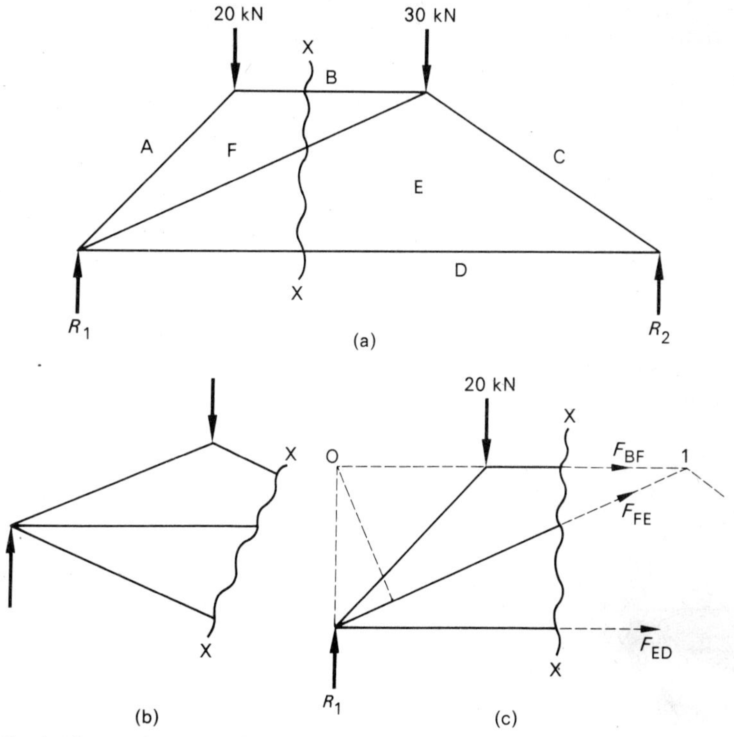

Fig. 1.21 Application of method of sections

This cutting plane divides the truss into two parts. Suppose the right-hand part of the frame is removed. This will cause the left-hand part to collapse about the joint at R_1, as shown at (*b*), unless external forces represented by F_{BF}, F_{FE}, and F_{ED} are applied to the cut members as indicated. Now, in practice, the left-hand section of the truss is kept in equilibrium by the internal forces in the members BF, FE, and ED. Therefore, the external forces applied in diagram (*c*) actually represent the internal forces which we wish to calculate.

If we consider the equilibrium of the left-hand section of the truss, the external forces applied to the cut section must either push or pull on the joints to the left of section XX. It is a good practice to assume that *the cut members are all in tension* as a positive final answer then confirms tension (TIE) while a negative answer implies compression (STRUT).

The left-hand section of the frame is now in equilibrium under the action of the reaction R_1, the load of 20 kN and the three unknown forces F_{BF}, F_{FE}, and F_{ED}. By taking moments of all the forces about the left-hand reaction – i.e., the point of intersection of two of the unknown forces F_{FE} and F_{ED} – the value of F_{BF} can be easily determined. Similarly, the magnitude of F_{ED} can be determined by taking moments about position 1. Finally, moments about a point such as O will give the magnitude of the force in FE.

FORCE IN BF (FIG. 1.22):
Cutting plane XX:

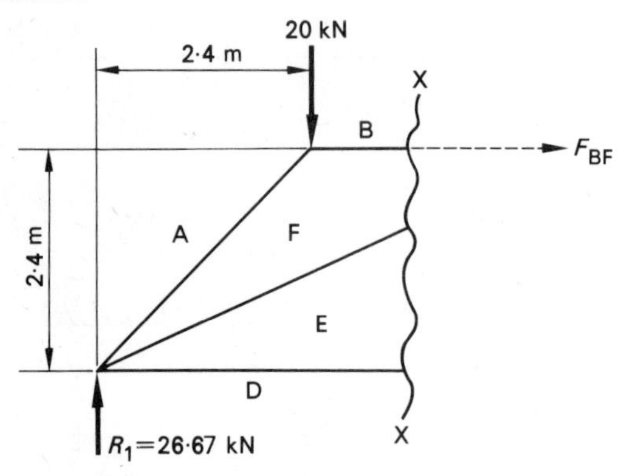

Fig. 1.22 Determination of force in BF

Taking moments about R_1.

$$20 \times 2{\cdot}4 + F_{BF} \times 2{\cdot}4 = 0$$

$$F_{BF} = -20 \text{ kN}$$

∴ BF is a strut.

FORCE IN ED (FIG. 1.23):
Cutting plane XX:

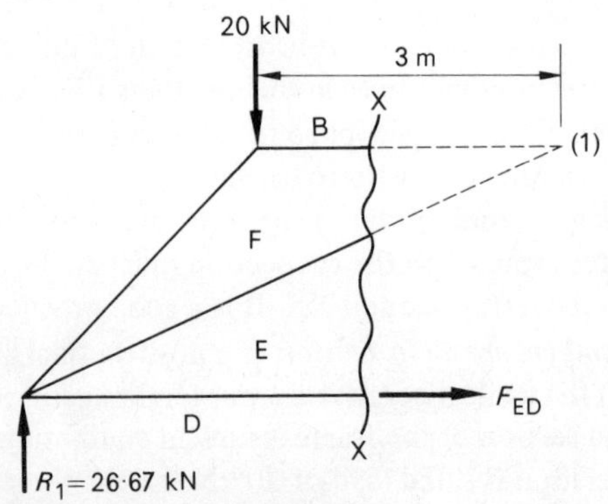

Fig. 1.23 Determination of force in ED

Taking moments about point 1.

$$26{\cdot}67 \times 5{\cdot}4 - 20 \times 3 - F_{ED} \times 2{\cdot}4 = 0$$

$$F_{ED} = 35 \text{ kN}$$

∴ ED is a tie.

FORCE IN FE (FIG. 1.24):
Cutting plane XX:

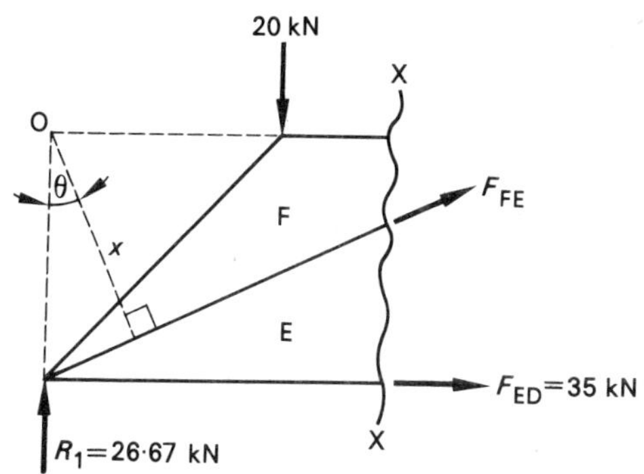

Fig. 1.24 Determination of force in FE

Taking moments about O.

$$20 \times 2{\cdot}4 - F_{FE} \times x - F_{ED} \times 2{\cdot}4 = 0$$
$$48 - F_{FE} \times 2{\cdot}4 \cos\theta - 35 \times 2{\cdot}4 = 0$$
$$F_{FE} = -16{\cdot}4 \text{ kN}$$

∴ FE is a strut.

FORCE IN AF (FIG. 1.25):
Cutting plane YY:

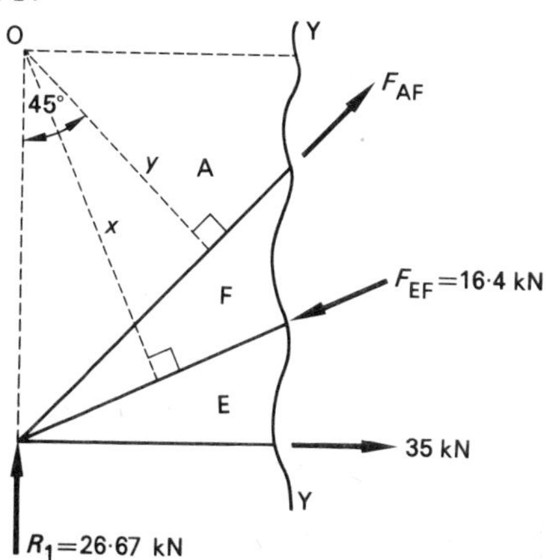

Fig. 1.25 Determination of force in AF

Taking moments about O.

$$16{\cdot}4 \times x - F_{AF} \times y - 35 \times 2{\cdot}4 = 0$$

$$F_{AF} = -\ 28{\cdot}3 \text{ kN}$$

∴ AF is a strut.

FORCE IN EC (FIG. 1.26):
Cutting plane ZZ:

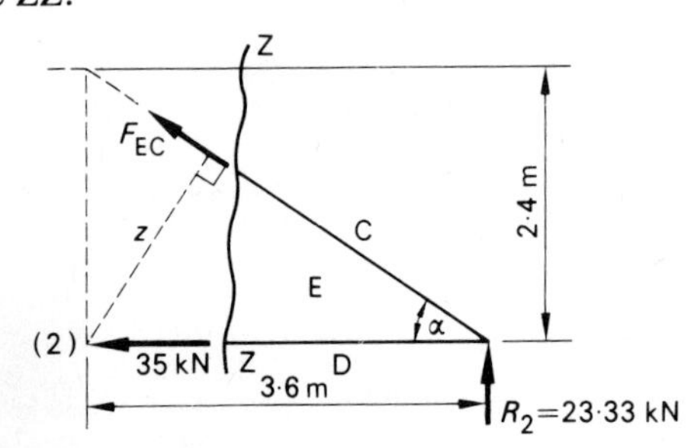

Fig. 1.26 Determination of force in EC

Taking moments about point 2.

$$23{\cdot}33 \times 3{\cdot}6 + F_{EC} \times z = 0$$

$$F_{EC} = -42 \text{ kN}$$

∴ EC is a strut.

Note that in all instances the force in the unknown member initially has been assumed to be in tension.

Example 1.4

A simple pin-jointed crane structure, shown in Fig. 1.27, supports a vertical load of 14 kN by means of a chain. Determine the forces in each member of the structure.

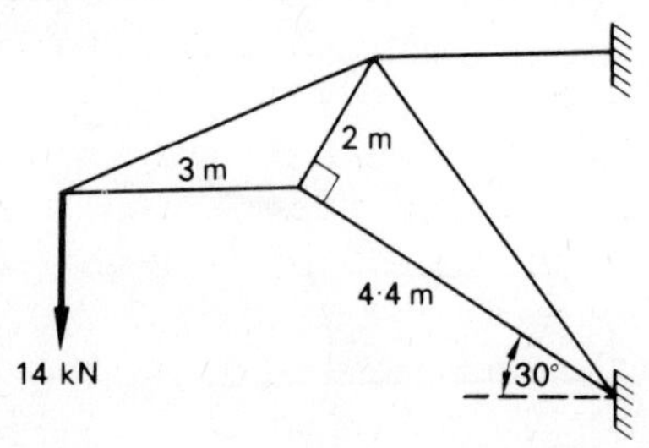

Fig. 1.27 Example 1.4

Solution

The structure is drawn to scale in Fig. 1.28(*a*) and lettered according to Bow's notation. The force vector diagram is drawn in Fig. 1.28(*b*). This

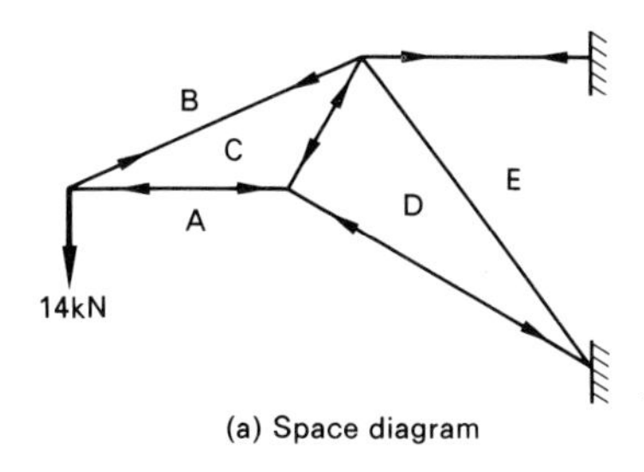

(a) Space diagram

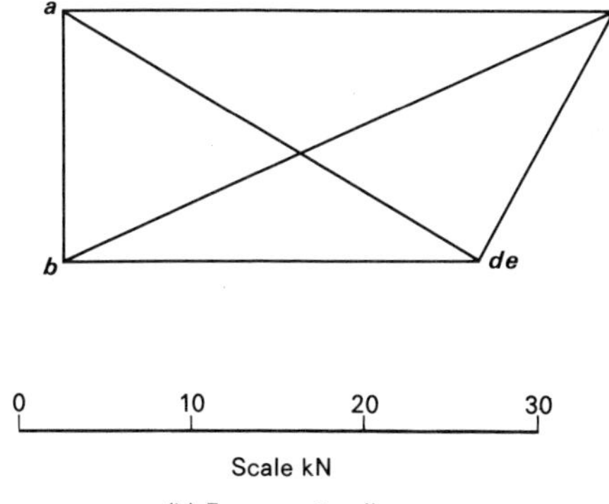

(b) Force vector diagram

Member	Force kN	Nature
BC	35	T
CA	32·2	S
CD	16·1	S
DA	28	S
DE	0	—
BE	24·2	T

Fig. 1.28 (a) Space diagram (b) Force vector diagram

diagram is commenced at the loaded joint, vector ***ab*** being drawn to an appropriate scale and the forces in BC and CA then obtained from the vector triangle ***abc.*** The directions of the forces at the joint are transferred to the space diagram. The diagram is completed by working round the two remaining joints in turn.

It is found that points ***d*** and ***e*** are coincident on the vector diagram. Thus, for the particular loading of this example, the member DE does not carry any load. This does not mean, however, that this member is unnecessary as if it were removed the structure would collapse due to the formation of a mechanism if the load were applied in any direction other than vertically.

From the vector diagram the results shown in the table are obtained.

Example 1.5

The pin-jointed cantilever framework shown in Fig. 1.29 is hinged at D and supported by rollers at E. Determine, for the loading shown, the forces in the members CD, BE and EF.

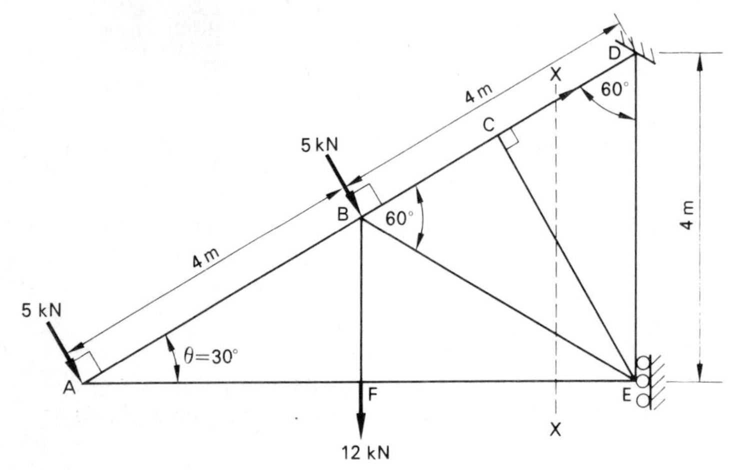

Fig. 1.29 Example 1.5

Solution

Referring to Fig. 1.29 it will be seen that

$$\text{Sin } \theta = \tfrac{4}{8} = 0{\cdot}5, \text{ i.e., } \theta = 30^\circ$$

Resolving perpendicular to AB at A, assuming AF is in tension, gives

$$F_{AF} \cos 60^\circ + 5 = 0$$

$$\therefore \qquad F_{AF} = -10 \text{ kN, i.e., AF is a strut}$$

Resolving horizontally at F gives

$$F_{EF} - F_{AF} = 0$$

$$\therefore \qquad F_{FE} = -10 \text{ kN, i.e., FE is a strut}$$

Resolving vertically at F gives

$$F_{BF} = 12 \text{ kN tie}$$

Referring to Fig. 1.30 it will be seen that the force in BE can be obtained by resolving perpendicular to ABC.

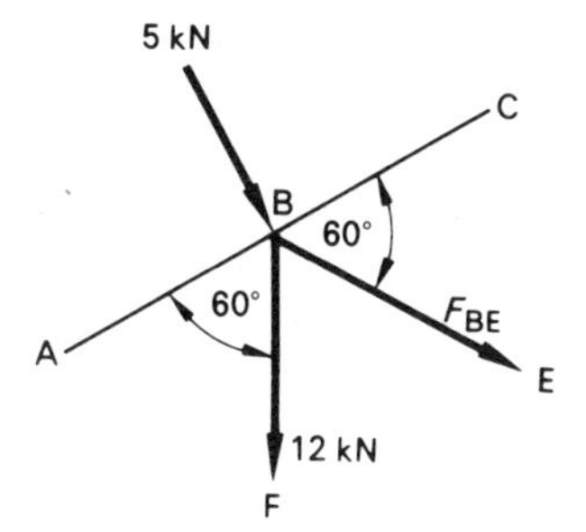

Fig. 1.30

Assume BE to be in tension.

Then $5 + 12 \sin 60^\circ + F_{BE} \sin 60^\circ = 0$

$$F_{BE} = -22 \text{ kN}$$

∴ BE is a strut

The force in BE could be obtained, for example, by resolving at joint E, knowing that the force in member CE must be zero (CE is perpendicular to members BCD) and that the reaction at E is perpendicular to the roller track.*

Although the force in CD could be obtained by resolution we can illustrate how the method of sections may be used to determine the force in a single member.

Considering the cutting plane XX (Fig. 1.29) and taking moments about E for the left-hand portion.

$$F_{CD} \times 4 \cos 30^\circ - 12 \times 4 \cos 30^\circ - 5 \times 4 \cos 60^\circ - 5\,(4 + 4 \cos 60^\circ) = 0$$

$$F_{CD} \times 2\sqrt{3} - 24\sqrt{3} - 10 - 30 = 0$$

$$F_{CD} = 20{\cdot}66 \text{ kN}$$

∴ CD is a tie

Example 1.6

The roof truss, shown in Fig. 1.31, is fixed at the right-hand support whilst the left-hand support rests on rollers. Determine the support reactions in magnitude and direction and also the force in each of the members CD, DO, and OP, indicating whether the member is a strut or a tie.

* See Jackson, A., *Mechanical Engineering Science for 01* (Chapter 2), Longman, London 1971.

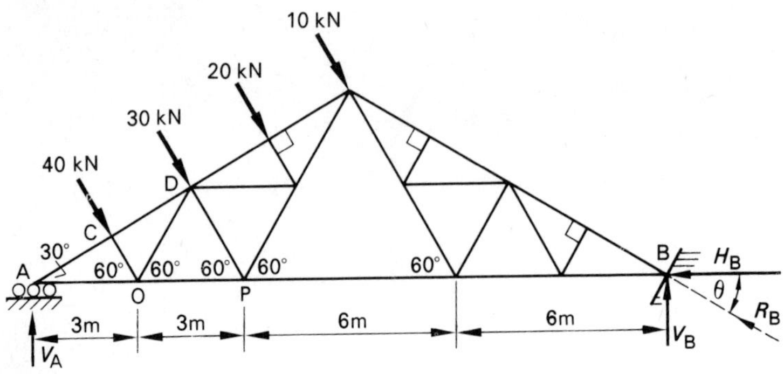

Fig. 1.31 Example 1.6

Solution

Let the vertical reaction at A be V_A (there cannot be any horizontal component of the reaction at A since this support rests on rollers) and the vertical and horizontal components of the reaction at B be V_B and H_B, respectively. These reactions are shown on Fig. 1.31.

The reaction H_B must therefore balance the horizontal components of the applied forces.

$$(40 + 30 + 20 + 10) \cos 60^\circ = H_B$$

$$H_B = 50 \text{ kN}$$

Taking moments about A

$$40 \times 3 \cos 30^\circ + 30 \times 6 \cos 30^\circ + 20 \times 9 \cos 30^\circ + 10 \times 12 \cos 30^\circ$$

$$= V_B \times 18$$

$$V_B = 28{\cdot}9 \text{ kN}$$

Then $R_B{}^2 = 50^2 + 28{\cdot}9^2$

$$R_B = 57{\cdot}7 \text{ kN}$$

and $\text{Tan}\ \theta = \dfrac{V_B}{H_B} = \dfrac{28{\cdot}9}{50} = 0{\cdot}578$

$$\theta = 30^\circ$$

For vertical equilibrium

$$V_A + V_B = (40 + 30 + 20 + 10) \cos 30^\circ$$

$$= 86{\cdot}6 \text{ kN}$$

$\therefore \qquad V_A = 57{\cdot}7 \text{ kN}$

Resolving vertically at A:

$$F_{AC} \cos 60^\circ + V_A = 0$$

$\therefore$ $F_{AC} = -115{\cdot}4$ kN AC is a strut

But for equilibrium at C

$$F_{AC} = F_{CD}$$

$\therefore$ $F_{CD} = -115{\cdot}4$ kN CD is a strut

Also $F_{CO} = -40$ kN CO is a strut

Resolving horizontally at A:

$$F_{AC} \cos 30^\circ + F_{AO} = 0$$

$$-115{\cdot}4 \cos 30^\circ + F_{AO} = 0$$

$\therefore$ $F_{AO} = 100$ kN AO is a tie

The forces now acting at O are shown in Fig. 1.32.

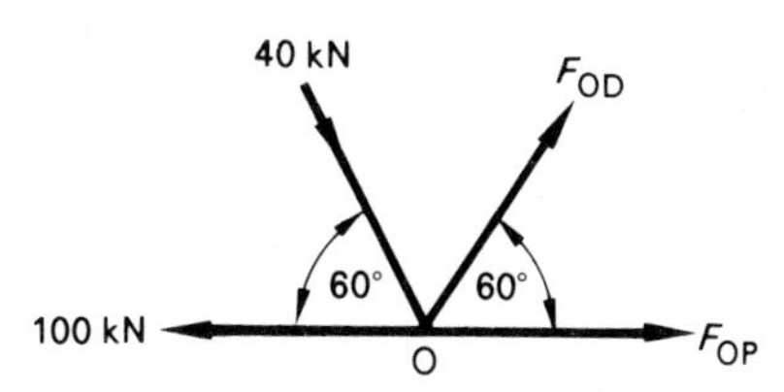

Fig. 1.32 Forces acting at O

Resolving vertically at O:

$$40 \sin 60^\circ - F_{OD} \sin 60^\circ = 0$$

$\therefore$ $F_{OD} = 40$ kN OD is a tie

Resolving horizontally at O:

$$40 \cos 60^\circ + F_{OD} \cos 60^\circ + F_{OP} - 100 = 0$$

$\therefore$ $F_{OP} = 60$ kN OP is a tie

Example 1.7

Figure 1.33 shows a simply supported truss. Determine:

(*a*) graphically, the forces in members (1), (2), (3), and (4);
(*b*) by calculation, the forces in members (5), (6), and (7).

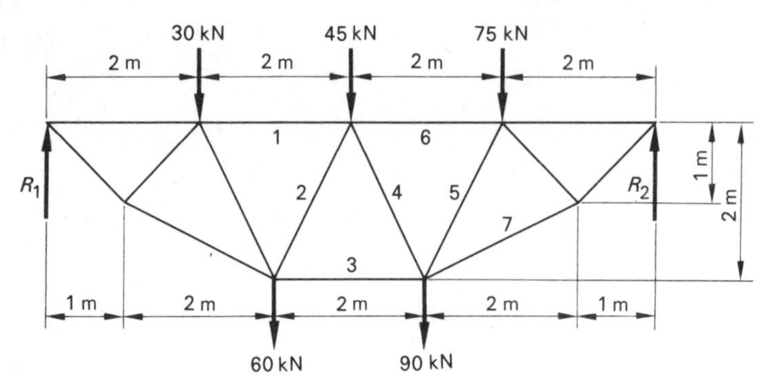

Fig. 1.33 Example 1.7

Solution

Let the support reactions be R_1 and R_2 as shown.
By moments about R_1:

$$30 \times 2 + 45 \times 4 + 75 \times 6 + 60 \times 3 + 90 \times 5 = 8R_2$$

$$R_2 = 165 \text{ kN}$$

$$\therefore \qquad R_1 = 135 \text{ kN}$$

(*a*) The part of the vector force diagram required to determine the forces in members (1), (2), (3), and (4) is given in Fig. 1.34. Note that the diagram, which is commenced with ***ab***, is centred upon the load line ***ed*** in the usual way.

The results obtained from the vector force diagram are:

Member	*Force* kN	*Nature*
1	187	Strut
2	50	Strut
3	210	Tie
4	0	–

(*b*) The forces in members (5), (6), and (7) will be obtained by application of the *method of sections.*

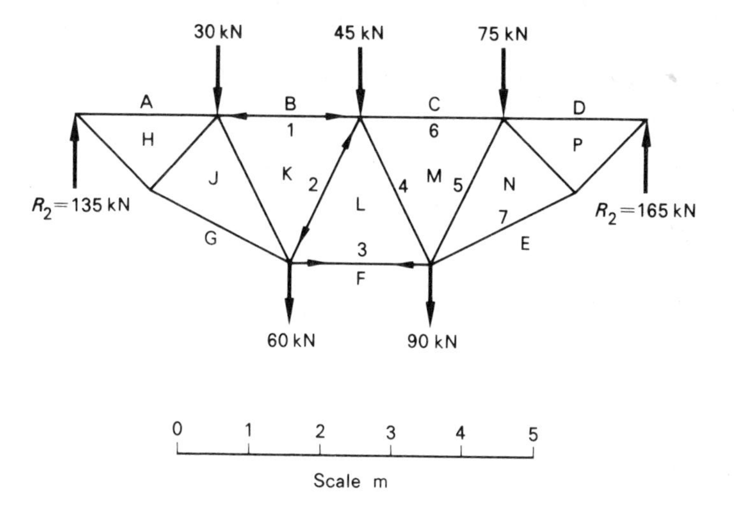

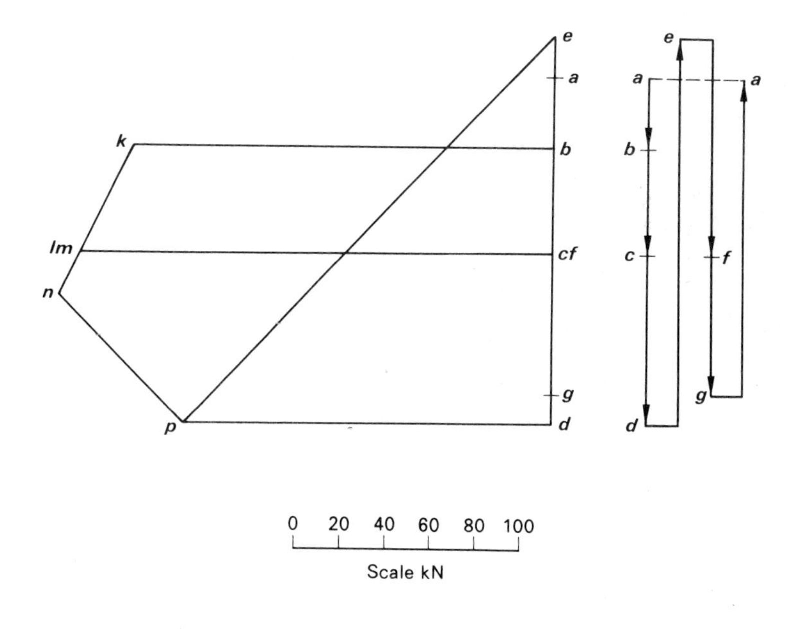

Fig. 1.34 Determination of forces by vector diagram

FORCE IN MN (5):
Consider cutting plane XX (Fig. 1.35) and take moments about O for the section to the right of XX.

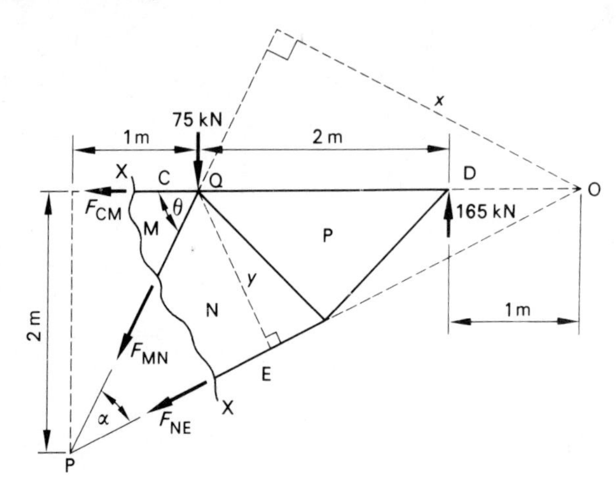

Fig. 1.35 Determination of forces by method of sections

Assuming MN to be a tie, then

$$165 \times 1 - 75 \times 3 - F_{MN} \cdot x = 0$$

Now $$x = 3 \sin \theta$$

But $$\text{Tan } \theta = \tfrac{2}{1}$$

∴ $$\text{Sin } \theta = \frac{2}{\sqrt{5}}$$

Therefore, $$x = \frac{6}{\sqrt{5}}$$

and $$F_{MN} \cdot \frac{6}{\sqrt{5}} = -60$$

$$F_{MN} = -22{\cdot}35 \text{ kN} \quad \text{MN is a strut}$$

FORCE IN CM (6):
Cutting plane XX:
Take moments about P, assuming CM is a tie.

$$75 \times 1 - 165 \times 3 - F_{CM} \cdot 2 = 0$$

∴ $$F_{CM} = -210 \text{ kN} \quad \text{CM is a strut}$$

FORCE IN EN (7):
Cutting plane XX:
Take moments about Q, assuming NE is a tie.

$$165 \times 2 = F_{NE} \cdot y$$

Now $\quad y = PQ \sin \alpha$

$$= \sqrt{5} \sin \alpha$$

But $\quad \sin \alpha = \dfrac{x}{OP}$

And $\quad OP = \sqrt{(2^2 + 4^2)} = 2\sqrt{5}$

$\therefore \quad \operatorname{Sin} \alpha = \dfrac{6/\sqrt{5}}{2\sqrt{5}} = \frac{3}{5}$

$\therefore \quad y = \dfrac{3}{\sqrt{5}}$

$\therefore \quad F_{NE} = \dfrac{330\sqrt{5}}{3} = 246$ kN $\quad$ NE is a tie

Problems

1. A beam AB, 3 m long, is simply supported at A and at a point C, 2·1 m from A. Loads of 20 kN, 35 kN and 15 kN are carried by the beam at distances of 1 m, 1·6 m, and 3 m, respectively, from A. Determine graphically the vertical reactions at A and C.

2. Determine, by means of the funicular polygon, the magnitude of the wall reaction applied to the loaded beam shown in Fig. 1.36.

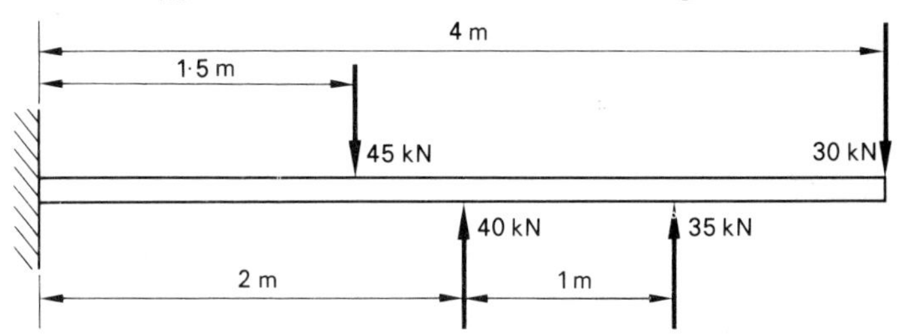

Fig. 1.36

3. A uniform bar AB, having a mass of 30 kg, is inclined at 20° to the horizontal plane and is supported by cords AC and BD. The cord BD is vertical and AC is inclined at 30° to the vertical, such that the angle

CAB is 100° and DBA is 110°. Determine the horizontal force P, applied at B, which is required to keep the bar in this position if P is in the same vertical plane as the bar and the cords.

4. Figure 1.37 shows a uniform horizontal beam AB, 6 m long, which is hinged at B and freely supported at C, 2 m from A. The beam has a

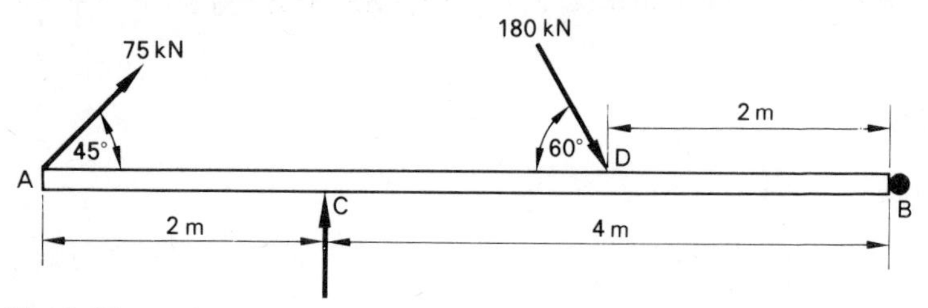

Fig. 1.37

mass of 4078 kg and carries loads of 75 kN and 180 kN at A and D, respectively, as shown. Determine the support reactions.

5. (*a*) State the conditions necessary for a system of co-planar forces to be in equilibrium.

(*b*) A horizontal beam XY, 7 m long, is hinged freely at X and rests at Y on a smooth surface inclined at 45° to the horizontal. The beam carries vertical loads of 50 kN and 65 kN at points 2 m and 5 m, respectively, from X.

Draw the beam to scale and determine, graphically, the magnitude and direction of the reaction at X.

6. A light beam, as shown in Fig. 1.38, carries a load of 300 N at D, and is supported in the horizontal position by three strings attached at A,

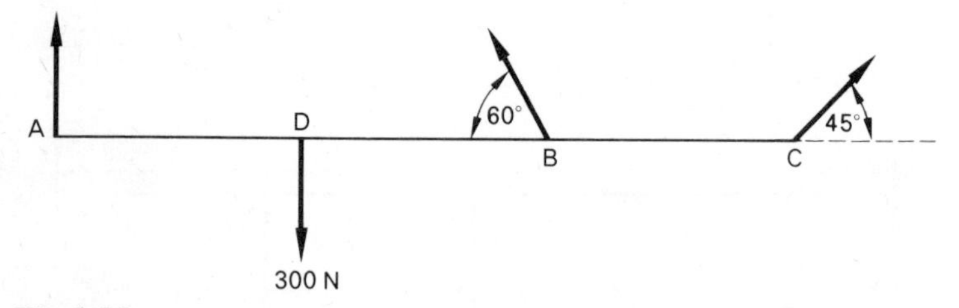

Fig. 1.38

B, and C, the angles which the strings make with the beam being 90°, 60°, and 45°, respectively. If the distances AD, DB, and BC are each 3 m determine, graphically, the tensions in the three strings.

7. A uniform beam AB has a mass of 25·5 kg and is 4 m long. It is hinged at A and supported in a horizontal position by a rope attached to a point C on the beam, 3·5 m from A. The rope is inclined at 30° to the beam and is fastened to the wall vertically above A. The beam carries a vertical load of 300 N at B. Determine the reaction at the hinge and the tension in the rope.

8. The pin-jointed structure shown in Fig. 1.39 carries vertical loads at the joints B, C and D and is simply supported at A and E.

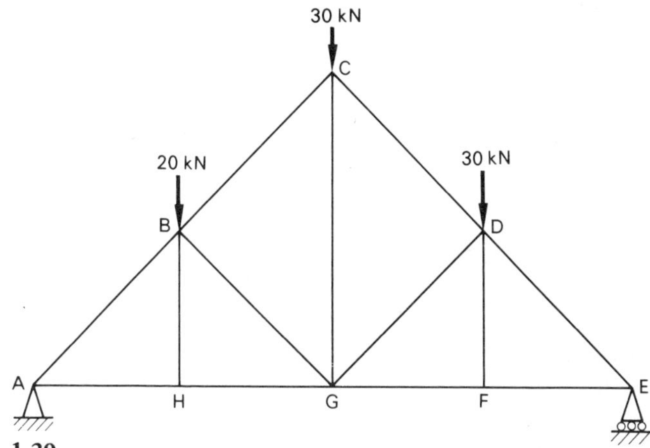

Fig. 1.39

CG = 3 m and AH = HG = GF = FE = 2 m

Determine the support reactions and draw the vector force diagram for the structure, so obtaining the forces in the members and their nature.

9. Figure 1.40 shows a pin-jointed frame carrying three vertical loads as shown.

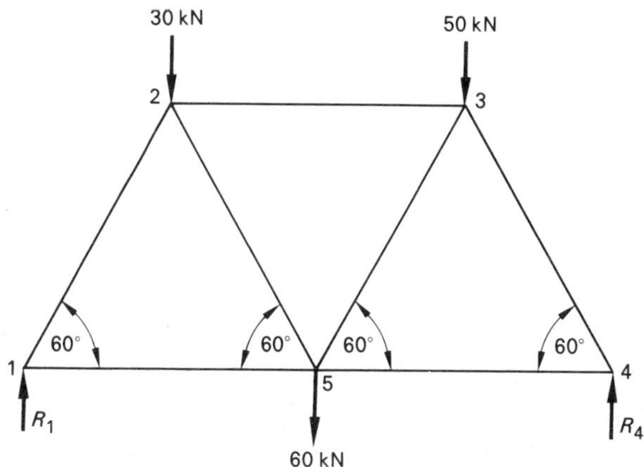

Fig. 1.40

(*i*) Determine 'by means of a funicular polygon' the supporting reactions R_1 and R_4;
(*ii*) Draw a force diagram for the frame and use it to determine the magnitude and nature of the forces in members 2–5 and 3–5.

10. A workshop has a north-light roof, the dimensions of a typical bay being given in Fig. 1.41, which also shows the design loads at the panel points. Points D and E are at the centre of their respective rafters.

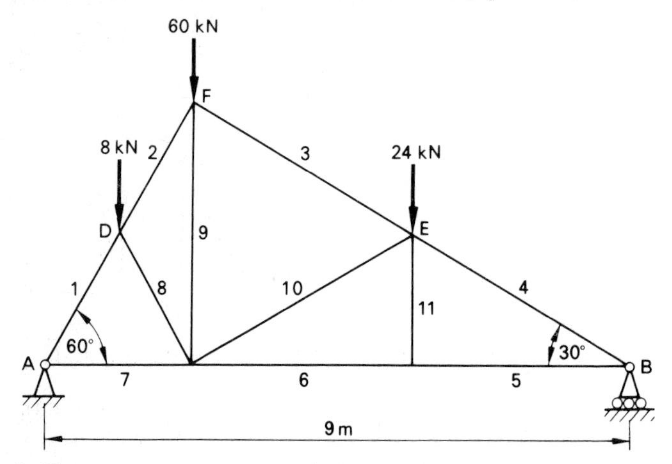

Fig. 1.41

(*a*) If the truss is pinned at A and B, calculate the reactions there.
(*b*) Draw the force diagram and hence determine the force in each member, stating its nature.

11. Due to a conveyor system, a roof truss carries loads of 30 kN and 20 kN as shown in Fig. 1.42. Determine the forces in the members and state their nature.

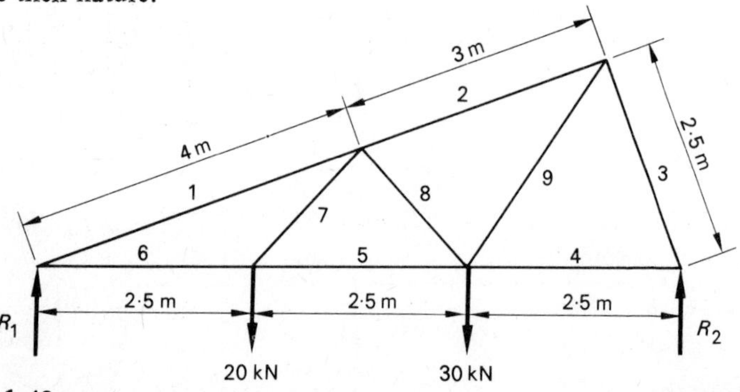

Fig. 1.42

12. A cantilever frame has horizontal upper and lower booms together with three other members which are inclined at 60° to the horizontal, as shown in Fig. 1.43.

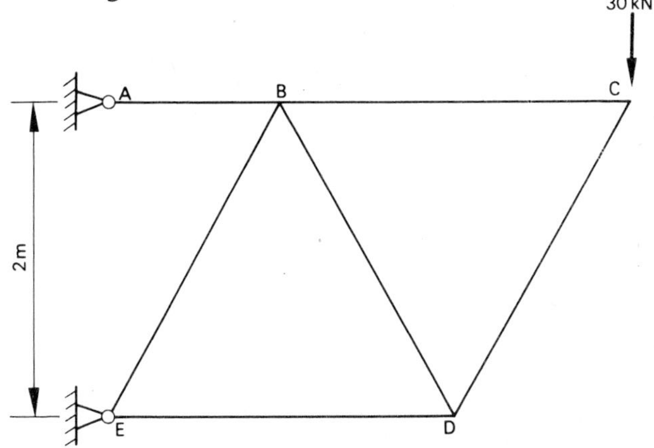

Fig. 1.43

When the structure supports a load of 30 kN at the end of the upper boom, find:

(*i*) the magnitude and direction of the reactions at the anchor pins;
(*ii*) the forces in the members of the frame.

13. Figure 1.44 represents a simply supported girder the joints of which may be regarded as pinned.

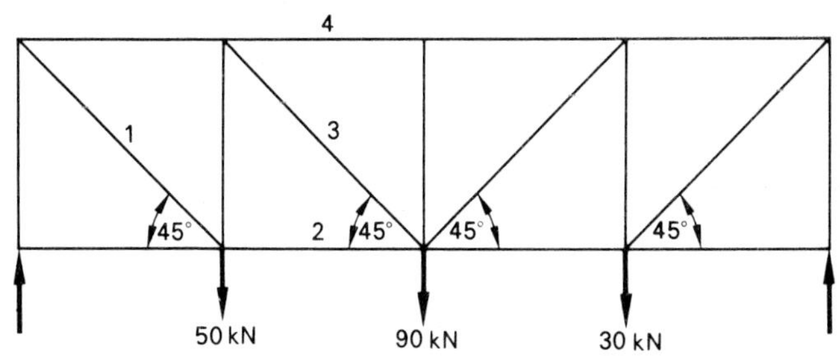

Fig. 1.44

Determine the forces in the members 1, 2, 3, and 4, stating whether the members are struts or ties.

14. Figure 1.45 shows a pin-jointed structure which is simply supported at its ends.

AJ = JI = IH = HG = 3·2 m BA = FG = 2·4 m

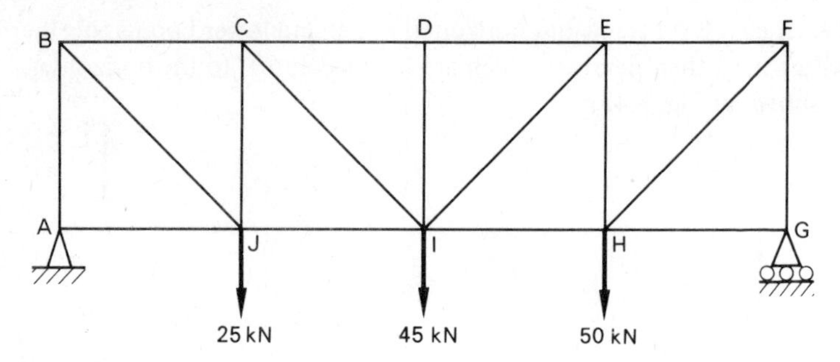

Fig. 1.45

Draw the force vector diagram for the structure and determine which members are struts and which members carry no load. State the force in member CI.

15. Figure 1.46 shows a loaded roof truss. Determine the values of the forces in the members BF, FG, and GD, stating whether they are struts or ties.

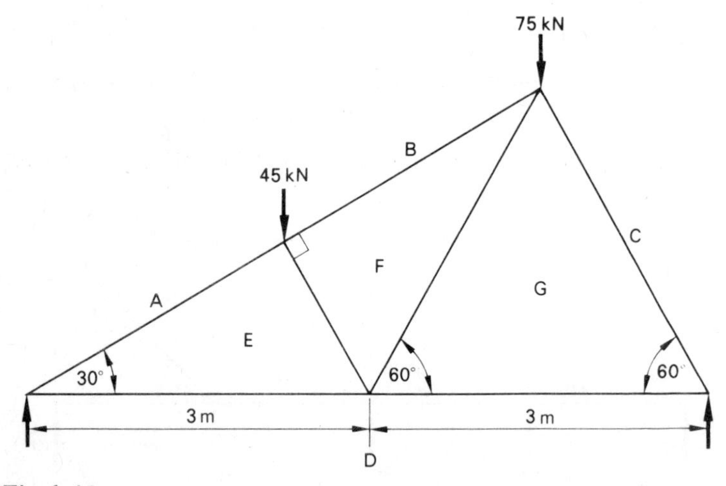

Fig. 1.46

16. The roof truss shown in Fig. 1.47 is hinged at B and supported on rollers at A. For the loading given, determine the values of the reactions at A and B in magnitude and direction. Also determine the forces in the members marked 1, 2, and 3, stating whether they are struts or ties.

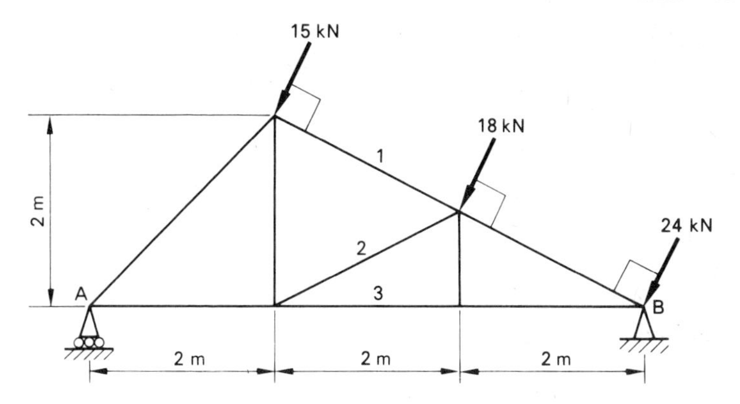

Fig. 1.47

17. Figure 1.48 gives details of a framework which is supported by a pin at Z and by a horizontal tie at Y. Determine the forces in the framework members arising from the given loading and the magnitude and direction of the reactions at Y and Z.

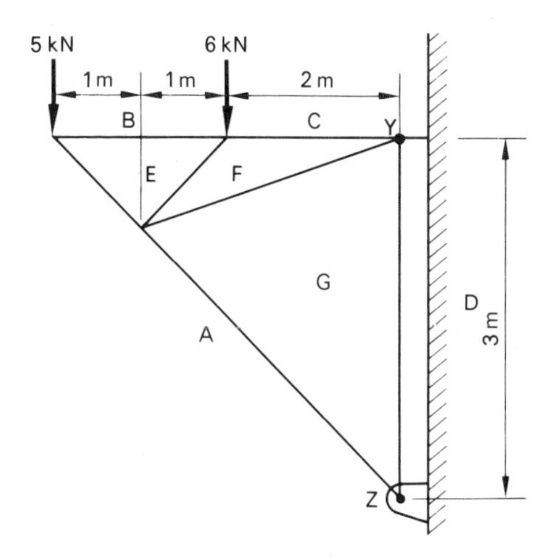

Fig. 1.48

18. A simple frame of six bars is connected to a rigid vertical wall by pins in brackets at A and at E, 2 m above A. There are two equal bars, AB and BC, each 1 m long, and pinned at B, in a horizontal line from A. Two other equal bars, CD and DE, pinned at D, form a straight-line connection between C and E. The two remaining members connect D with A and with B. Vertical forces act downwards at D, C, and B of

respective magnitudes, 20 kN, 8 kN, and 4 kN. Determine the forces in the bars, stating their nature, and the magnitude and direction of the reaction at A.

19. The central panel ABCD of a pin-jointed truss is rectangular in form, 4 m long by 3 m deep. The bottom member is AD, with A on the left; the top member is BC, with B on the left; and there is a diagonal member AC. Loads of 200 kN and 60 kN act vertically downwards at A and D, respectively. The members of the adjacent panels act at the corners in such a way as to set up a horizontal force at A of 150 kN directed to the left; a horizontal force of 70 kN at D directed to the right; and an inclined force at B, the components of which are 140 kN vertically upwards and 210 kN horizontally to the right. Determine the nature and magnitude of the force in each member of the central panel and state the components of the force at C due to the connection with the adjacent panel.

2

Shear force and bending moment

2.1 Definitions of shear force and bending moment

A beam which is resting on two simple supports,* as shown in Fig. 2.1, is a particular case of a body at rest under the action of parallel forces. This follows from our previous work on statics.

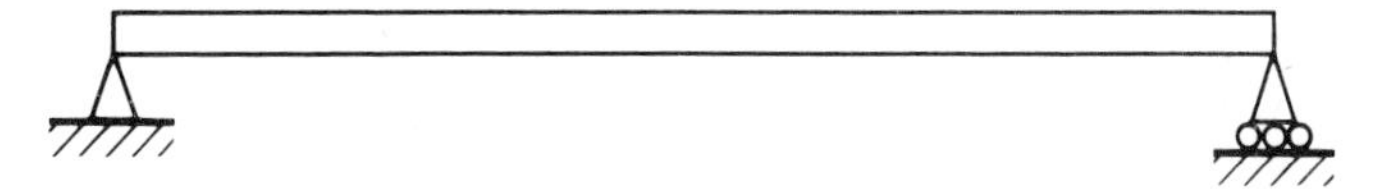

Fig. 2.1 A beam on two simple supports

If the beam has a uniform cross-section then its weight (the gravitational force acting upon its mass) can be considered as uniformly distributed over the beam length as indicated in Fig. 2.2.

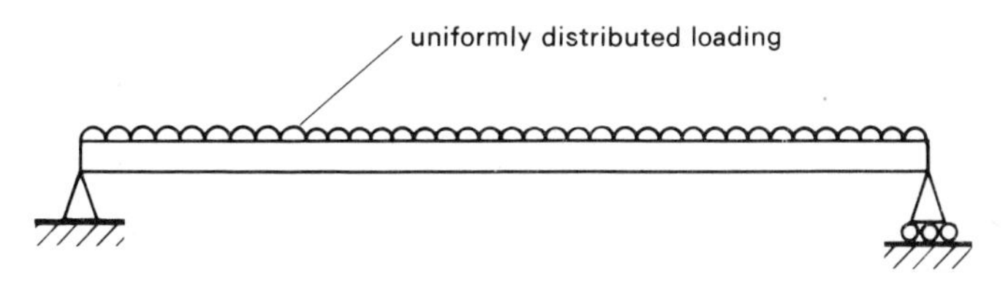

Fig. 2.2 Beam weight considered as a uniformly distributed loading

The evaluation of the shear force and bending moment throughout a beam, or other structure, is necessary for the determination of the stresses within the material and the definitions of these quantities can

* A simple support is one which offers no resistance to any rotation of the beam at the support.

be obtained by considering the simply supported beam shown in Fig. 2.3.

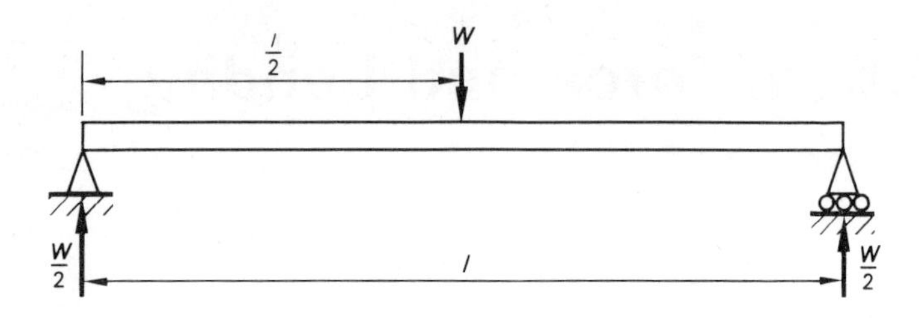

Fig. 2.3

For simplicity the beam is assumed to be of negligible mass, and due to the single concentrated load acting at the centre the support reactions will each be $W/2$ as shown.

Let us now consider what is happening at a section ZZ, distance z from the left hand support ($0 \leqslant z \leqslant l/2$). It will be seen from Fig. 2.4

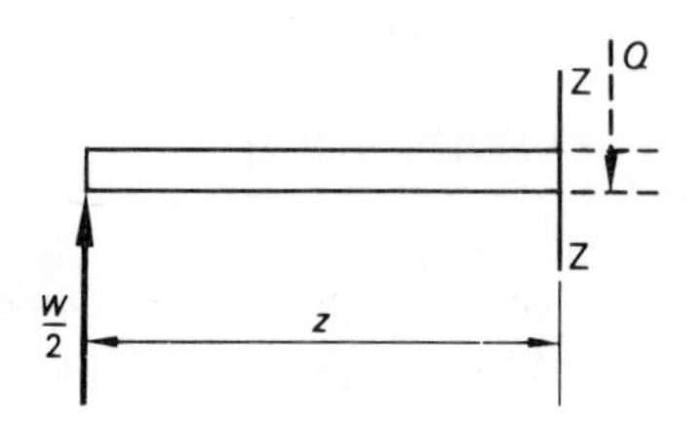

Fig. 2.4 Shear force at beam section

that the only force acting on the section is $W/2$ upwards. Thus for equilibrium the right-hand section of the beam must exert a force Q (shown dotted) on the left-hand section such that

$$Q = W/2$$

The force Q is referred to as the shearing force (commonly called 'shear force') acting at the section.

The tendency of these forces is to produce the action shown in Fig. 2.5(a). This is defined in this text as positive shearing action while negative shearing action is that shown in Fig. 2.5(b).

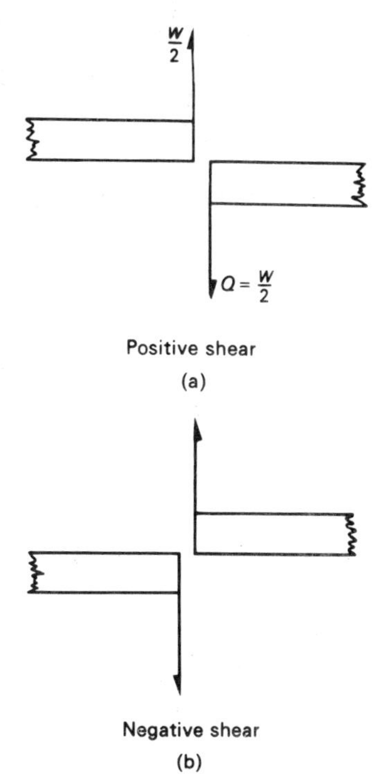

Fig. 2.5 Shear force notations

Let us now return to our simple beam and consider the case when $l/2 \leqslant z \leqslant l$, as shown in Fig. 2.6.

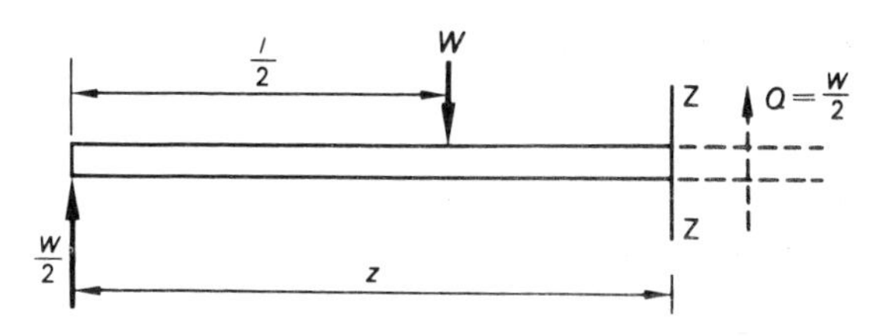

Fig. 2.6 Shear force at beam section

The resultant force to the left of the section is now $W/2$ downwards and therefore the right-hand section of the beam must exert an upward force of $Q = W/2$ on the left hand side if equilibrium is to be maintained. This therefore constitutes a negative shearing action. Obviously the change from positive shear to negative shear occurs at the applied load and the shear force diagram for the beam is Fig. 2.7(*a*). From this diagram, it would appear that the centre of the beam is subjected to two shear forces – one of + $W/2$ and another of $-W/2$. This is because the load W was considered to be a concentrated load acting at a single

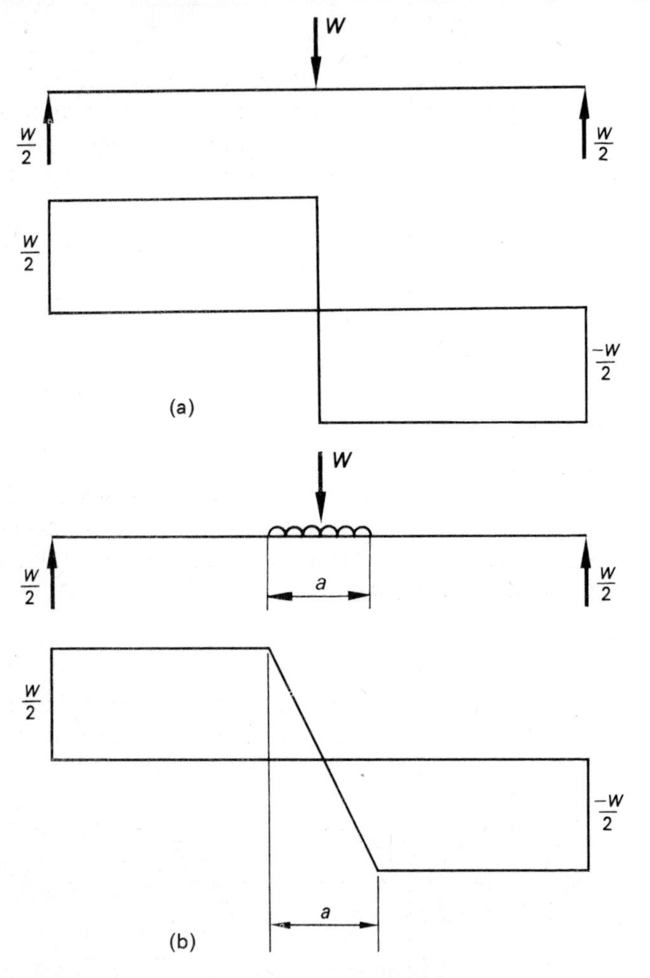

Fig. 2.7 Shear force diagram at a concentrated load

point whereas in reality the load is most likely to be applied over a short length *a* so that the true shear force diagram would be as shown in Fig. 2.7(*b*). The student should appreciate at this stage that true point loadings rarely occur.

Thus we may state:

> The shear force acting at a section is the algebraic sum of the forces acting on either side of that section. Positive shear is when the resultant of the forces to the left of the section is vertically upwards.

Let us now look at the left-hand section of the beam again. Having examined the equilibrium of vertical forces acting at any section of the beam we can now turn our attention to moment equilibrium. At section ZZ (ref. Fig. 2.8) there will be a clockwise moment of $Wz/2$ as

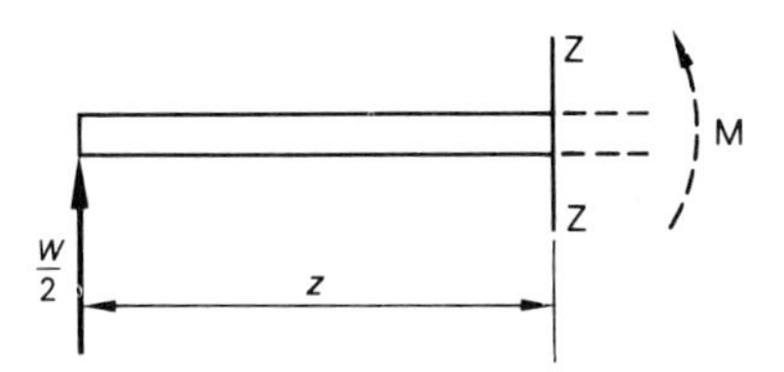

Fig. 2.8 Moment equilibrium of beam section

a result of the left-hand reaction and therefore for equilibrium of the section there must be a moment of equal magnitude applied by the right-hand section of the beam onto the left-hand section. This moment is shown as M (dotted) on the diagram.

Thus $M = Wz/2$ for $0 \leqslant z \leqslant l/2$

For a section of the beam to the right of the applied load the forces acting are as in Fig. 2.9. In this instance there is a clockwise moment

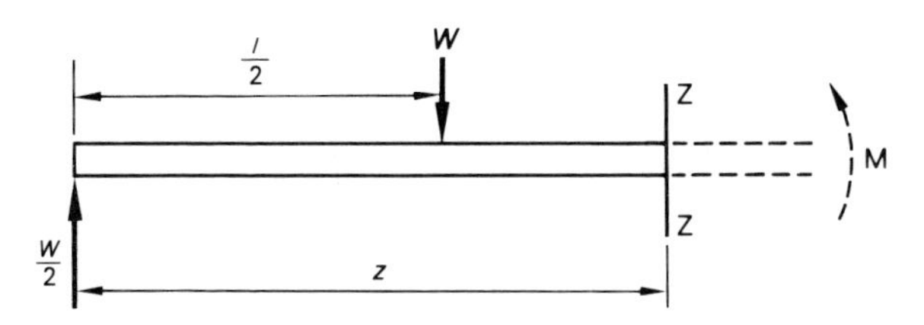

Fig. 2.9 Moment equilibrium of beam section

at section ZZ, resulting from the forces to the left of ZZ, of

$$\frac{Wz}{2} - W\left(z - \frac{l}{2}\right), \text{ i.e., } \frac{W}{2}(l - z)$$

Again for equilibrium there must be a moment M, shown dotted, which is applied from the right-hand section to the left-hand section of the beam. Thus

$$M = \frac{W}{2}(l - z) \quad \text{for} \quad \frac{l}{2} \leqslant z \leqslant l$$

The moment M is referred to as the bending moment.

It will be seen that these expressions for the *internal* moment in the beam are dependent upon the magnitude of z. Before a diagram can be drawn to show the variation of internal bending moment along the length of the beam it is necessary to specify a sign convention.

A bending moment which tends to produce sagging of a beam, see Fig. 2.10(*a*), will be taken throughout this text as positive while one which tends to produce hogging, see (*b*) will be taken as negative.

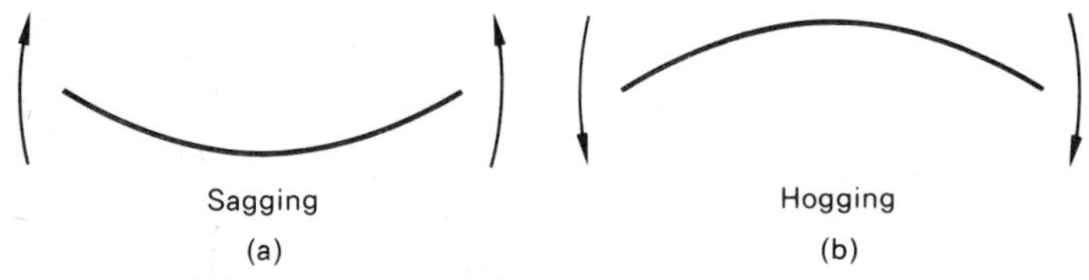

Fig. 2.10 Bending moment notation

It should be appreciated that although the choice of a particular sign convention is purely arbitrary it is important to specify the one which is being used.

The bending moment diagram for our simple beam example can now be drawn. It will be seen that the beam is subjected to a sagging moment throughout its entire length, and by substituting values for z into the expressions given, the bending moment diagram shown in Fig. 2.11 is obtained.

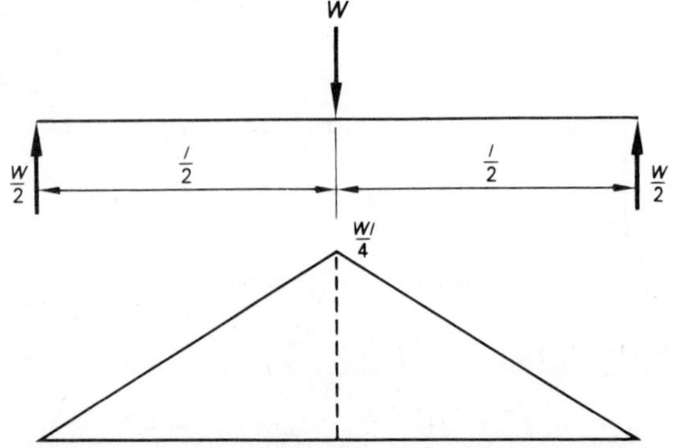

Fig. 2.11 Bending moment diagram for beam with central load

Note: Beams forming parts of structures will not always be horizontal so the terms sagging and hogging cannot be used in general.

Example 2.1

A beam which is simply supported at its ends has a length of 6 m and carries loads of 15 kN, 25 kN, and 40 kN at distances of 1 m, 3 m, and 5 m respectively from the left-hand support.

Determine the bending moment and shear force at sections 2.5 m and 4 m from the left-hand support.

The mass of the beam is to be neglected.

Solution

A diagram is given in Fig. 2.12.

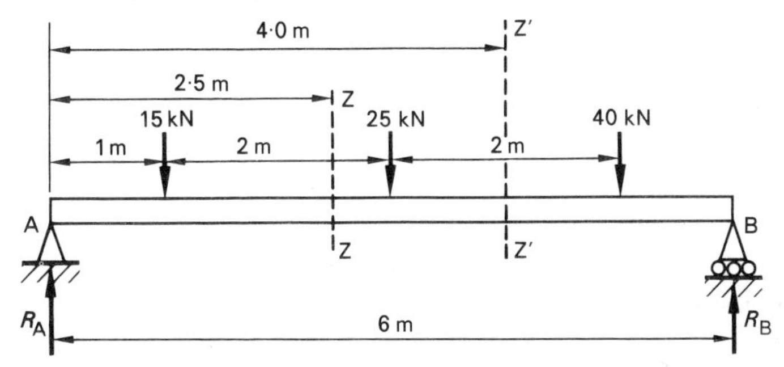

Fig. 2.12 Example 2.1

It is first necessary to determine the support reactions R_A and R_B. Taking moments about A:

$$15 \times 1 + 25 \times 3 + 40 \times 5 - 6R_B = 0$$

$$\therefore \qquad R_B = \frac{290}{6} = 48{\cdot}3 \text{ kN}$$

For vertical equilibrium:

$$R_A = 15 + 25 + 40 - R_B$$
$$= 31{\cdot}7 \text{ kN}$$

SECTION ZZ, 2.5 m FROM A:

Considering equilibrium of the section to the left of ZZ gives:

$$\text{Shear force at ZZ} = R_A - 15$$
$$= 16{\cdot}7 \text{ kN}$$

$$\text{Bending moment at ZZ} = R_A \times 2{\cdot}5 - 15 \times 1{\cdot}5$$
$$= 56{\cdot}75 \text{ kN m}$$

SECTION Z′Z′, 4.0 m FROM A:
Considering equilibrium of the section to the left of Z′Z′ gives:

$$\text{Shear force at } Z'Z' = R_A - 15 - 25 = -8{\cdot}3 \text{ kN}$$

$$\text{Bending moment at } Z'Z' = R_A \times 4{\cdot}0 - 15 \times 3 - 25 \times 1 = 56{\cdot}8 \text{ kN m}$$

2.2 Uniformly distributed loading

We can now progress to the consideration of a uniformly distributed loading or to the inclusion of the weight of a beam, the two being virtually the same as regards the shear force and bending moment diagrams.

For the purposes of the determination of support reactions and bending moments only, a uniformly distributed loading may be replaced by an equivalent imaginary concentrated load acting at the centre span of the uniformly distributed loading. This substitution must not, however, be used when determining the shear force at any section of the beam. The following examples indicate the method of dealing with this type of loading.

Example 2.2

A uniform cantilever is 5 m long and has a weight of 12 kN. A load of 50 kN is applied at the free end of the cantilever. Determine the bending moment and shear force at the centre of beam and at the fixed end.

Solution

A diagram is given in Fig. 2.13.

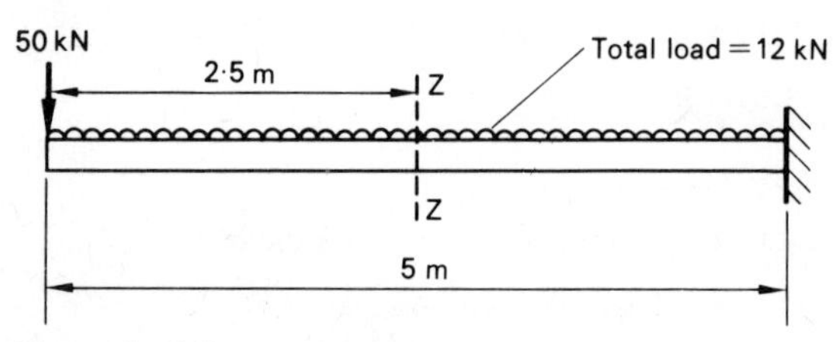

Fig. 2.13 Example 2.2

As the beam is uniform its weight per metre length = 12/5 = 2·4 kN/m

AT SECTION ZZ:

Shear force at ZZ $= -50 - 2{\cdot}5 \times 2{\cdot}4$

$= -56$ kN

Bending moment at ZZ $= -50 \times 2{\cdot}5 - 2{\cdot}4 \times 2{\cdot}5 \times \frac{2{\cdot}5}{2}$

$= -132{\cdot}5$ kN m

AT THE BUILT-IN END:

Shear force $= -50 - 12$

$= -62$ kN

Bending moment $= -50 \times 5 - 12 \times \frac{5}{2}$

$= -280$ kN m

Example 2.3

Plot the bending moment and shearing force diagrams for the cantilever beam shown in Fig. 2.14(*a*). The mass of the beam is to be neglected.

Solution

SHEAR FORCE:

From A to B: Shear Force $= -10$ kN

From B to C: Shear Force $= -10 - 18 = -28$ kN

From C to D: Shear Force $= -10 - 18 - 18 = -46$ kN

The shear force diagram is plotted as Fig. 2.14(*b*).

BENDING MOMENT:

Bending moment at A = 0

From A to B: Bending moment $= -10\,z$

where z is the distance from A.

This is a linear relationship giving a straight line.

At B ($z = 1$): Bending moment $= -10$ kN m

From B to C: Bending moment $= -10\,z - 18\,(z - 1)$

$= -28\,z + 18$

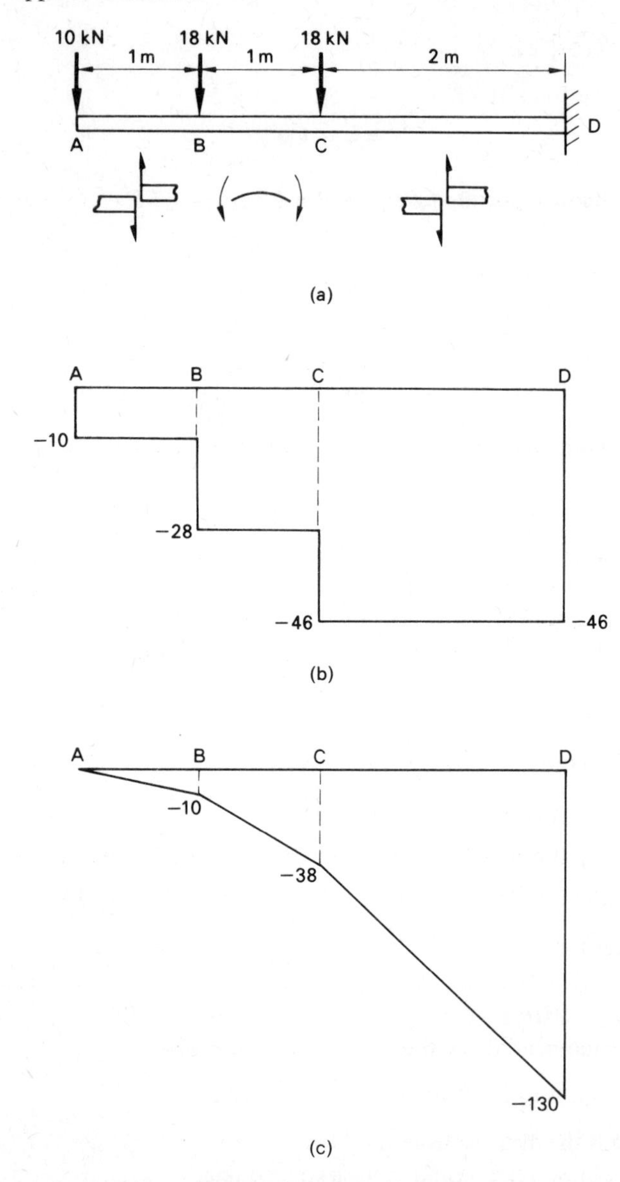

Fig. 2.14 (a) Loading diagram (b) Shear force diagram (c) Bending moment diagram–Example 2.3

Substituting $z = 1$ into this expression will produce the same value for the moment at B as obtained above.

At C ($z = 2$): Bending moment $= -56 + 18$

$= -38$ kN m

From C to D: Bending moment $= -10\,z - 18\,(z-1) - 18\,(z-2)$

$= -46\,z + 54$

This expression similarly gives the same value for the moment at C as previously obtained.

At D ($z = 4$): Bending moment $= -184 + 54$

$= -130$ kN m

The bending moment diagram is therefore a series of straight lines and is drawn as Fig. 2.14(*c*).

Example 2.4

A uniform cantilever, 5 m long, has a weight of 30 kN/m and carries a load of 40 kN applied 2 m from the built-in end.

Draw the shear force and bending moment diagrams.

Solution

A diagram is given in Fig. 2.15(*a*).

SHEAR FORCE:

Consider a section ZZ on the length AB which is at a distance z from the free end of the cantilever A.

Shear force at ZZ $= -30\,z$

This relationship only applies from A to B (i.e., for $0 \leqslant z \leqslant 3$) since there is a concentrated load at B.

Substituting in the appropriate values for z gives:

Shear force at A $= 0$

Shear force just to the left of B $= -30 \times 3 = -90$ kN

Between B and C a section Z_1Z_1 at a distance z_1 from A is considered.

Shear force at $Z_1Z_1 = -30\,z_1 - 40$

This only applies from B to C (i.e., for $3 \leqslant z_1 \leqslant 5$)

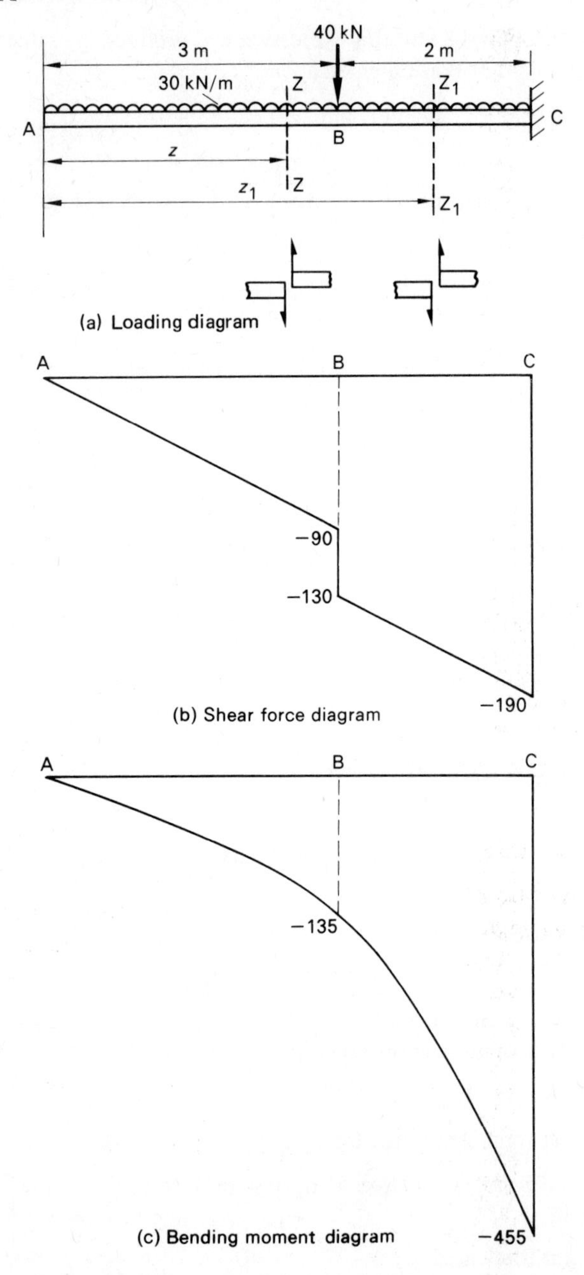

Fig. 2.15 Example 2.4

Then,

Shear force just to the right of B, $(z_1 = 3) = -30 \times 3 - 40 = -130\,\text{kN}$

Shear force at C, $(z_1 = 5) = -30 \times 5 - 40 = -190\,\text{kN}$

BENDING MOMENT:

For section ZZ.

$$\text{Bending moment at ZZ} = -30 \,.\, z \,.\, \frac{z}{2} \text{ for } 0 \leqslant z \leqslant 3$$

This equation gives a parabolic curve.

Bending moment at A, $(z = 0) = 0$

$$\text{Bending moment at B, } (z = 3) = \frac{-30 \times 3^2}{2}$$

$$= -135 \text{ kN m}$$

For section Z_1Z_1:

$$\text{Bending moment} = -30 \,.\, z_1 \,.\, \frac{z_1}{2} - 40\,(z_1 - 3) \text{ for } 3 \leqslant z_1 \leqslant 5$$

$$\text{Bending moment at C, } (z_1 = 5) = \frac{-30 \times 5^2}{2} - 40 \times 2$$

$$= -455 \text{ kN m}$$

The shear force and bending moment diagrams are shown as Figs. 2.15(*b*) and (*c*). An accurate bending moment diagram may be plotted by evaluating the expression for bending moment at intermediate values to those indicated above.

Example 2.5

A beam 8 m long is simply supported at its ends and symmetrically loaded by two 12 kN loads which are applied at a distance of 2 m from each support. Draw the bending moment and shear force diagrams.

Solution

A loading diagram is given in Fig. 2.16(*a*).

By symmetry $R_A = R_D = 12$ kN

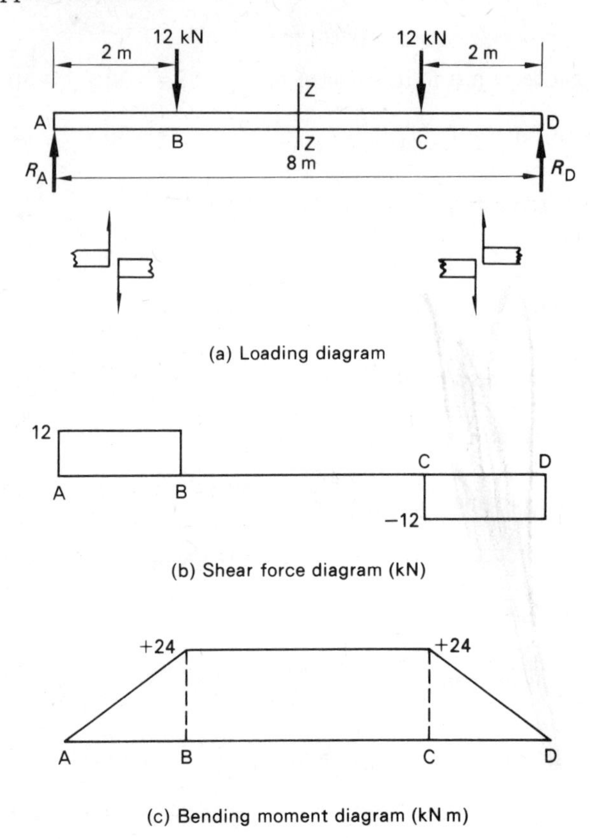

Fig. 2.16 Example 2.5

SHEAR FORCE:

From A to B: Shear force $= +12$ kN

From B to C: Shear force $= 0$

From C to D: Shear force $= -12$ kN

The shear force diagram is drawn as Fig. 2.16(*b*).

BENDING MOMENT:

At A: Bending moment $= 0$

At B: Bending moment $= +R_A \times 2 = +24$ kN m

At any section ZZ between B and C:

$$\text{Bending moment} = R_A \times z - 12\,(z - 2)$$

$$= +24 \text{ kN m}$$

Thus, Bending moment $= +24$ kN m from B to C.

At D, Bending moment = 0

The bending moment diagram is drawn as Fig. 2.16(*c*).

From these diagrams some important observations can be made.

1. Between B and C the beam is subjected to a constant bending moment and over this length the shear force is zero.
2. The area of the shear force diagram between A and B is equal to the change in bending moment between A and B.
3. The maximum bending moment occurs where the shear force is zero.

Example 2.6

A simply supported beam of span l carries a uniformly distributed load of intensity w. Draw the shear force and bending moment diagrams.

Solution

The loading diagram is shown in Fig. 2.17(*a*). By symmetry the support reactions are each equal to $wl/2$.

Consider a section ZZ at a distance z from the left-hand support.

$$\text{Shear force at ZZ} = \frac{wl}{2} - wz$$

This is the equation of a straight line and the shear force diagram is drawn as Fig. 2.17(*b*).

$$\text{Bending moment at ZZ} = \frac{wl}{2} \cdot z - w \cdot z \cdot \frac{z}{2}$$

i.e.,

$$M_z = \frac{w}{2}(lz - z^2) \qquad (1)$$

This is the equation of a parabola.

The maximum bending moment occurs where $\dfrac{dM_z}{dz} = 0$

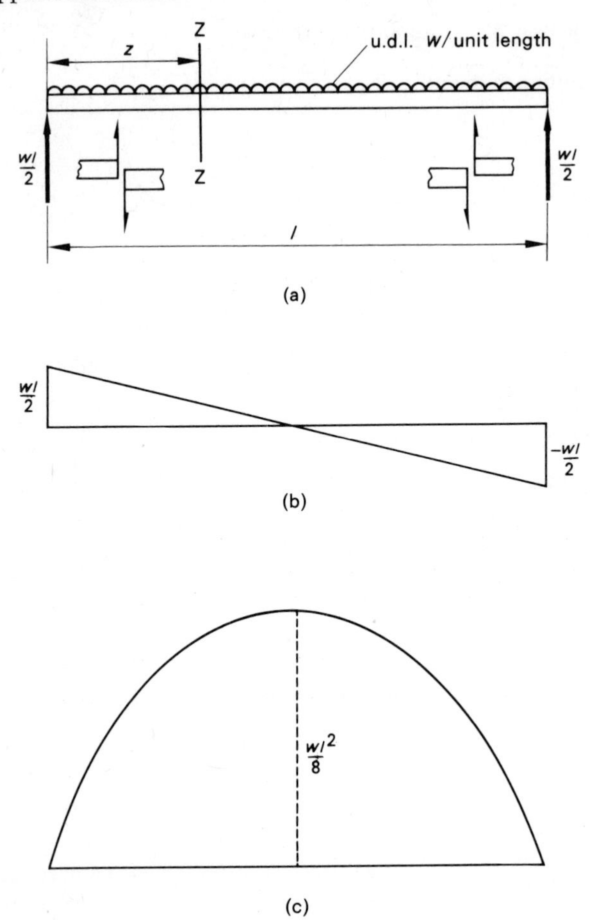

Fig. 2.17 (a) Loading diagram (b) Shear force diagram (c) Bending moment diagram–Example 2.6

Thus, differentiating eq. (1):

$$\frac{dM_z}{dz} = \frac{w}{2}(l - 2z)$$

$$= 0$$

when $\quad z = l/2$

The maximum bending moment therefore occurs at the centre span, a result to be expected because of the symmetry of loading.

Substituting $z = l/2$ into eq. (1) gives:

$$M_{max} = \frac{w}{2}\left(\frac{l^2}{2} - \frac{l^2}{4}\right)$$

$$= \frac{wl^2}{8}$$

The bending diagram is drawn as Fig. 2.17(*c*).

Note: The maximum bending moment occurs when the shear force is zero.

2.3 Contraflexure or inflexion

The examples considered so far have produced either hogging or sagging bending but not both in the same example. However, a beam loaded as in Fig. 2.18(*a*) will produce both sagging and hogging and the position B at which the sign of the bending moment changes is known as a position of contraflexure, or inflexion. Thus, a position of contraflexure is one of zero bending moment.

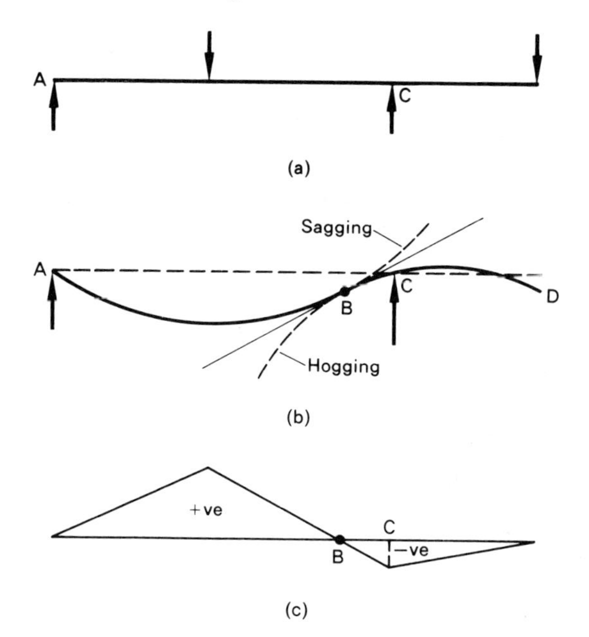

Fig. 2.18 (a) Loading diagram (b) Deflection ABCD (exaggerated) (c) Bending moment diagram–Beam contraflexure or inflexion

An exaggerated deflection diagram and a bending moment diagram are shown in Figs. 2.18(*b*) and (*c*) respectively.

Example 2.7

A beam AD has a length of 12 m and is simply supported at A and at C, which is 3 m from D. The beam carries a uniformly distributed load of 12 kN/m between the supports, a load of 45 kN at a position B, 2 m from A and a load of 24 kN at D. Draw the shear force and bending moment diagrams and determine the magnitude and position of the maximum bending moment and the position of contraflexure.

Solution

A loading diagram is given in Fig. 2.19(*a*).

REACTIONS:
Taking moments about A:

$$45 \times 2 + (12 \times 9) \times 4{\cdot}5 + 24 \times 12 - 9R_C = 0$$

$$R_C = 96 \text{ kN}$$

$$\therefore \quad R_A = 81 \text{ kN}$$

SHEAR FORCE:
Measuring z from A then:

For $0 \leqslant z \leqslant 2$, shear force $= (81 - 12z)$ [kN]

For $2 \leqslant z \leqslant 9$, shear force $= 81 - 12z - 45$

$$= (36 - 12z) \text{ [kN]} \qquad (1)$$

For $9 \leqslant z \leqslant 12$, shear force $= 81 - 12 \times 9 - 45 + 96$

$$= +24 \text{ [kN]}$$

(This result could obviously have been obtained by working from D.)

These relationships give straight lines and the shear force diagram is drawn as Fig. 2.19(*b*).

BENDING MOMENT:

For $0 \leqslant z \leqslant 2$, $M = 81z - 12\dfrac{z^2}{2}$

$$= 81z - 6z^2 \text{ [kN m]}$$

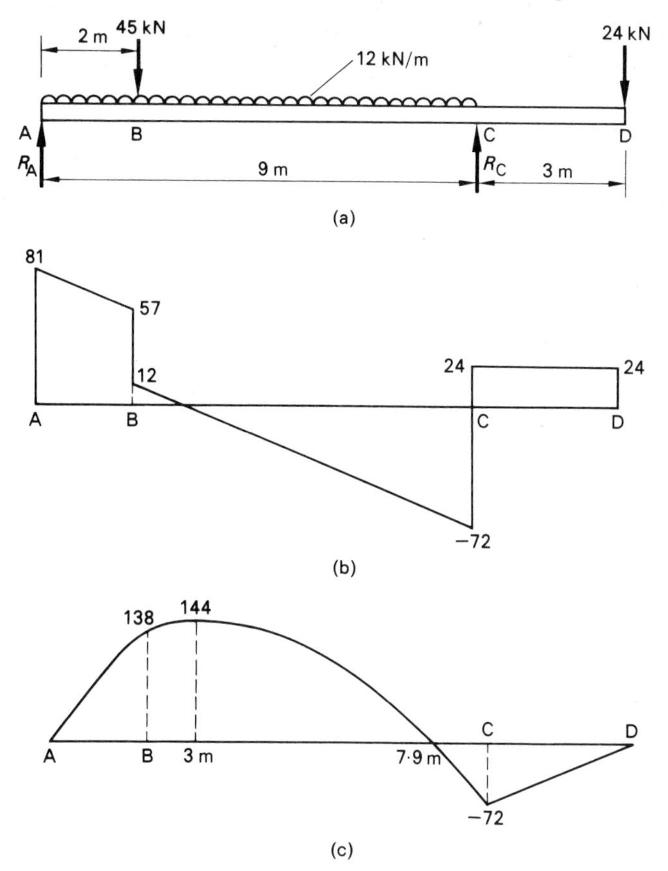

Fig. 2.19 (a) Loading diagram (b) Shear force diagram (c) Bending moment diagram–Example 2.7

Thus, at A, $z = 0$ and $M_A = 0$

at B, $z = 2$ $\quad M_B = 162 - 24 = 138$ kN m

For $2 \leqslant z \leqslant 9$,

$$M = 81\,z - \frac{12\,z^2}{2} - 45\,(z-2)$$

$$= 36\,z - 6\,z^2 + 90 \text{ kN m} \qquad (2)$$

At $\quad z = 2$, $\quad M_B = 72 - 24 + 90 = 138$ kN m $\quad$ (as above)

At $\quad z = 9$, $\quad M_C = 324 - 486 + 90 = -72$ kN m

Although the variation of bending moment between C and D can be obtained by working from the left hand side it is easier to work from D to C. Due to the single concentrated load at D the bending moment varies linearly from zero at D to –72 kN m at C.

The position of maximum bending moment occurs between B and C, where $dM/dz = 0$, or where the shear force is zero.
From eq. (2):

$$\frac{dM}{dz} = 36 - 12z \qquad \text{[cf eq. (1)]}$$

$$= 0$$

when $z = 3$ m

Substituting $z = 3$ into eq. (2) gives:

$$M_{max} = 36 \times 3 - 6 \times 3^2 + 90$$

$$= 144 \text{ kN m}$$

The maximum bending moment is 144 kN m and occurs at a position 3 m from A.

The position of contraflexure will occur where the bending moment is zero and is between B and C, since there is a change in the sign of the moment between these positions.

Thus, equating the moment eq. (2) for this section of the beam to zero gives:

$$0 = 36z - 6z^2 + 90$$

or $$0 = z^2 - 6z - 15$$

$$z = \frac{+6 \pm \sqrt{((-6)^2 - 4(-15))}}{2}$$

$$= \frac{6 \pm 9{\cdot}8}{2}$$

$z = 7{\cdot}9$ is the only positive solution

The position of contraflexure occurs 7·9 m from A.

The bending moment diagram is drawn in Fig. 2.19(*c*), sufficient points being obtained by evaluating M at intermediate values of z, as indicated.

Example 2.8

A uniform beam ABCD is 9 m long and is simply supported at B and C, where AB = 2 m and BC = 6 m. The beam carries a uniformly distributed load of 15 kN/m over its complete span.

Calculate:

(*a*) the position and magnitude of the maximum bending moment;
(*b*) the positions of the points of contraflexure.

Sketch the shear force and bending moment diagrams indicating the principal values.

Solution

A loading diagram is drawn in Fig. 2.20(*a*).

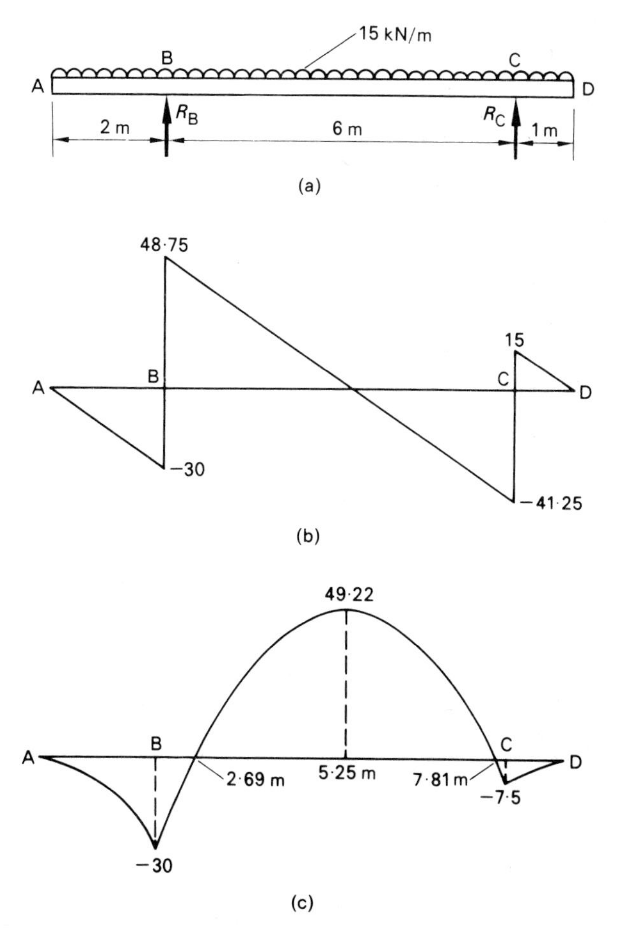

Fig. 2.20 (a) Loading diagram (b) Shear force diagram (c) Bending moment diagram–Example 2.8

REACTIONS:

Taking moments about B:

$$(15 \times 9)\left(\frac{9}{2} - 2\right) - 6R_C = 0$$

$$R_C = 56{\cdot}25 \text{ kN}$$

$$\therefore \quad R_B = (15 \times 9) - 56{\cdot}25$$

$$= 78{\cdot}75 \text{ kN}$$

SHEAR FORCE:

Measuring z from A then:

For $0 \leqslant z \leqslant 2$ Shear force $= -15z$

For $2 \leqslant z \leqslant 8$ Shear force $= -15z + R_B$

$$= -15z + 78{\cdot}75 \qquad (1)$$

For $8 \leqslant z \leqslant 9$ Shear force $= -15z + R_B + R_C$

$$= -15z + 135$$

From these relationships the shear force diagram, Fig. 2.20(*b*), is drawn.

BENDING MOMENT:

From A to B $\quad 0 \leqslant z \leqslant 2$

$$M = -15\frac{z^2}{2}$$

At A, $\quad z = 0$ and $M_A = 0$

At B, $\quad z = 2$ and $M_B = -30$ kN m

From B to C $\quad 2 \leqslant z \leqslant 8$

$$M = -15\frac{z^2}{2} + 78{\cdot}75(z - 2) \qquad (2)$$

At B, $\quad z = 2$ and $M_B = -30$ kN m (as above)

At C, $\quad z = 8$ and $M_C = \dfrac{-15 \times 8^2}{2} + 78{\cdot}75 \times 6$

$$= -7{\cdot}5 \text{ kN m}$$

The maximum bending moment will occur between B and C, where

$$\frac{dM}{dz} = 0.$$

From eq. (2)

$$\frac{dM}{dz} = -15\,z + 78{\cdot}75 \qquad \text{[cf eq. (1)]}$$

$$= 0$$

when $z = \dfrac{78{\cdot}75}{15} = 5{\cdot}25$ m

Substituting $z = 5{\cdot}25$ m into eq. (2) gives

$$M_{max} = \frac{-15 \times 5{\cdot}25^2}{2} + 78{\cdot}75 \times 3{\cdot}25$$

$$= +\,49{\cdot}22 \text{ kN m}$$

The maximum bending moment is 49·22 kN m and occurs at a distance of 5·25 m from A.

The points of contraflexure occur where $M = 0$ and as the sign of M changes twice there are two points of contraflexure, both within the length BC.

Thus, putting $M = 0$ in eq. (2) gives

$$-7{\cdot}5\,z^2 + 78{\cdot}75\,z - 157{\cdot}5 = 0$$

$$z^2 - 10{\cdot}5\,z + 21 = 0$$

$$z = \frac{10{\cdot}5 \pm \surd(10{\cdot}5^2 - 4 \times 21)}{2}$$

$$= 2{\cdot}69 \text{ m or } 7{\cdot}81 \text{ m}$$

The positions of contraflexure are 2·69 m and 7·81 m from A. A sketch of the bending moment diagram is given in Fig. 2.20(*c*).

Example 2.9

A horizontal steel girder having a span of 4 m is built-in at one end and freely supported at the other. The girder carries a uniformly distributed load of 30 kN/m over the whole span and a concentrated load of 20 kN at a position 2·75 m from the built-in end. The supporting conditions are such that the reaction at the built-in end is 65 kN. Determine:

(*a*) the bending moment at the built-in end;
(*b*) the maximum bending moment and the position at which this occurs.

Solution

A diagram is given in Fig. 2.21.

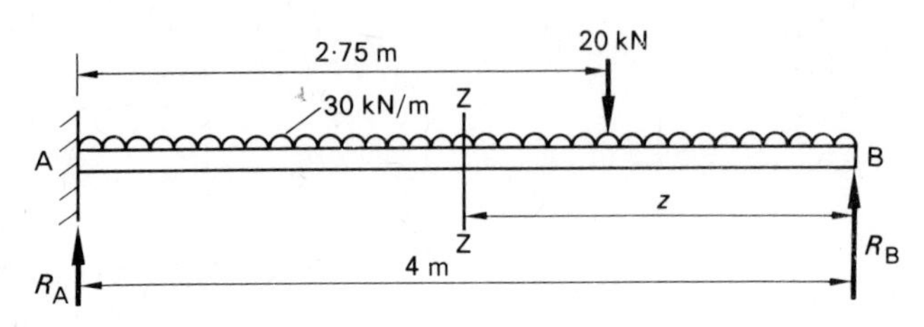

Fig. 2.21 Example 2.9

(*a*) For vertical equilibrium:

$$R_A + R_B = 30 \times 4 + 20$$

$$\therefore \quad R_B = 75 \text{ kN}$$

The moment at the built-in end is obtained by taking moments about A; i.e.,

$$M_A = 75 \times 4 - 20 \times 2{\cdot}75 - (30 \times 4) \times 2$$
$$= 5 \text{ kN m}$$

The moment at the built-in end is 5 kN m.

(*b*) Assume the maximum bending moment to occur between the 20 kN load and the built-in end and consider a section ZZ at a distance z from the right-hand support B, where $1{\cdot}25 \leqslant z \leqslant 4$.

Then $M_z = 75\,z - 20\,(z - 1{\cdot}25) - 30z\,.\,\dfrac{z}{2}$

The maximum bending moment will occur where $dM_z/dz = 0$; i.e.,

$$\frac{dM_z}{dz} = 75 - 20 - 30z$$
$$= 0$$

when $\quad z = \dfrac{55}{30} = 1{\cdot}833 \text{ m}$

$$\therefore \quad M_{max} = 75 \times 1{\cdot}833 - 20 \times 0{\cdot}583 - \frac{30 \times 1{\cdot}833^2}{2}$$
$$= 75{\cdot}42 \text{ kN m}$$

The maximum bending moment is 75·42 kN m and occurs at a distance of 1·833 m from the freely-supported end.

Note: The value of z where

$$\frac{dM_z}{dz} = 0$$

must lie within the limits for z applicable to the equation for M_z. If this is not the case then the wrong section of the beam has been selected for the maximum bending moment and the solution is not valid. This may be illustrated as follows.

Suppose the assumption had been made that the maximum bending moment occurred between B and the 20 kN load.

Then $M_z = 75z - 30z \cdot \frac{z}{2}$ for $0 \leqslant z \leqslant 1{\cdot}25$

$$\frac{dM_z}{dz} = 75 - 30z$$

$$= 0$$

when $z = 2{\cdot}5$ m

The value of 2·5 m is outside the limits for which M_z is applicable, so indicating that the maximum moment occurs in another section of the beam.

It is worth noting that the section of the beam where the maximum moment occurs can frequently be predicted by inspection.

Problems

Draw the shear force and bending moment diagrams for the loaded beams shown in problems 1–6 (diagrams 2.22–2.27). Also determine the maximum bending moment and position of contraflexure if applicable.

1.

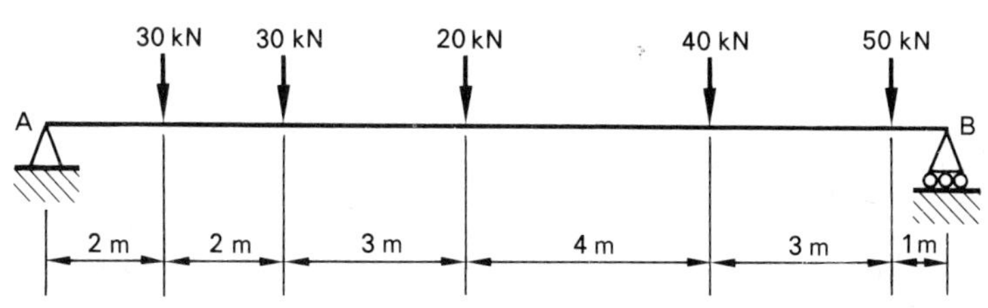

Fig. 2.22

2.

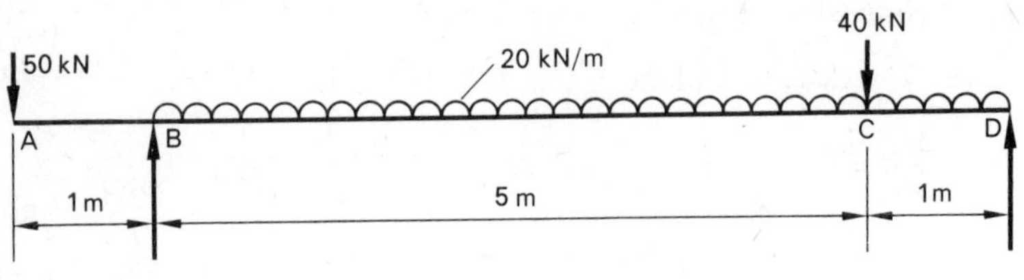

Fig. 2.23

3.

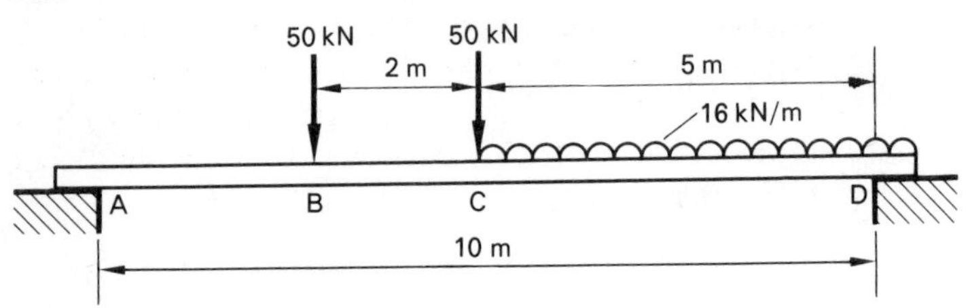

Fig. 2.24

4.

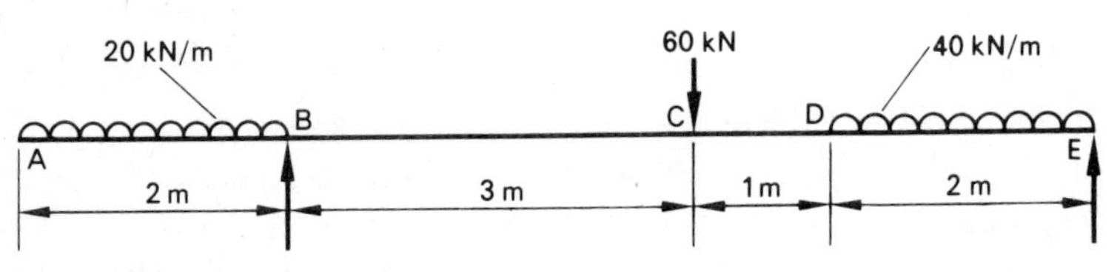

Fig. 2.25

5.

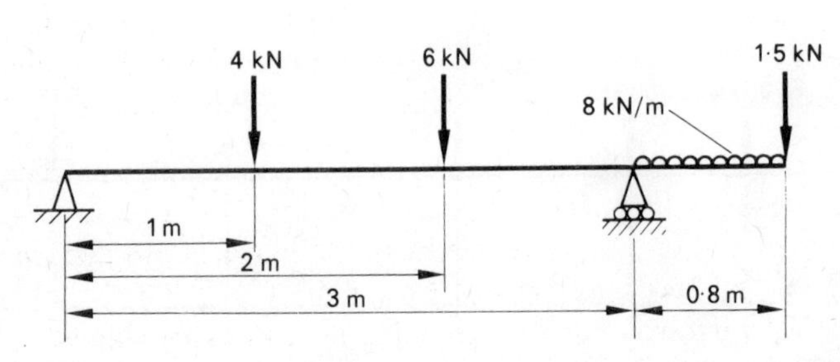

Fig. 2.26

6.

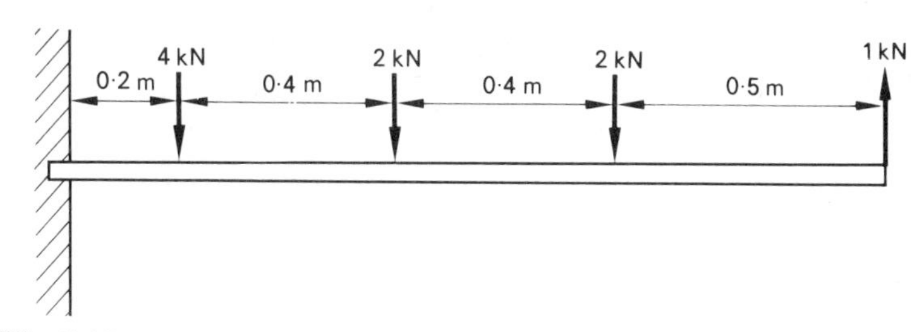

Fig. 2.27

7. A horizontal beam is 14 m long, has supports 9 m apart, the overhangs being 2 m and 3 m at the left- and right-hand ends respectively. The beam supports a vertical load of 45 kN at its left-hand end and a load of 70 kN at its right-hand end. The beam also carries a uniformly distributed load of 15 kN/m over the whole span. Draw the shear force and bending moment diagrams, indicating the principal values.

8. A horizontal beam with overhanging ends is shown in Fig. 2.28.

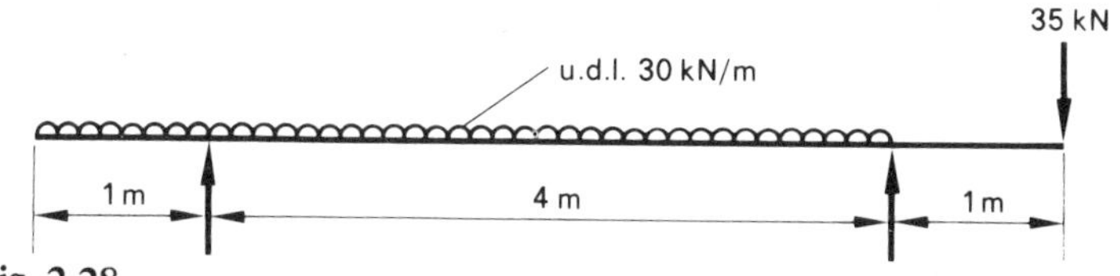

Fig. 2.28

Draw the shear force diagram to scale for the loaded beam. Calculate the maximum bending moment on the beam and the value of the bending moment at five other sections between the two ends.

Hence sketch the bending moment diagram for the loaded beam, approximately to scale.

(N.C.T.E.C.)

9. A horizontal beam is simply supported at its ends which are 6 m apart, and carries a uniformly distributed load of 6 kN/m over the whole span together with a concentrated load of 2 kN at the middle of the span. Draw the shear force and bending moment diagrams.

10. A beam ABCDEF is simply supported at A and at E, 8 m from A. AB = 3 m; BC = 3 m; CD = 1 m; DE = 1 m; EF = 2 m. There is a uniform load of 25 kN/m over the length AB and concentrated loads of 40, 20, and 30 kN at C, D, and F respectively.

Calculate the shear force and bending moment at the lettered positions and draw the shear force and bending moment diagrams.

Determine the position of zero shear force and the bending moment at this position.

11. A beam RS, of length 6 m, rests on two supports, one at each end of the beam. A load of 8 kN is applied at a position 2 m from R and another load of 10 kN is applied 5 m from R. The reactions are known to be 14·5 kN and 21·5 kN at R and S respectively. Calculate the position and magnitude of the load which maintains the beam in equilibrium. Draw the shear force and bending moment diagrams for the beam and determine the position and magnitude of the maximum bending moment.

12. Sketch, in good proportion, the shear force and bending moment diagrams for the beam, loaded as shown in Fig. 2.29. Determine the position and magnitude of the maximum bending moment and include these and other values as salient points on the diagrams. (U.E.I.)

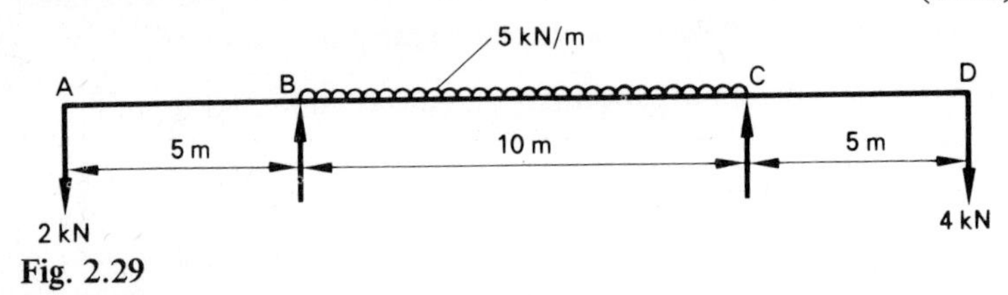

Fig. 2.29

13. A beam ABCD, 9 m long, is simply supported at A and C, 7 m apart. The beam carries a uniformly distributed load of 20 kN/m only between the supports, a vertical concentrated load of 29 kN at B, which is 1 m from A, and a vertical concentrated load of 24 kN at D. Draw the shear force and bending moment diagrams and indicate the principal values and the position of the point of contraflexure. Also determine the position and value of the maximum bending moment.

14. ABCD is a beam of length 6 m and is simply supported at B and C, where AB = 1 m, CD = 0·5 m, and BC = 4·5 m. Concentrated loads of 30 kN and 60 kN act at A and D respectively. The span BC carries a uniformly distributed load of 40 kN/m. Draw the shear force and bending moment diagrams and determine the positions of the points of contraflexure.

15. A simply supported beam ABCDE is loaded as shown in Fig. 2.30. Sketch the shear force and bending moment diagrams indicating the principal values. Calculate the magnitude of the maximum bending moment and state where it occurs. (U.L.C.I.)

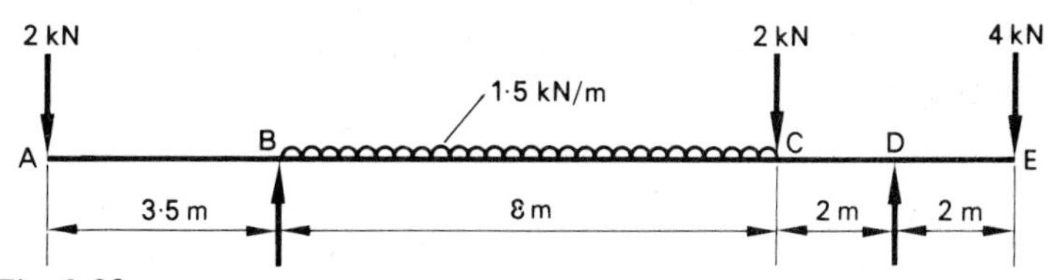

Fig. 2.30

16. A beam, 8 m long, carries a uniformly distributed load of 15 kN/m over its whole length. The beam rests on two supports, each 1 m from an end. Draw the shear force and bending moment diagrams. Determine the maximum bending moment and positions of the points of contraflexure. Determine also the position of the supports which will reduce the magnitude of the maximum bending moment on the beam to a minimum.

17. A simply supported horizontal beam is loaded as shown in Fig. 2.31.

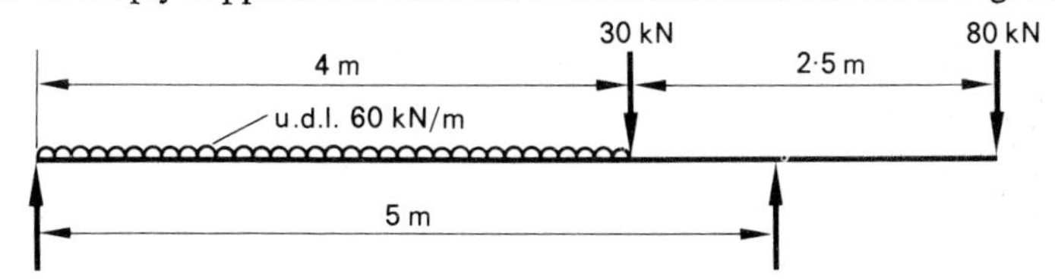

Fig. 2.31

Draw to scale the shear force diagram for the beam. Calculate the value of the maximum bending moment on the beam.

Also sketch to scale the bending moment diagram for the beam showing the necessary calculations of the bending moments.

(N.C.T.E.C.)

18. Figure 2.32 shows a simply supported beam with equal overhangs. The beam carries a concentrated load of 80 kN at the centre and three uniformly distributed loads of w kN/m in the positions shown. If the maximum bending moment at any position on the beam is not to exceed 125 kN m, find the value of w and draw the shear force and bending moment diagrams, indicating principal values and positions of contraflexure.

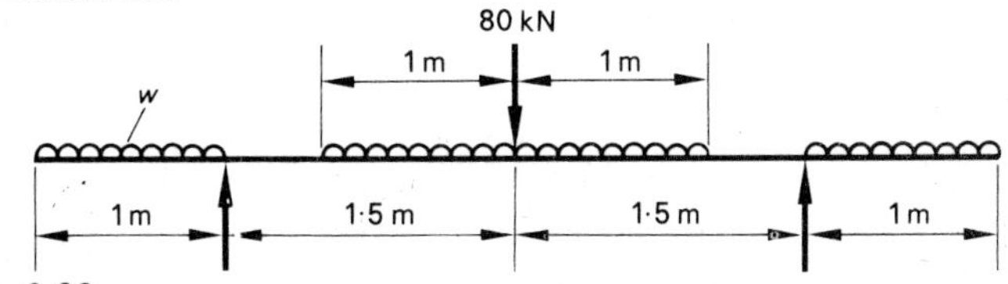

Fig. 2.32

3

Stress and strain

3.1 Transmission of forces

So far we have dealt with the forces arising in the members of a pin-jointed framework from the application of forces to the structure. The distorting effect caused by the application of a force to a body must now be considered.

The effect produced by a force will depend upon the manner in which the force is applied to the structure and also upon the material from which the structure is made. For design purposes it is important to know how a load is transmitted through a structure so that a high concentration of loading, which can lead to cracking or excessive distortion, can be avoided. The path which a load takes through a structure is obtained by analytical methods and frequently confirmed by experimental work.

3.2 Direct stress

When an external force is applied to a body, as shown in Fig. 3.1, distortion occurs and the body is said to be strained. The body will only

Fig. 3.1 Direct stress

remain in equilibrium under the action of this force if the internal forces can resist the enforced deformation. A knowledge of the internal forces which the material of a member can create is therefore necessary.

Consider a bar to be subjected to a force F such that a state of tension is produced, as shown in Fig. 3.2(a). If we imagine the bar to be divided into two parts across a section XX, at right angles to its axis, the equilibrium of the left-hand section can now be considered. Obviously, for equilibrium, the external force F acting to the left must be balanced by an equal internal force acting to the right [Fig. 3.2(*b*)]. This internal force is the result of inter-molecular forces created within the material, i.e., at XX the molecules on the right-hand side pull on the molecules of the left-hand side with a total force that is equal to F, so preventing the left-hand portion from moving in the direction of the externally applied force. A similar argument exists for the action of the left-hand side on the right-hand side, and for all other sections throughout the length of the bar.

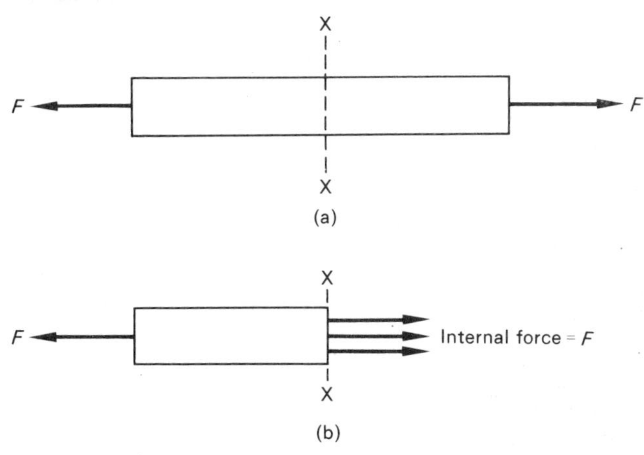

Fig. 3.2 Internal force in a member

The stress in a material is a measure of the force transmitted per unit cross-sectional area. Thus, for any cross-section such as XX having an area A,

$$\text{Direct stress} = \frac{\text{applied force}}{\text{cross-sectional area}}$$

$$\sigma = \frac{F}{A}$$

The units of stress are therefore N/m^2, kN/m^2, etc.

This type of stress, which may be tensile or compressive, is known as direct stress because the area transmitting the stress is normal to the axis of the loading. The above formula gives the average stress within the material, but providing the section considered is some distance from the point of application of the load it may be assumed that the stress is uniformly distributed over the section.

3.3 Direct strain

Whenever a load is applied to a member some deformation will occur resulting in a change in its dimensions. This deformation may not be perceptible to the human eye but the use of accurate measuring instruments will show that it is always present.

Consider the circular bar shown in Fig. 3.3.

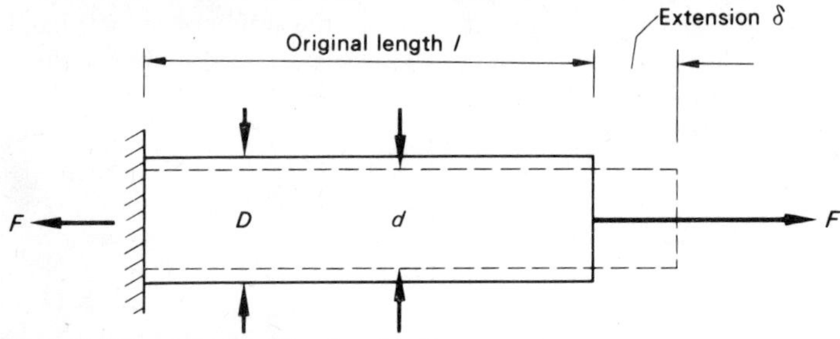

Fig. 3.3 Direct strain of a circular bar

Under the action of a force F an extension will occur in the direction of the loading which will be accompanied by a decrease in diameter – the reverse occurring in the case of compression. Then,

$$\text{Direct strain} = \frac{\text{change in length}}{\text{original length}}$$

$$\epsilon = \frac{\delta}{l}$$

As strain is a ratio of lengths it is dimensionless. Also,

$$\text{Lateral strain} = \frac{\text{change in diameter}}{\text{original diameter}} = \left(\frac{d - D}{D}\right)$$

The lateral strain takes place in the opposite sense to the direct strain, i.e. a tensile direct strain results in a compressive lateral strain.

3.4 Elasticity and Hooke's Law

It has been stated that when a material is subjected to a stress a corresponding strain must occur. If the strain is totally removed upon the removal of the load and the material returns to its original dimensions, the material is said to be perfectly elastic. This property of the material is known as its elasticity and, in the case when the strain is totally removed, it is apparent that the limit of elasticity (the elastic limit) has not been exceeded.

Our elementary theories on the behaviour of materials under different loading conditions are based upon the assumption that the material remains perfectly elastic and obeys Hooke's law, which states that:

Providing the limit of proportionality of a material is not exceeded the stress is directly proportional to the strain produced.

The validity of Hooke's law for most engineering materials can be demonstrated by experiment.

Thus, $\text{stress} \propto \text{strain}$

$$\frac{\text{stress}}{\text{strain}} = \text{constant (for any given material)}$$

For direct stresses this gives

$$\frac{\text{direct stress}}{\text{direct strain}} = \text{modulus of elasticity, } E$$

and for shear stresses

$$\frac{\text{shear stress}}{\text{shear strain}} = \text{modulus of rigidity, } G$$

If the strain is not totally removed when the load is removed the material is only partially elastic. In this case a permanent strain will remain and the material is said to have gained a permanent set.

3.5 The tensile test

A tensile test on a material is carried out on a machine which is designed to apply a true axial load. The test specimen is of uniform cross-section and should conform to dimensions recommended by the British Standards Institution (BS18). The strain is determined by making accurate measurements over a known length of the material, the gauge length, which is some distance from the ends of the test specimen, as illustrated in Fig. 3.4.

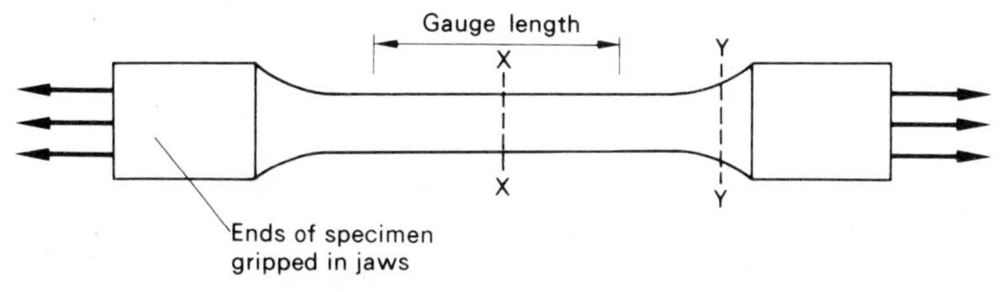

Fig. 3.4 A tensile test specimen

If recordings of load and extension are taken during a tensile test to destruction a load- extension or stress-strain graph can be plotted. A typical stress-strain diagram for irons and low carbon steels is shown in Fig. 3.5.

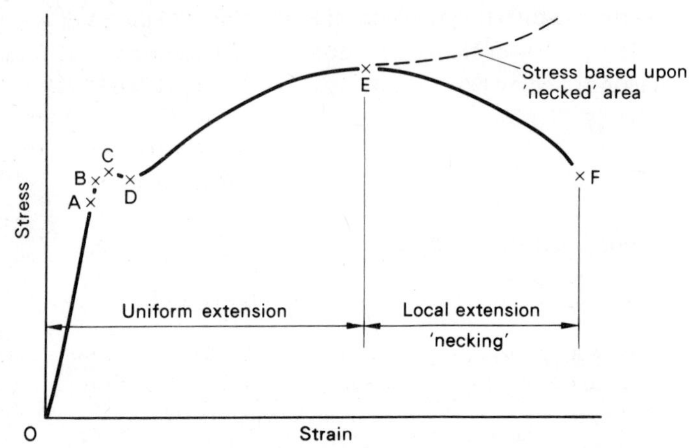

Fig. 3.5 Typical stress-strain graph for irons and low carbon steels

From O to A the strain is proportional to the stress and point A is therefore known as the limit of proportionality. Point B marks the elastic limit of the material and providing that this is not exceeded the material will return to its original length upon removal of the load, i.e., the strain is elastic up to this point. It is worth noting at this stage that for an iron or steel the limit of proportionality and the elastic limit are virtually coincident but for other materials, in particular non-ferrous materials, this is not the case.

Further loading beyond B will cause yielding of the material to occur at C. At this stage a small decrease in load occurs while the material undergoes a considerable strain to point D. The stress at C is known as the yield stress. Recovery of the material then occurs at D and further loading is required to strain the material until the maximum load point is reached at E. The stress at E is known as the tensile strength of the material. Up to the maximum load position elongation of the specimen occurs with hardly any visible reduction in diameter, but from E a local reduction in diameter, known as 'necking', occurs and this becomes more clearly defined as the load falls off to the fracture point F.

Although the stress-strain diagram is based upon the original cross-sectional area of the material, it is worth noting that whilst the load falls from E to F the stress over this region, if based upon the actual necked cross-sectional area of the specimen, would rise considerably above E as shown on Fig. 3.5.

As the determination of the limit of proportionality and elastic limit requires the use of very accurate instruments it is frequently convenient to take the more easily obtainable yield point as the limit of the elastic range when determining the maximum working stress of a material. However, in the case of non-ferrous materials and high carbon steels no definite yield point is apparent and another criteria must be used to assess the maximum working stress of these materials (see Proof stress, section 3.7).

3.6 Results from a tensile test

The following information may be obtained from a tensile test:

(*i*) The modulus of elasticity of a material. This is obtained from the initial straight part of the graph up to the limit of proportionality. Thus,

$$\text{modulus of elasticity } E = \frac{\text{stress}}{\text{strain}}$$

(*ii*) The yield stress, or proof stress (section 3.7)

(*iii*) $\text{Percentage elongation} = \dfrac{\text{increase in gauge length}}{\text{original gauge length}} \times 100$

(*iv*) $\text{Percentage reduction in area} = \dfrac{\text{original area} - \text{area at fracture}}{\text{original area}} \times 100$

(*v*) $\text{Tensile strength} = \dfrac{\text{maximum load}}{\text{original cross-sectional area}}$

The percentage elongation and percentage reduction in area give an indication of the ductility of the material, which is its ability to withstand strain without fracture occurring.

The total extension of the test specimen comprises the uniform extension which occurs up to the maximum load plus the local extension which occurs on either side of the fracture, i.e., over the necked portion. As the local extension is dependent upon the cross-sectional area and not the gauge length it follows that the total extension is more dependent on the local extension in the case of a short gauge length than for a longer gauge length. Hence the importance of using test specimens that conform to the recommendations of the British Standards Institution (ref. BS18). Otherwise the gauge length must be quoted.

The shape of a test specimen at fracture can often give useful information regarding the material. Brittle materials, such as cast iron, show little extension and produce a rather flat, ragged surface at the fracture. Ductile materials, on the other hand, produce a cup and cone type fracture. Examples of these fractures are shown in Fig. 3.6.

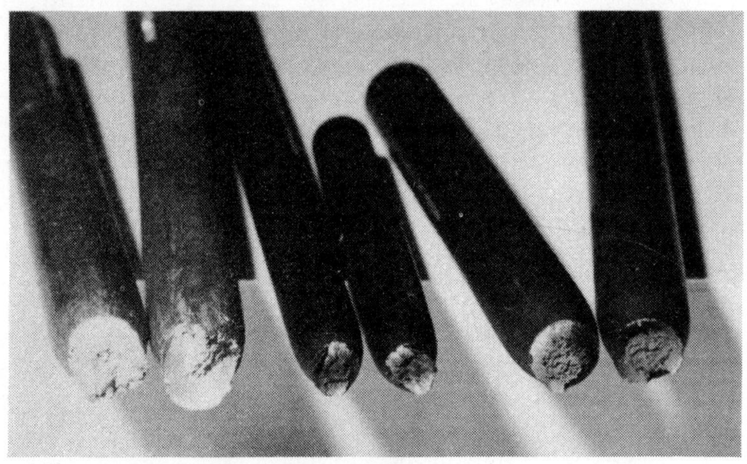

Fig. 3.6 Fractures of aluminium, copper and mild steel

Example 3.1

A tensile test on two specimens of the same mild steel, one cold rolled and the other annealed, gave the following results:

	Cold rolled	*Annealed*
Yield load	79 kN	44·7 kN
Maximum load	95·6 kN	64·2 kN
Final length between gauge points	88·9 mm	93·2 mm
Diameter at fracture	9·11 mm	7·09 mm

Both test specimens had an original diameter of 13·82 mm and a gauge length of 69 mm.

Determine the yield stress, tensile strength, percentage elongation and percentage reduction in area for each specimen and tabulate the results to show the effect of the heat treatment upon the properties of the material.

Solution

	Cold Rolled	*Annealed*
Yield stress	$= \dfrac{\text{yield load}}{\text{cross-sectional area}}$	
	$= \dfrac{79 \times 10^3}{(\pi/4) \times 13{\cdot}82^2 \times 10^{-6}} \dfrac{[\text{N}]}{[\text{m}^2]}$	$= \dfrac{44{\cdot}7 \times 10^3}{(\pi/4) \times 13{\cdot}82^2 \times 10^{-6}}$
	$= 526{\cdot}7 \times 10^6 \text{ N/m}^2$	
	$= 526{\cdot}7 \text{ MN/m}^2$	$= 298 \text{ MN/m}^2$
Tensile strength	$= \dfrac{\text{maximum load}}{\text{cross-sectional area}}$	
	$= \dfrac{95{\cdot}6 \times 10^3}{(\pi/4) \times 13{\cdot}82^2 \times 10^{-6}} \dfrac{[\text{N}]}{[\text{m}^2]}$	$= \dfrac{64{\cdot}2 \times 10^3}{(\pi/4) \times 13{\cdot}82^2 \times 10^{-6}}$
	$= 637{\cdot}3 \text{ MN/m}^2$	$= 428 \text{ MN/m}^2$
Percentage elongation	$= \dfrac{\text{increase in gauge length}}{\text{original gauge length}} \times 100$	
	$= \left(\dfrac{88{\cdot}9 - 69}{69}\right) \times 100$	$= \left(\dfrac{93{\cdot}2 - 69}{69}\right) \times 100$
	$= 28{\cdot}85\%$	$= 35{\cdot}1\%$
Percentage reduction in area	$= \dfrac{\text{original area} - \text{final area}}{\text{original area}} \times 100$	
	$= \left\{\dfrac{\pi/4\,(13{\cdot}82^2 - 9{\cdot}11^2)}{(\pi/4) \times 13{\cdot}82^2}\right\} \times 100$	$= \left\{\dfrac{(\pi/4)\,(13{\cdot}82^2 - 7{\cdot}09^2)}{(\pi/4) \times 13{\cdot}82^2}\right\} \times 100$
	$= 58{\cdot}46\%$	$= 74{\cdot}84\%$

The effect of the heat treatment upon the material is evident from the above results.

3.7 Proof stress

High carbon steels, cast iron, and most of the non-ferrous alloys do not exhibit a well defined yield point as is the case with mild steel. The form of the stress-strain graph for such materials is shown in Fig. 3.7.

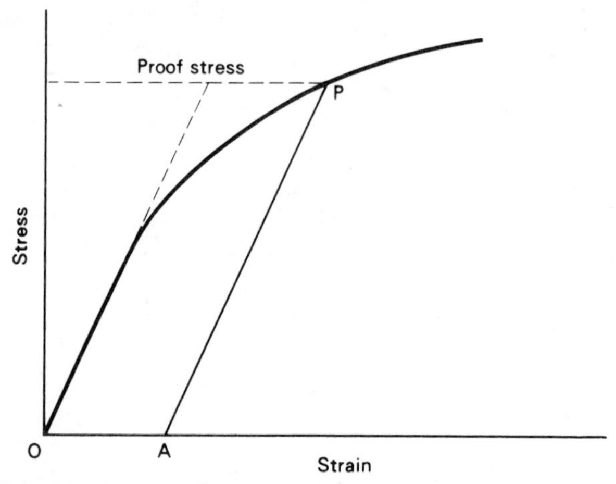

Fig. 3.7 Determination of proof stress

In the case of these materials it is usual to specify a limiting stress, known as the proof stress, corresponding to a specified non-proportional extension. This non-proportional extension is a specified percentage of the original gauge length, usually 0·05%, 0·10%, 0·20%, or 0·50% and when quoting a proof stress the non-proportional extension should be specified. The proof stress is obtained by drawing AP parallel to the initial slope of the stress-strain graph, the distance OA being the strain corresponding to the required non-proportional extension. The stress at P, based upon the original cross-sectional area of the test specimen, is the required proof stress.

Example 3.2

A tensile test on a non-ferrous alloy test specimen, having a diameter of 15·96 mm and a gauge length of 80 mm produced the following results:

Load (kN)	10	20	30	40	50
Extension (mm)	0·0312	0·064	0·0945	0·128	0·1624
Load (kN)	60	70	80	90	
Extension (mm)	0·1942	0·2304	0·276	0·3183	
Load (kN)	96	102	108		
Extension (mm)	0·3545	0·4057	0·464		

Fracture of the test specimen occurred at a maximum load of 119·2 kN.

Draw a stress-strain graph from these results and determine:

(*a*) the modulus of elasticity;
(*b*) the 0·05% and 0·1% proof stress;
(*c*) the tensile strength of the material.

Solution

$$\text{Cross-sectional area} = (\pi/4) \times 15{\cdot}96^2 = 200 \text{ mm}^2$$

Then, $\quad$ direct stress, $\sigma = \dfrac{\text{load}}{\text{area}}$

Thus, for a load P [kN] the stress is given by

$$\sigma = \frac{P \times 10^3}{200 \times 10^{-6}} \frac{[\text{N}]}{[\text{m}^2]}$$
$$= 5P \times 10^6 \text{ N/m}^2$$
$$= 5P \text{ MN/m}^2$$

Also, $\quad$ direct strain, $\epsilon = \dfrac{\text{change in length}}{\text{original length}}$

and as the extension and original gauge length are both given in mm units it follows for any extension δ [mm] that

$$\epsilon = \frac{\delta}{80}$$

Applying these relationships to the test results gives the following values.

Stress σ (MN/m^2)	50	100	150	200	250	300	350
Strain ($\times 10^{-3}$)	0·39	0·8	1·181	1·60	2·03	2·427	2·88
Stress σ (MN/m^2)	400	450	480	510	540		
Strain (10^{-3})	3·45	3·979	4·431	5·071	5·8		

A stress-strain graph is plotted as Fig. 3.8.

(*a*) Modulus of elasticity $= \dfrac{\text{stress}}{\text{strain}}$

$$= \text{gradient of stress-strain graph}$$
$$= \frac{247{\cdot}5 \times 10^6 \text{ [N/m}^2\text{]}}{2 \times 10^{-3}}$$
$$= 123{\cdot}75 \text{ GN/m}^2$$

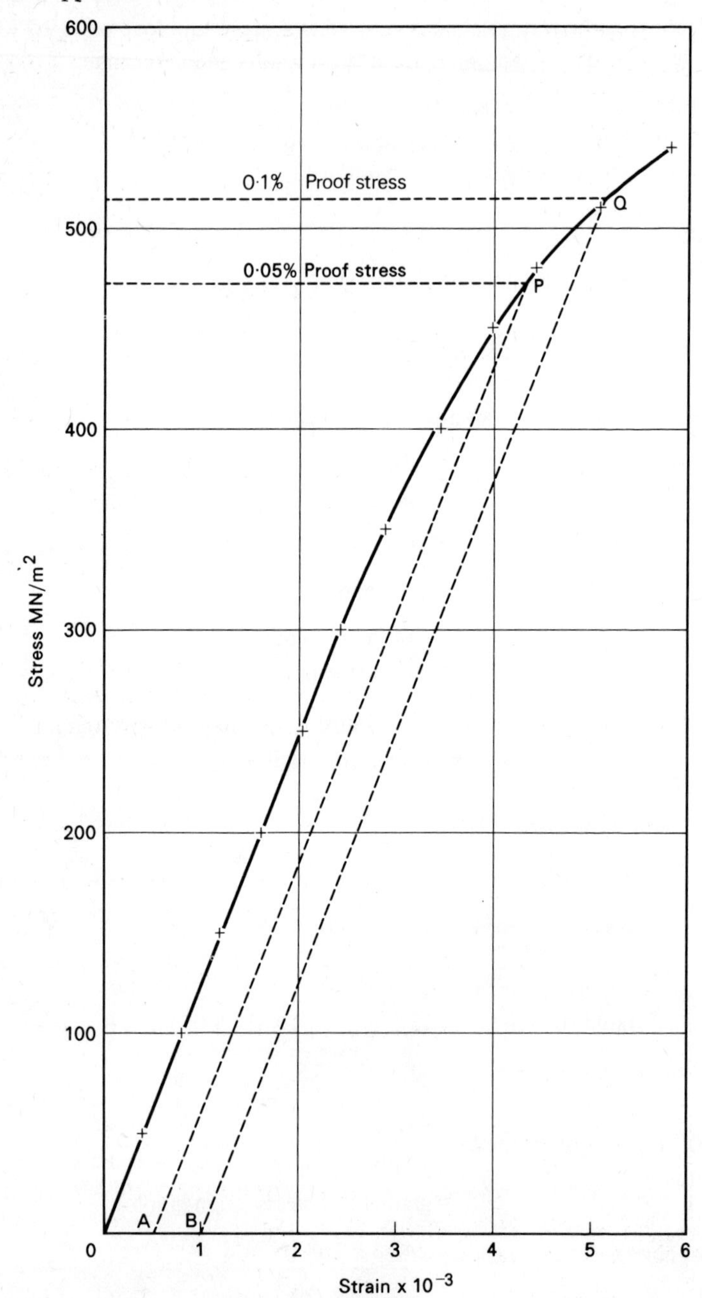

Fig. 3.8 Stress-strain graph—Example 3.2

(*b*) To obtain the 0·05% proof stress, A is marked on the strain axis at a strain of 0·0005 and a line AP drawn parallel to the initial linear part of the diagram.

This gives a 0·05% proof stress of 472·5 MN/m^2.

Similarly, the 0·1% proof stress is found from line BQ to be 515 MN/m^2.

(*c*) $\text{Tensile strength} = \dfrac{\text{maximum load}}{\text{original cross-sectional area}}$

$$= \frac{119{\cdot}2 \times 10^3}{200 \times 10^{-6}} \; \frac{[\text{N}]}{[\text{m}^2]}$$

$$= 596 \text{ MN/m}^2$$

The modulus of elasticity is 123·75 GN/m^2, the 0·05% proof stress 472·5 MN/m^2, the 0·1% proof stress 515 MN/m^2, and the tensile strength 596 MN/m^2.

3.8 Compound bars

A compound bar is one comprising two or more parallel elements, of different materials, which are rigidly fixed together at their ends. The compound bar may be loaded in tension or compression but if the bar is not symmetrical then bending stresses will be produced in addition to any direct stresses.

A section through a typical compound bar consisting of a circular bar (1) surrounded by a tube (2) is shown in Fig. 3.9.

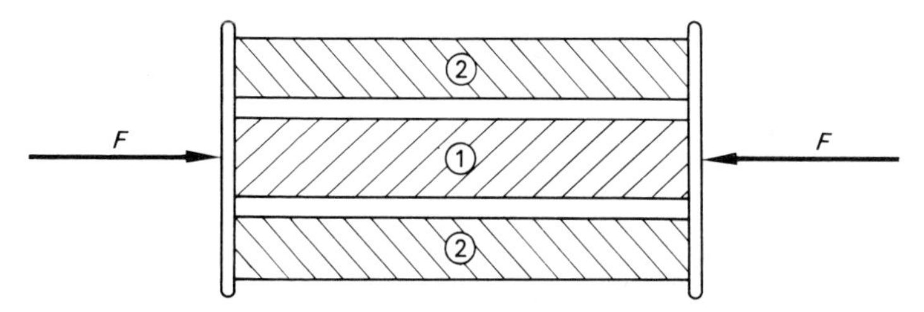

Fig. 3.9 A compound bar

If such a bar is loaded in compression by a force F it is evident that each material must be compressed by the same amount and hence the two materials have the same strain, i.e.,

$$\epsilon_1 = \epsilon_2$$

But, $\text{strain} = \dfrac{\text{stress}}{E}$

$$\therefore \qquad \frac{\sigma_1}{E_1} = \frac{\sigma_2}{E_2} \qquad (3.1)$$

where E_1 and E_2 are the elastic moduli of materials 1 and 2 respectively.

Also, the sum of the loads carried by the individual elements must be equal to the applied load F, i.e.,

$$F = F_1 + F_2$$

F_1 and F_2 being the loads in the individual elements.

Now, as force = stress x area

Then, $F = \sigma_1 A_1 + \sigma_2 A_2$

where A_1 and A_2 are the areas of 1 and 2 respectively.
Substituting for σ_2 from eq. (3.1) gives

$$\therefore \qquad F = \sigma_1 A_1 + \frac{\sigma_1 E_2 A_2}{E_1}$$

or

$$\sigma_1 = \frac{F E_1}{(E_1 A_1 + E_2 A_2)} \qquad (3.2)$$

Similarly,

$$\sigma_2 = \frac{F E_2}{(E_1 A_1 + E_2 A_2)} \qquad (3.3)$$

Note that the ratio of the stresses in the two materials is constant, being equal to E_1/E_2, irrespective of the individual areas of the two elements.

Example 3.3

A composite bar 0·6 m long comprises a steel bar 0·2 m long and 40 mm diameter which is fixed at one end to a copper bar having a length of 0·4 m. Determine the necessary diameter of the copper bar in order that the extension of each material shall be the same when the composite bar is subjected to an axial load.

What will be the stresses in the steel and copper when the bar is subjected to an axial tensile loading of 30 kN?
(For steel, $E = 210\ \text{GN/m}^2$; for copper, $E = 110\ \text{GN/m}^2$.)

Solution

A diagram is given in Fig. 3.10.

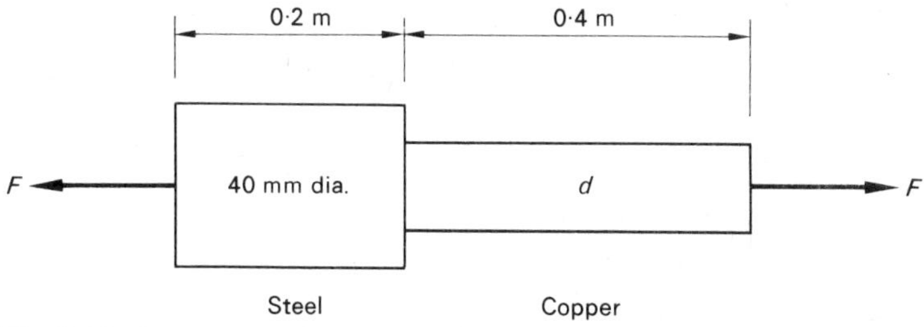

Fig. 3.10 A composite bar–Example 3.3

Let the diameter of the copper bar be d [mm]. The specified condition is that the copper and steel have equal extensions; i.e.,

$$\delta_c = \delta_s$$

But

$$\delta = \epsilon l = \frac{\sigma l}{E} = \frac{Fl}{AE}$$

Thus, the condition becomes

$$\frac{F_c l_c}{A_c E_c} = \frac{F_s l_s}{A_s E_s}$$

But under an applied axial load, the total load F is transmitted by both the copper and steel, i.e.,

$$F_c = F_s = F$$

$$\therefore \quad \frac{l_c}{A_c E_c} = \frac{l_s}{A_s E_s}$$

For the values given,

$$\frac{0{\cdot}4}{(\pi d^2/4) \times 110 \times 10^9} = \frac{0{\cdot}2}{(\pi \times 40^2/4) \times 210 \times 10^9}$$

$$d^2 = \frac{2 \times 210 \times 40^2}{110} \text{ mm}^2$$

$$d = 78{\cdot}16 \text{ mm}$$

The required diameter of the copper bar for equal extensions of the copper and steel is 78·16 mm.

Then, for a loading of 30 kN,

$$\text{Stress in steel, } \sigma_s = \frac{30 \times 10^3}{(\pi \times 40^2/4) \times 10^{-6}} = 23{\cdot}87 \text{ MN/m}^2$$

$$\text{Stress in copper, } \sigma_c = \frac{30 \times 10^3}{(\pi \times 78{\cdot}16^2/4) \times 10^{-6}} = 9 \text{ MN/m}^2$$

Example 3.4

A compound bar is made up from a brass bar 30 mm in diameter fitting into a cylindrical steel tube having an internal diameter of 35 mm. Determine the external diameter of the steel tube if:

(*i*) an axial load is to be carried equally by the two components;
(*ii*) the stress in the brass is not to exceed 80 MN/m^2 when the compound bar is subjected to an axial load of 200 kN.

$E_{steel} = 210 \text{ GN/m}^2$; $E_{brass} = 120 \text{ GN/m}^2$

Solution

(*i*) Using suffixes s and b to represent the steel and the brass respectively and substituting $\sigma = F/A$ into eqs. (3.2) and (3.3) gives

$$F_s = \frac{FE_sA_s}{(E_sA_s + E_bA_b)} \quad \text{and} \quad F_b = \frac{FE_bA_b}{(E_sA_s + E_bA_b)}$$

If the load is to be equally shared between the steel and the brass then

$$F_s = F_b$$

i.e., $E_sA_s = E_bA_b$

Let the required external diameter of the steel tube be d_1 [mm].

$$\text{Then,} \quad 210 \times 10^9 \times \frac{\pi(d_1^2 - 35^2)}{4 \times 10^6} = 120 \times 10^9 \times \frac{\pi \times 30^2}{4 \times 10^6}$$

$$d_1^2 - 35^2 = \frac{120}{210} \times 30^2$$

$$d_1 = 41{\cdot}7 \text{ mm}$$

For equal load distribution between the steel and brass the external diameter of the steel tube should be 41·7 mm.

(*ii*) Now $\sigma_b = \dfrac{FE_b}{(E_sA_s + E_bA_b)}$ [from eq. (3.3)]

If d_2 [mm] is the required external diameter of the steel tube then,

$$80 \times 10^6 = \frac{200 \times 10^3 \times 120 \times 10^9}{\left(210 \times 10^9 \times \frac{\pi(d_2^2 - 35^2)}{4 \times 10^6} + 120 \times 10^9 \times \frac{\pi \times 30^2}{4 \times 10^6}\right)}$$

Re-arranging,

$$\frac{10^9}{10^6} \times \frac{\pi}{4}(210\,(d_2^2 - 35^2) + 120 \times 30^2) = \frac{200 \times 10^3 \times 120 \times 10^9}{80 \times 10^6}$$

$$210\,(d_2^2 - 35^2) + 120 \times 30^2 = \frac{200 \times 10^3 \times 120}{80} \times \frac{4}{\pi}$$

$$210\,(d_2^2 - 35^2) + 108\,000 = 381\,970$$

$$d_2^2 - 35^2 = \frac{273\,970}{210}$$

$$d_2^2 - 1225 = 1305$$

$$d_2 = 50{\cdot}3 \text{ mm}$$

The required external diameter is 50·3 mm.

Note: The determination of d_2 from the first equation may appear to be extremely difficult. The layout is intended to show that this can be easily achieved if a systematic approach is adopted.

3.9 Temperature stresses

When a component is subjected to a change in temperature there will be a natural change in its length. Providing that the ends of the component are free to move then no force, and hence stress, is induced in the component. However, if the ends of the component are fixed, or if the free expansion of the component is in any way restricted, then a stress will be created. Previously, we have considered stresses arising from the application of external forces but it must also be appreciated that stresses can occur without external forces being applied. This does not mean that forces are not involved, merely that they are internally created within a structure or series of components. The internal forces, and hence stresses, which arise from a variation in temperature can be quite large, as is indicated, for example, by the occasional buckling of railway lines in hot weather and the bursting of a glass bottle by the expansion of ice. It is to allow for this natural change in length, or size, with temperature variation that expansion joints are incorporated in piping systems and bridges for example.

The magnitude of the stresses arising from temperature variation will now be examined for a single member and for a compound bar.

Consider a member of length l whose ends are rigidly fixed. This condition is illustrated in Fig. 3.11(a).

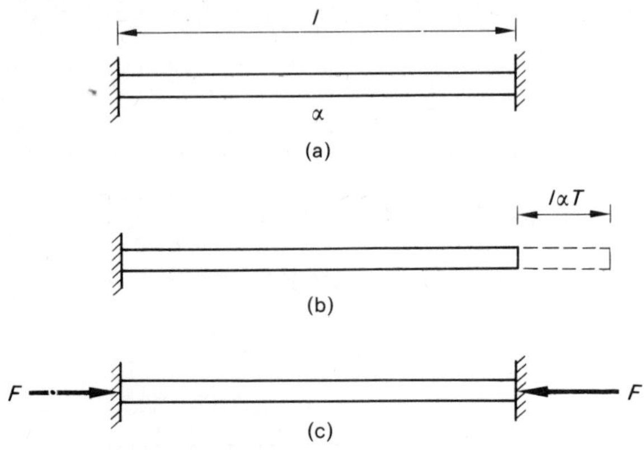

Fig. 3.11 Stress due to temperature rise

If the member is subjected to a uniform temperature rise T then, if one end of the member is free there would be an expansion of $l\alpha T$ where α is the coefficient of linear expansion of the material. This is illustrated in Fig. 3.11(b).

However, because the ends of the member are rigidly built in this expansion is suppressed and a force F is created at the built-in ends to reduce the length by $l\alpha T$, as indicated in Fig. 3.11(c).

Then, strain $= \dfrac{\text{change in length}}{\text{original length}}$

$$\epsilon = \alpha T$$

and, stress, $\sigma = \epsilon E$

$$= E\alpha T$$

where E = modulus of elasticity

If the built-in ends are not completely rigid then only part of the temperature expansion would be suppressed and a correspondingly reduced stress would be created in the member.

The effect of temperature variation upon a compound bar will now be examined.

Consider a compound bar of length l consisting of two different materials (1) and (2), having coefficients of expansion α_1 and α_2

respectively, where α_1 is greater than α_2. The initial conditions are shown in Fig. 3.12(*a*). If the bar is now subjected to a uniform temperature rise T and the right-hand fixing released then the two materials would expand independently to take up the positions shown in Fig. 3.12(*b*). However, this difference in free expansion cannot occur

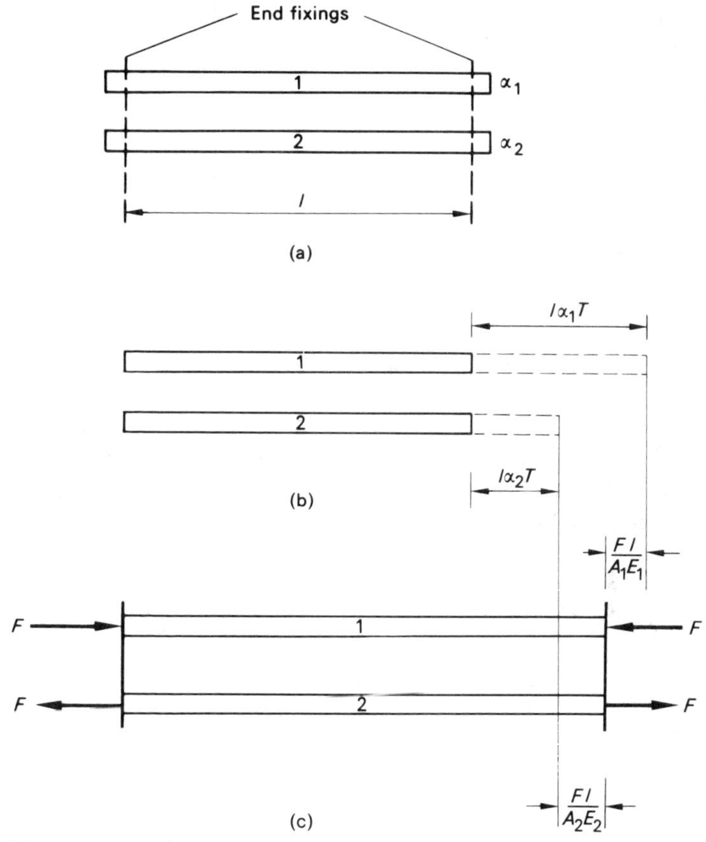

Fig. 3.12 A compound bar subjected to temperature rise

due to the end fixing. Thus, the end fixing must supply a force which decreases the length of bar 1 and increases the length of bar 2 until equilibrium is achieved at a common length as indicated in Fig. 3.12(*c*). As no external forces are involved it follows that a self equilibrating (balancing) force system is created, as shown in Fig. (*c*), but if the end fixing is not sufficiently strong to withstand the force involved then fracture of the end fixings will occur.

The free expansions of the two component members are $l\alpha_1 T$ and $l\alpha_2 T$ respectively. Due to force F: the decrease in length of (1) is Fl/A_1E_1, and the increase in length of (2) is Fl/A_2E_2, where A and E are the cross-sectional area and modulus of elasticity respectively.

By examining Figs. (*b*) and (*c*) it is seen that the condition for the two component bars to have the same final lengths is

$$l\alpha_1 T - \frac{Fl}{A_1E_1} = l\alpha_2 T + \frac{Fl}{A_2E_2}$$

(Note that the expansions shown in Fig. 3.12 are exaggerated and that the error involved in calculating the changes of length, due to force F, on the original length instead of the freely expanded length is negligible.)

Rearranging the above equation and dividing throughout by l gives

$$(\alpha_1 - \alpha_2)\,T = F\left(\frac{1}{A_1E_1} + \frac{1}{A_2E_2}\right) \qquad (3.4)$$

or Difference of temperature strains = sum of fixing strains.

In all instances the stresses arising in a component due to temperature effects will be in addition to those present or arising from other effects.

Example 3.5

A steel bolt of 20 mm diameter passes through a copper tube having an inside diameter of 25 mm and an external diameter of 45 mm. Washers are placed at each end of the tube and the length of the bolt and tube is 0·75 m when the nut is just tight. If the pitch of the bolt thread is 2·5 mm what stresses will be produced in the bolt and tube if the nut is tightened by half a turn?

If the temperature of the assembly is now increased by 60°C what will be the resulting stresses in each material?

Modulus of elasticity: Steel $210\ \text{GN/m}^2$
Copper $100\ \text{GN/m}^2$

Coefficients of Linear Expansion: Steel $11 \times 10^{-6}/°\text{C}$
Copper $18 \times 10^{-6}/°\text{C}$

Solution

STRESSES DUE TO TIGHTENING OF NUT

Tightening the nut will put the bolt into tension and the tube into compression. Let F be the force created in each.

Although no change in temperature occurs during the tightening of the nut the solution is similar, and after the nut is tightened, the displacement relationship is:

Extension of bolt + compression of tube = displacement of nut

This is illustrated in Fig. 3.13.

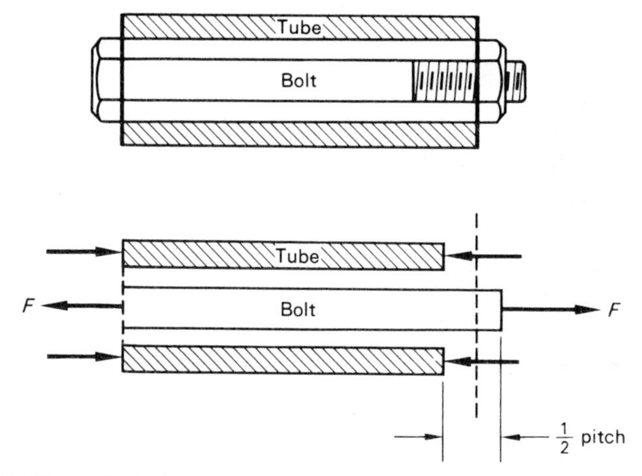

Fig. 3.13 Example 3.5

i.e. $$\frac{Fl}{A_s E_s} + \frac{Fl}{A_c E_c} = \tfrac{1}{2} \text{ pitch}$$

$$A_s = \frac{\pi}{4} \times 20^2 = 100\pi \text{ mm}^2 \qquad A_c = \frac{\pi}{4}(45^2 - 25^2) = 350\pi \text{ mm}^2$$

$$\therefore \quad F \times 0{\cdot}75 \left(\frac{1}{100\pi \times 10^{-6} \times 210 \times 10^9} + \frac{1}{350\pi \times 10^{-6} \times 100 \times 10^9} \right)$$

$$= \tfrac{1}{2} \cdot \frac{2{\cdot}5}{10^3}$$

$$F = 68\,723 \text{ N}$$

$$\text{Tensile stress in the steel bolt} = \frac{68\,723}{100\pi \times 10^{-6}} \quad \frac{[\text{N}]}{[\text{m}^2]}$$

$$= 218{\cdot}8 \text{ MN/m}^2$$

$$\text{Compressive stress in copper tube} = \frac{68\,723}{350\pi \times 10^{-6}}$$

$$= 62{\cdot}5 \text{ MN/m}^2$$

STRESSES DUE TO INCREASE IN TEMPERATURE

The stresses set up as a result of the temperature increase will be in addition to those induced by tightening the nut.

Let F' be the force induced in the bolt and tube due to the temperature rise of 60°C.

Then applying eq. (3.4),

$$(\alpha_c - \alpha_s)T = F'\left(\frac{1}{A_s E_s} + \frac{1}{A_c E_c}\right)$$

$$(18 \times 10^{-6} - 11 \times 10^{-6})\,60 = F'\left(\frac{1}{100\pi \times 10^{-6} \times 210 \times 10^9} + \frac{1}{350\pi \times 10^{-6} \times 100 \times 10^9}\right)$$

$$F' = 24\,245 \text{ N}$$

Tensile stress induced in steel bolt $= \dfrac{24\,245}{100\pi \times 10^{-6}}$

$= 77{\cdot}2 \text{ MN/m}^2$

Compressive stress induced in copper tube $= \dfrac{24\,245}{350\pi \times 10^{-6}}$

$= 22{\cdot}05 \text{ MN/m}^2$

∴ Resultant tensile stress in steel bolt $= 218{\cdot}8 + 77{\cdot}2$

$= 296 \text{ MN/m}^2$

and Resultant compressive stress in copper tube $= 62{\cdot}5 + 22{\cdot}05$

$= 84{\cdot}5 \text{ MN/m}^2$

3.10 Stresses in thin cylindrical shells

When a thin cylindrical shell, such as a steam boiler, large pipe or other similar vessel, is subjected to an internal pressure then stresses will be set up. Arising from the internal pressure there is a variation of stress in the radial direction from a value equal to the internal pressure (gauge) at the internal surface to zero at the outside surface, as indicated in Fig. 3.14(*a*). However, this stress is usually small, and so can be neglected in comparison with the stresses that are created in the longitudinal and circumferential directions. These two stresses and the possible resulting failure of the cylinder will now be examined.

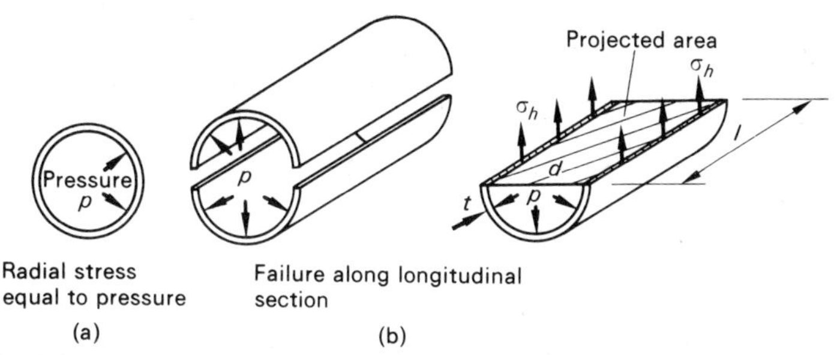

Fig. 3.14 Hoop stress in a thin cylindrical shell

As a result of the internal pressure there is a tendency for the diameter of the cylinder to be increased. This produces a circumferential or hoop stress (tensile), as shown in Fig. 3.14(*b*), and in the event of failure a longitudinal burst would occur. The equilibrium condition necessary is most conveniently obtained by considering half the cylinder, as shown.

The force due to the internal pressure, p, is balanced by the force due to the hoop stress, σ_h.

Thus, hoop stress x area = pressure x projected area

$$\sigma_h \times 2lt = p \times dl$$

or

$$\sigma_h = \frac{pd}{2t} \tag{3.5}$$

where d = internal diameter of cylinder

t = wall thickness of cylinder

The internal pressure also produces a tensile stress in the longitudinal direction as indicated in Fig. 3.15(*a*). The pressure p acting over an area $\pi d^2/4$ is balanced by the longitudinal stress σ_l acting over an approximate area πdt (the mean diameter should strictly be used).

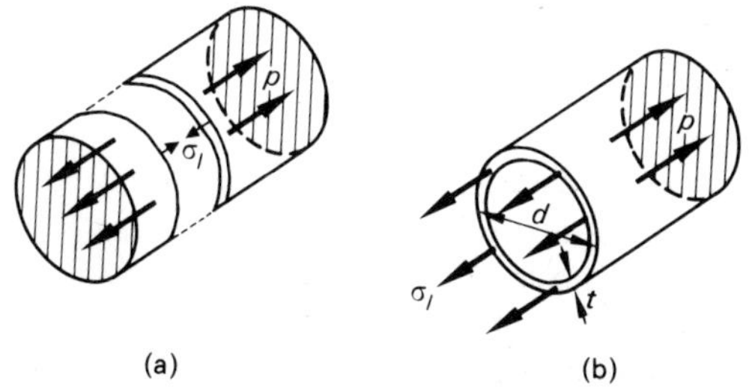

Fig. 3.15 Longitudina! stress in a thin cylindrical shell

Equating forces

$$\sigma_1 \times \pi d t = p \times \frac{\pi d^2}{4}$$

or $$\sigma_1 = \frac{pd}{4t} \qquad (3.6)$$

It will be seen from eqs. (3.5) and (3.6) that the hoop stress (acting on a longitudinal section) is twice the longitudinal stress (acting on a circumferential section). Thus, the longitudinal joints should be stronger than the circumferential (ref. Fig. 3.16) and if riveted joints are involved an appropriate joint efficiency factor should be used (see section 3.17).

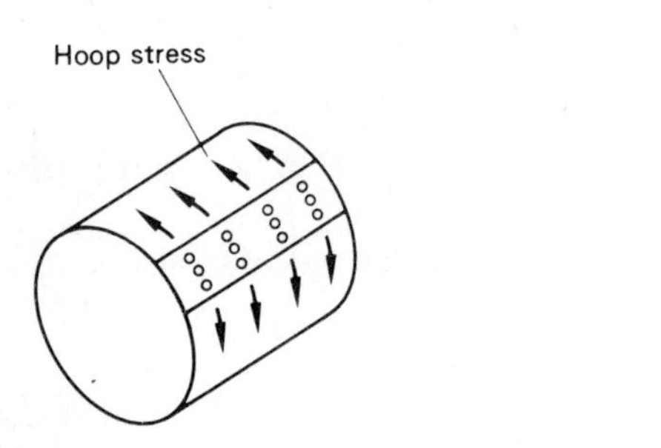

Fig. 3.16 Riveted joints in a thin shell

Note: The expressions (3.5) and (3.6) for the hoop and longitudinal stresses in a thin cylinder are developed on the assumption that these stresses are constant over the thickness of the material. This assumption is only valid, to any reasonable accuracy, if the ratio of thickness to internal diameter is less than 1/20. Improved accuracy is obtained by using the mean diameter in relationships (3.5) and (3.6).

Example 3.6

A thin-walled seamless steel cylinder is 4 m long, has an internal diameter of 0·8 m and a wall thickness of 30 mm. Determine the hoop and longitudinal stresses in the cylinder when it is subjected to an internal pressure of 2 MN/m^2.

At what pressure should a safely valve be set to operate if there is to be a safety factor of 12, based upon a tensile yield strength of 348 MN/m^2?

Solution

$$\text{Hoop stress } \sigma_h = \frac{pd}{2t} \quad \text{[eq. (3.5)]}$$

$$= \frac{2 \times 10^6 \times 0{\cdot}8}{2 \times 0{\cdot}03}$$

$$= 26{\cdot}67 \text{ MN/m}^2$$

$$\text{Longitudinal stress } \sigma_l = \frac{pd}{4t} \quad \text{[eq. (3.6)]}$$

$$= 13{\cdot}33 \text{ MN/m}^2$$

Now, $$\text{factor of safety} = \frac{\text{yield stress}}{\text{working stress}}$$

Therefore, $$\text{maximum working stress} = \frac{348}{12}$$

$$= 29 \text{ MN/m}^2$$

As the maximum stress in the cylinder is directly proportional to the internal pressure, then

$$\text{Maximum allowable pressure} = \frac{29}{26{\cdot}67} \times 2$$

$$= 2{\cdot}175 \text{ MN/m}^2$$

3.11 Stresses in thin rotating rings

If a thin circular ring, or cylinder, is rotated about its centre there will be a natural tendency for the diameter of the ring to be increased. To maintain a body in a circular motion a centripetal force is required (see section 8.8) and in the case of a rotating ring this force can only arise from the circumferential (or hoop) stress that is created in the ring.

Consider a thin ring of mean radius r, density ρ, and having a cross-sectional area A, to be rotating about centre O with an angular velocity ω [rad/s]. For an elemental length which subtends an angle $\delta\theta$ at O, as shown in Fig. 3.17(*a*):

Circumferential length of element $= r \,.\, \delta\theta$

Volume of element $= rA \,.\, \delta\theta$

Mass of element $= \rho rA \,.\, \delta\theta$

Therefore, centripetal force to maintain circular motion

$$= (\rho rA \,.\, \delta\theta) \,.\, \omega^2 r$$

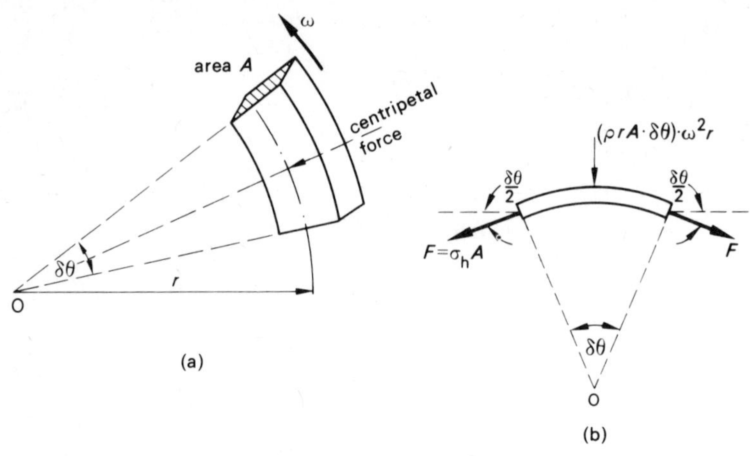

Fig. 3.17 Thin rotating ring

If the hoop stress created in the ring is σ_h, then:

Force F acting on cross-section $= \sigma_h \,.\, A$

This force is shown in Fig. 3.17(*b*), from which it will be seen that,

Radial component of the forces $F = 2\,(\sigma_h \,.\, A) \sin \delta\theta/2$

It is this radial component of the forces F that supplies the necessary centripetal force to maintain the element in circular motion. Thus, equating expressions and putting $\sin \delta\theta/2 = \delta\theta/2$ (for small angles of $\delta\theta$) gives

$$2(\sigma_h \,.\, A) \,.\, \frac{\delta\theta}{2} = (\rho r A \,.\, \delta\theta) \,.\, \omega^2 r$$

or $$\sigma_h = \rho\omega^2 r^2 \tag{3.7}$$

Putting $$v = \omega r$$

gives $$\sigma_h = \rho v^2 \tag{3.8}$$

Hence, the hoop stress created in a thin rotating ring, or cylinder, is independent of the cross-sectional area and furthermore, for a given peripheral speed, this stress is independent of the radius of the ring.

Example 3.7

What is the maximum speed with which a thin brass cylinder can be rotated if the hoop stress is not to exceed 50 MN/m^2?

The mean diameter of the cylinder is 0·4 m.

Density of brass = 8·4 Mg/m^3.

Solution

From eq. (3.7) the maximum hoop stress created in a rotating cylinder is

$$\sigma_h = \rho\omega^2 r^2$$

$$\therefore \quad \omega^2 = \frac{\sigma_h}{\rho r^2} = \frac{50 \times 10^6 \ [\text{N/m}^2]}{8400 \ [\text{kg/m}^3] \times 0{\cdot}2^2 \ [\text{m}^2]}$$

$$\omega = 385{\cdot}7 \text{ rad/s}$$

$$= 3686 \text{ rev/min}$$

The maximum speed of rotation is 3686 rev/min.

3.12 Strain energy

If a material is strained by a gradually applied load then work is done on the material by the applied load. This work is stored in the material in the form of strain energy and providing that the strain is kept within the elastic range of the material this energy is not retained by the material upon the removal of the load.

Figure 3.18 shows the load-extension graph for a uniform bar, the extension δ being associated with a gradually applied load F which is

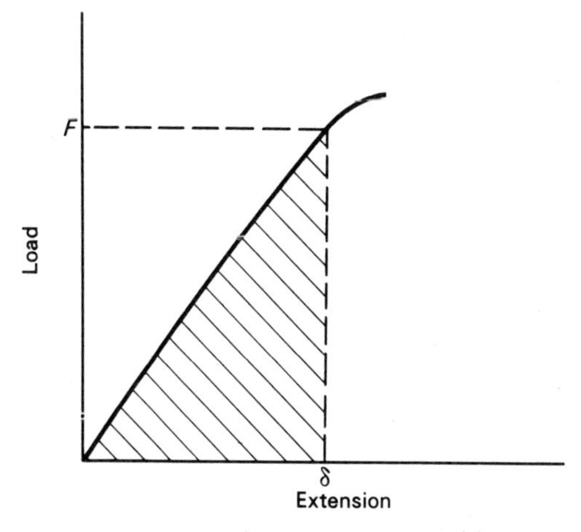

Fig. 3.18 Strain energy

within the elastic range of the material. The shaded area represents the work done in increasing the load from zero to its final value F. Thus,

$$\text{Work done} = \text{strain energy of bar} = \text{shaded area}$$

$$W = U = \tfrac{1}{2}F\delta$$

Now, stress in material, $\sigma = \dfrac{F}{A}$

where A is the cross-sectional area

and strain = stress/E

$$\therefore \quad \frac{\delta}{l} = \frac{\sigma}{E}$$

where l is the original length.
Substituting for P and δ gives

$$W = U = \tfrac{1}{2} \, . \, \sigma A \, . \, \frac{\sigma l}{E}$$

$$= \frac{\sigma^2}{2E} \times Al$$

But Al is the volume of the bar

$$\therefore \quad U = \frac{\sigma^2}{2E} \times \text{volume} \qquad (3.9)$$

The units of strain energy are the same of those of work, i.e., joules.

The strain energy per unit volume, $\sigma^2/2E$, is known as the resilience. It follows that the greatest amount of strain energy that can be stored in a material without permanent set occurring will be when σ is equal to the elastic limit stress (or proof stress, if applicable). This particular value of the strain energy is known as the proof resilience.

Example 3.8

A steel bar has a rectangular section 50 mm x 30 mm and is 0·6 m long. Determine the strain energy stored in the bar when it is subjected to an axial load of 200 kN.

If the bar reaches its elastic limit when the axial load is 465 kN what is the proof resilience of the steel?

Modulus of elasticity = 210 GN/m^2.

Solution

$$\text{Strain energy } U = \frac{\sigma^2}{2E} \times \text{volume}$$

Now $$\sigma = \frac{P}{A} = \frac{200 \times 10^3}{(50 \times 30)/10^6} = 133{\cdot}3 \times 10^6 \text{ N/m}^2$$

and $$\text{volume} = \frac{50}{10^3} \times \frac{30}{10^3} \times 0{\cdot}6 = \frac{0{\cdot}9}{10^3} \text{ m}^3$$

$$\therefore \quad U = \frac{(133{\cdot}3 \times 10^6)^2}{2 \times 210 \times 10^9} \times \frac{0{\cdot}9}{10^3}$$

$$= 30{\cdot}2 \text{ J}$$

The strain energy stored in the bar is 30·2 J.

$$\text{The elastic limit stress} = \frac{465 \times 10^3}{1500/10^6}$$

$$= 310 \times 10^6 \text{ N/m}^2$$

$$\therefore \quad \text{Proof resilience} = \frac{(310 \times 10^6)^2}{2 \times 210 \times 10^9}$$

$$= 228{\cdot}8 \times 10^3 \text{ J}$$

$$= 228{\cdot}8 \text{ kJ}$$

The proof resilience of the steel is 228·8 kJ.

3.13 Shear stress

When a material is loaded such that one layer of the material tends to slide over its adjacent layer then the material is said to be in a state of shear. Two common examples of shearing action are illustrated in

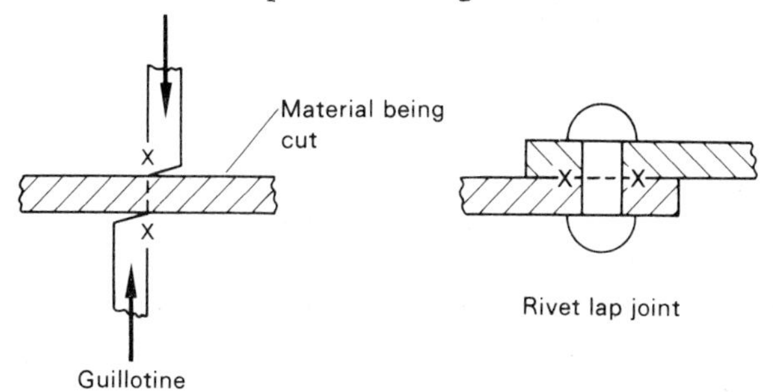

Fig. 3.19 Examples of shearing action

Fig. 3.19 where the material on either side of XX moves, or has a tendency to move, in opposite directions.

The magnitude and variation of shear forces in simply loaded beams has been dealt with in chapter 2 and we must now investigate the stresses that arise from shearing action. Although the shear stress varies over the area under shear its mean value can be obtained from

$$\text{Average shear stress} = \frac{\text{shear force}}{\text{area resisting shear}}$$

$$\tau = \frac{Q}{A}$$

[τ Greek tau (pronounced 'tor')].

3.14 Shear strain

When a shear stress is created within a material it must be accompanied by a shear strain. The meaning of shear strain can be best explained by considering the shearing of a block of material, as shown in Fig. 3.20.

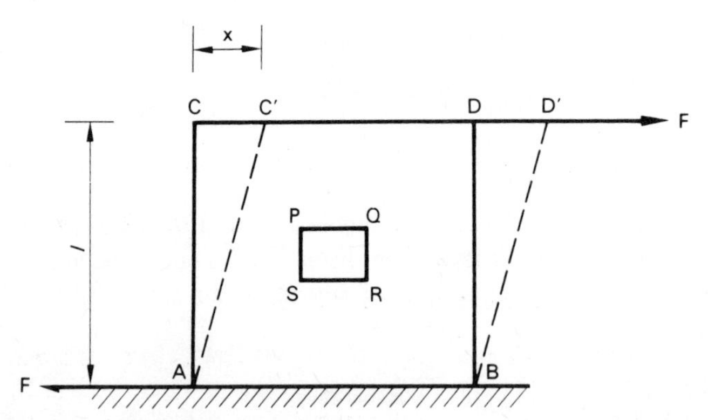

Fig. 3.20 Shearing of a block of material

Under the action of forces F the block is strained from ABCD to ABC′D′. The shear strain ϕ (phi) is then given by:

$$\phi = \frac{x}{l}$$

3.15 Modulus of rigidity

Providing that the limit of proportionality of a material is not exceeded, the shear stress is proportional to the shear strain. Then,

$$\text{Modulus of rigidity} = \frac{\text{shear stress}}{\text{shear strain}}$$

$$G = \frac{\tau}{\phi}$$

3.16 Complementary shear stress

The action of the forces F on the block of material (ref. Fig. 3.20) will cause shear stresses to be set up throughout the material. Consider a small element PQRS of this material and let the shear stresses created on faces PQ and RS be τ_1, as shown in Fig. 3.21. The element is therefore subjected to a couple and for equilibrium a balancing couple must

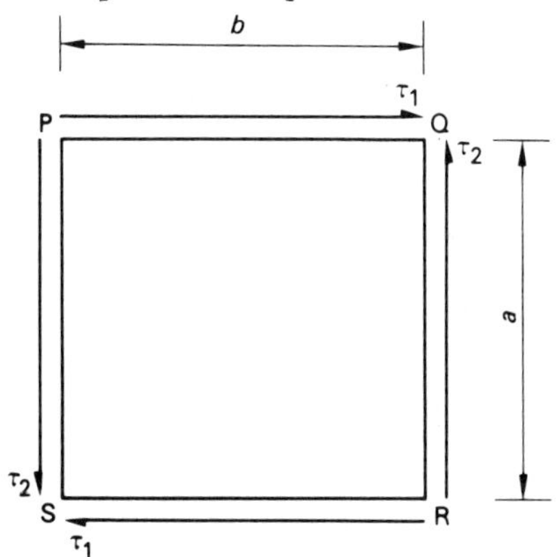

Fig. 3.21 Complementary shear stresses

be brought into action. This can only arise from the shear stresses arising on faces QR and PS. Let the stresses on these faces be τ_2. If the thickness of the material (at right angles to the paper) is t, and the lengths of the element sides are a and b, as shown, then for equilibrium:

$$\text{Clockwise couple} = \text{anticlockwise couple}$$

$$\text{Force on PQ (or RS)} \times a = \text{force on QR (or PS)} \times b$$

$$\tau_1 \times bt \times a = \tau_2 \times at \times b$$

$$\tau_1 = \tau_2$$

Thus, whenever a shear stress occurs on a plane within a material it is automatically accompanied by an equal shear stress on the perpendicular plane. The direction of the complementary shear stresses is always such that their couple opposes that of the original shear stresses.

3.17 Riveted joints

Whenever rivets, or bolts, are used as the means of joint fastening they are designed to be in shear. There are two main types of joints: the lap joint, and the butt joint. These are illustrated in Fig. 3.22.

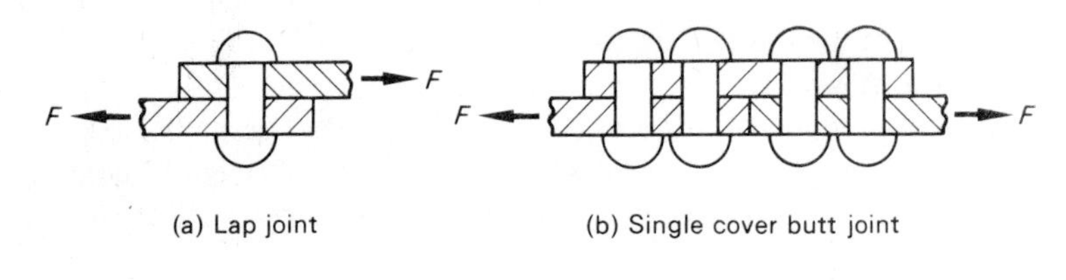

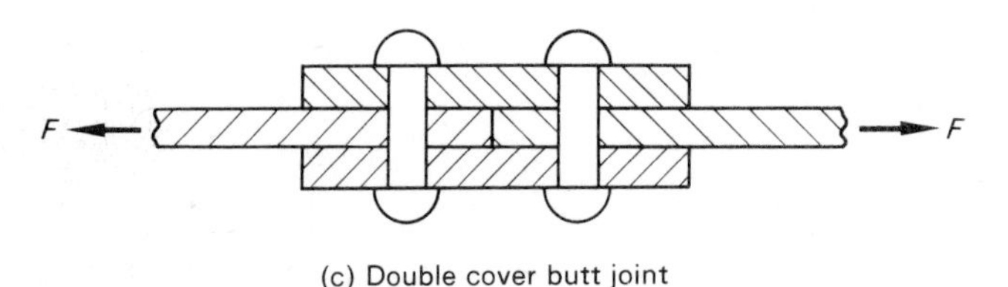

Fig. 3.22 Riveted joints

The lap joint (*a*) has overlapping plates which are connected by one or more rows of rivets in single shear. The single cover butt joint (*b*) consists of a lap joint between the cover plate and each member. It will be seen that the rivets are again in single shear. However, the double butt joint (*c*), which is most frequently used for a butt joint, has two cover plates and therefore each rivet is in double shear, i.e., two planes of the rivet are each subjected to shear and the area resisting shear is $2 \times \pi d^2/4$, where d is the diameter of the rivet. To allow for uneven bearing of the rivets against the two cover plates the resisting shear area is frequently taken as $1{\cdot}75 \times \pi d^2/4$.

When designing riveted joints various possible modes of failure must be examined.

Consider one pitch-width of the single riveted lap joint shown in Fig. 3.23. The joint may fail due to:

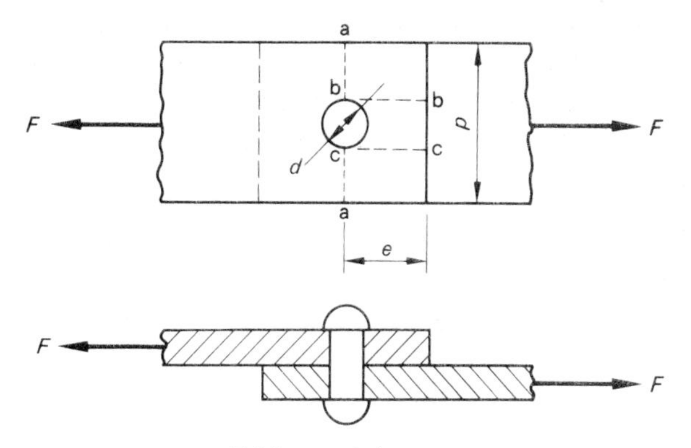

Fig. 3.23 Single riveted tap joint

1. SHEAR OF THE RIVET

Force F to cause rivet shear is

$$F = \tau \,.\, \frac{\pi d^2}{4} \tag{3.10}$$

where τ is the allowable shear stress of the rivet.

2. TENSION IN PLATE ACROSS aa

Force F to cause failure is

$$F = \sigma_t \,.\, (p - d)\, t \tag{3.11}$$

where σ_t is the allowable tensile stress of the plate.

3. SHEAR OUT OF THE PLATE ALONG bb AND cc

If the distance of the rivet from the edge of the plate is e then the area resisting shear is $2et$. The force F to cause failure is then

$$F = \tau_p \,.\, 2et \tag{3.12}$$

where τ_p is the allowable shear stress of the plate.

This mode of failure is usually eliminated by pitching the rivets at a distance of $1{\cdot}5d$ from the edge of the plate.

4. CRUSHING OF THE PLATE BY THE RIVET

The projected area of the rivet is dt and the force F to cause failure of the plate is therefore

$$F = \sigma_b \,.\, dt \tag{3.13}$$

where σ_b is the allowable bearing stress of the rivet.

The plate is deemed to have failed in bearing when the elongation of the hole reaches 10%.

The joint is then designed so as to be able to withstand the applied load in accordance with the eqs. (3.10), (3.11), (3.12) and (3.13).

The strength of a riveted joint must always be less than that of the original plate simply because material is removed from the plate for the rivet holes and the efficiency of a riveted joint is defined as

$$\text{Joint efficiency} = \frac{\text{strength of joint}}{\text{strength of original plate}} \times 100\%$$

The strength and efficiency of joints with more than one row of rivets may be obtained in a similar manner. In industry it is usual practice to refer to the company data sheets, the above being an empirical treatment only.

Example 3.9

A cylindrical boiler shell is constructed from 15 mm thick mild steel plate. The shell is 1·2 m in diameter and has a single riveted longitudinal lap joint. The rivets are 20 mm in diameter and are pitched at 75-mm intervals. The distance of the rivets from the edge of the plate is 30 mm. If the maximum allowable working stresses are 100 MN/m^2 in tension, 80 MN/m^2 in shear, and 200 MN/m^2 in bearing determine.

(*a*) the efficiency of the joint;

(*b*) the maximum internal pressure that the boiler shell can safely withstand.

Solution

(*a*) To determine the efficiency of the joint the maximum load that can be carried by the joint must be obtained.

From eq. (3.10),

$$\text{Load to shear rivet} = \tau . \frac{\pi d^2}{4}$$

$$= 80 \times 10^6 \times \frac{\pi}{4} \times \frac{20^2}{10^6}$$

$$= 25\,132 \text{ N}$$

From eq. (3.11),

Load to cause tension failure of plate across rivet hole

$$= \sigma_t(p - d)t$$

$$= 100 \times 10^6 \times \frac{(75 - 20)}{10^3} \times \frac{15}{10^3}$$

$$= 82\,500 \text{ N}$$

From eq. (3.12),

Load to cause shear out of plate at rivet holes

$$= \tau \times 2et$$

$$= 80 \times 10^6 \times 2 \times \frac{30}{10^3} \times \frac{15}{10^3}$$

$$= 72\,000 \text{ N}$$

From eq. (3.13),

Load to cause bearing failure of plate

$$= \sigma_b \,.\, dt$$

$$= 200 \times 10^6 \times \frac{20}{10^3} \times \frac{15}{10^3}$$

$$= 60\,000 \text{ N}$$

Thus, the maximum load that can be carried by the joint is 25 132 N per rivet pitch.

Strength of original plate per rivet pitch

$$= \sigma_t \,.\, pt$$

$$= 100 \times 10^6 \times \frac{75}{10^3} \times \frac{15}{10^3}$$

$$= 112\,500 \text{ N}$$

$$\therefore \quad \text{Joint efficiency} = \frac{25\,132}{112\,500} \times 100\%$$

$$= 22{\cdot}34\%$$

This efficiency is very low and could be increased by reducing the rivet pitch.

(*b*) The hoop stress in a thin cylinder is given by

$$\sigma_h = \frac{pd}{2t} \qquad \text{[eq. (3.5)]}$$

where d is now the internal diameter of the cylinder.

In this instance the allowable hoop stress, which is tensile, is limited by the efficiency of the joint and is given by

$$\sigma_h = \eta\sigma_t$$

$$= \frac{22{\cdot}34}{100} \times 100 \times 10^6$$

$$= 22{\cdot}34 \times 10^6 \text{ N/m}^2$$

Then $p = \frac{2t\sigma_h}{d}$

$$= \frac{2 \times 15/10^3 \times 22{\cdot}34 \times 10^6}{1{\cdot}2}$$

$$= 558\,500 \text{ N/m}^2$$

$$= 558{\cdot}5 \text{ kN/m}^2$$

The maximum pressure that the cylinder can safely withstand is $558{\cdot}5$ kN/m^2.

Example 3.10

Two mild steel plates 0·24 m wide and 18 mm thick are to be joined by a double butt joint using mild steel rivets. Determine the number of rivets required and indicate a suitable joint design. The following criteria are to be applied:

Rivet diameter, $d = 25$ mm

Minimum rivet pitch $= 3d$

Double shear factor $= 1{\cdot}75$

Also determine the maximum load which can be carried by the joint and the joint efficiency.
The allowable stresses are: Tension, 110 MN/m^2; Shear 85 MN/m^2; Bearing 180 MN/m^2.

Solution

Assuming that the full load will be carried by the width of plate less one rivet diameter then the force F that could be carried in tension is

$$F = \sigma_t(p - d)t \qquad \text{[eq. (3.11)]}$$

$$= 110 \times 10^6 \left(\frac{240 - 25}{10^3}\right) \times \frac{18}{10^3}$$

$$= 425\,700 \text{ N}$$

Strength of one rivet in double shear $= \tau \times \frac{\pi d^2}{4} \times 1{\cdot}75$

$$= 85 \times 10^6 \times \frac{\pi \times 25^2}{4 \times 10^6} \times 1{\cdot}75$$

$$= 73\,020 \text{ N}$$

Strength of plate in bearing = $\sigma_b \, . \, dt$

$$= 180 \times 10^6 \times \frac{25}{10^3} \times \frac{18}{10^3}$$

$$= 81\,000 \text{ N}$$

As the bearing strength is greater than the shear strength then the number of rivets required is determined on the basis of the shear strength, i.e.,

$$\text{number of rivets} = \frac{\text{tensile strength}}{\text{shear strength of 1 rivet}}$$

$$= \frac{425\,700}{73\,020}$$

$$= 5{\cdot}83$$

Therefore six rivets will be used and a suitable arrangement is indicated in Fig. 3.24.

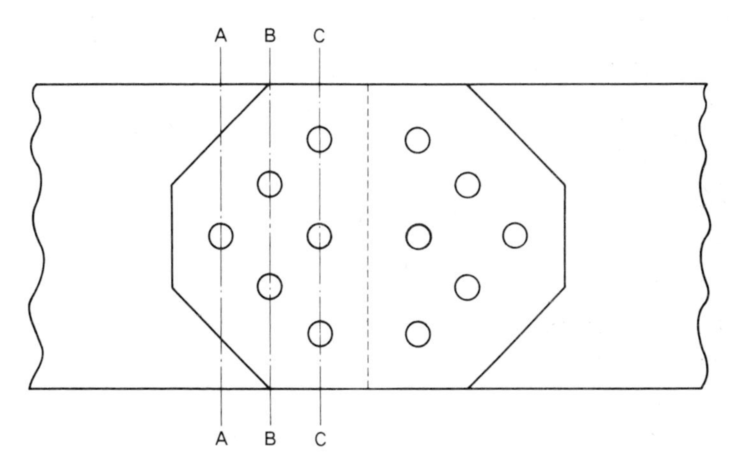

Fig. 3.24 A double butt rivet joint–Example 3.10

To determine the maximum load that can be carried by the proposed joint various modes of failure will be examined.

1. Shear of all the rivets.

$$F = 6 \times 73\,020$$

$$= 428\,120 \text{ N}$$

2. Tension across AA.

$$F = 425\,700 \text{ N}$$

as calculated previously

3. Tension of plate across BB plus the shear of the single rivet to the left of BB.

$$F = 110 \times 10^6 \frac{(240 - 2 \times 25)}{10^3} \times \frac{18}{10^3} + 73\,020$$

$$= 449\,220 \text{ N}$$

4. Tension of plate across CC plus the shear of the three rivets to the left of CC.

$$F = 110 \times 10^6 \frac{(240 - 3 \times 25)}{10^3} \times \frac{18}{10^3} + 3 \times 73\,020$$

$$= 545\,760 \text{ N}$$

5. Shear out of plate due to rivets on CC plus the shear of the three rivets to left of CC.

Let the distance of the rivets at CC be $1{\cdot}5 \times 25 = 37{\cdot}5$ mm or 40 mm, say, from the edge of the plate.
Then, load to cause shear out of one rivet

$$= \tau \,.\, 2et$$

$$= 85 \times 10^6 \times 2 \times \frac{40}{10^3} \times \frac{18}{10^3}$$

$$= 122\,400 \text{ N}$$

$$F = 3 \times 122\,400 + 3 \times 73\,020$$

$$= 586\,260 \text{ N}$$

6. Shear out of plate due to rivets on CC together with crushing of the three rivets to the left of CC.

As the crushing strength of a rivet is greater than its shear strength the load to cause failure will be greater than for case 5.

7. Tension of cover plates across CC.
It is usual to make the thickness of cover plates for a double butt joint equal to $\frac{5}{8}$ x thickness of main plate.

Thus,

$$\text{thickness of cover plates} = \tfrac{5}{8} \times 18$$

$$= 11{\cdot}25 \text{ mm (say 12 mm)}$$

Then $F = 2 \times \sigma_t \times$ area of each cover plate

$$= 2 \times 110 \times 10^6 \times \left(\frac{240 - 3 \times 25}{10^3} \right) \times \frac{12}{10^3}$$

$$= 435\,600 \text{ N}$$

It will be seen that the failure load in this instance is only just greater than that for tension of the plate across AA.

Then, maximum strength of joint = 425 700 N

$$\text{Strength of original plate} = 110 \times 10^6 \times \frac{240}{10^3} \times \frac{18}{10^3}$$

$$= 475\,200 \text{ N}$$

$$\therefore \quad \text{Joint efficiency} = \frac{425\,700}{475\,200} \times 100$$

$$= 92{\cdot}85\%$$

Suitable final dimensions for the joint are given in Fig. 3.25.

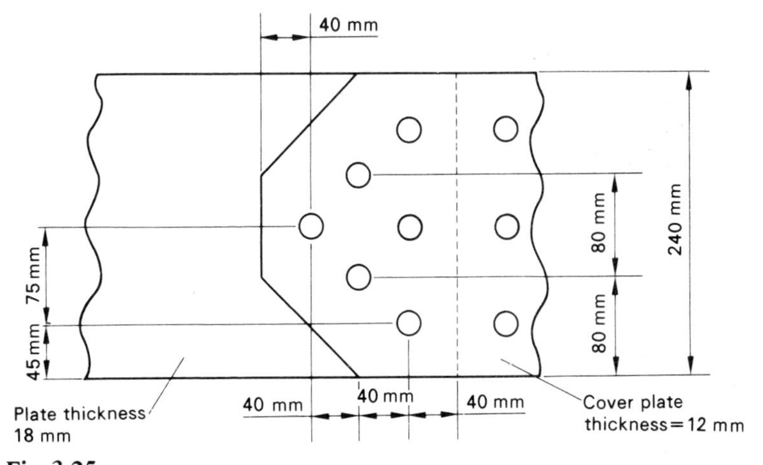

Fig. 3.25

Problems

1. A connecting link in a machine drive has a cross-sectional area of 400 mm^2 and a length of 1·6 m. It is subjected to a maximum tensile load of 28 kN. Determine:

(*a*) the elongation at the maximum load;
(*b*) the factor of safety if the tensile yield strength is 336 MN/m^2.

Modulus of elasticity = 200 GN/m^2.

2. A tensile test on a steel test specimen having a diameter of 20 mm and a gauge length of 100 mm produced the following results:

Load [kN]	20	40	60	80	100	120	130	143
Extension [mm]	0·032	0·065	0·097	0·129	0·162	0·194	0·212	0·50

Maximum load applied during the test = 208 kN

Final gauge length at fracture = 128·7 mm

Diameter at fracture = 15·82 mm

Determine:

(*a*) the yield stress;
(*b*) the tensile strength;
(*c*) the percentage reduction in area;
(*d*) the modulus of elasticity.

3. A test on an aluminium alloy test specimen having a diameter of 10 mm and a gauge length of 50 mm produced the following results:

Load [kN]	2·5	5·0	7·5	10·0	12·5	15·0	17·5
Strain [$\times 10^{-3}$]	0·375	0·8	1·25	1·65	2·06	2·5	2·92
Load [kN]	20·0	22·5	25·0	27·5	30·0	32·5	35·0
Strain [$\times 10^{-3}$]	3·36	3·8	4·25	4·7	5·15	5·63	6·3
Load [kN]	37·5	38·5	39·5	40·0	40·5		
Strain [$\times 10^{-3}$]	7·0	7·8	10·25	13·2	17·5		

From these results determine:

(*a*) the 0·1% and 0·2% proof stresses;
(*b*) the modulus of elasticity.

4. The conductor of a transmission system consists of a steel wire 5 mm in diameter which is covered with copper so that the outer diameter is 7 mm. The copper and steel adhere together under all conditions. When the conductor is subjected to a tension of 2·6 kN what will be the stresses in the two materials and the extension over a length of 80 m?

For steel, E = 205 GN/m^2
For copper, E = 123 GN/m^2

5. A brass rod, 20 mm in diameter is fitted into a steel tube, 20 mm internal diameter and 30 mm external diameter. The rod and tube are of the same length. If the allowable compressive stresses are 100 MN/m^2

for the steel and 60 MN/m^2 for the brass what is the maximum compressive load that can be applied to the compound bar?

For brass, modulus of elasticity = 85 GN/m^2
For steel, modulus of elasticity = 210 GN/m^2

6. A steel rod, 20 mm diameter and threaded at each end, is placed symmetrically inside a brass tube 250 mm long whose internal and external diameters are 25 mm and 40 mm respectively. The tube is closed by rigid washers of negligible thickness and nuts adjusted on the rod so as to produce a tensile stress in the steel of 28 MN/m^2.

Assuming no buckling occurs, calculate:

(*a*) the intensity of stress produced in the brass;
(*b*) the reduction in length of the brass tube.

If one nut is then tightened by 0·05 of a turn relative to the other and the pitch of the thread is 2·5 mm, calculate the increase in the intensity of stress of the rod.

The moduli of elasticity for steel and brass are 212 GN/m^2 and 110 GN/m^2 respectively. (N.C.T.E.C.)

7. A compound bar consists of a copper bar having an area of 800 mm^2 and a length 0·6 m connected at its ends to a steel bar of equal length. If the load carried by the steel bar is twice that carried by the copper determine:

(*a*) the area of the steel bar;
(*b*) the extension of the bar for a tensile load of 100 kN.

The moduli of elasticity of copper and steel are 120 GN/m^2 and 210 GN/m^2 respectively.

8. A composite tie bar 1·5 m long is made of an inner steel core 30 mm diameter surrounded by a close fitting copper tube 40 mm external diameter. The metals are bonded together so that relative movement cannot occur. If an axial load of 80 kN is applied to the tie bar calculate:

(*a*) the stress in each material;
(*b*) the extension of the bar;
(*c*) the strain energy stored in the steel core.

Modulus of elasticity of steel = 200 GN/m^2
Modulus of elasticity of copper = 110 GN/m^2

(E.M.E.U.)

9. A 20 mm steel bolt is passed through a brass sleeve of 25 mm internal and 30 mm external diameter. A nut and rigid washer are put on the bolt and the nut tightened until the brass sleeve has shortened by 0·040 mm. Determine the extension of the bolt and the ratio of the stresses in the two materials.

For steel, $E = 200\ \text{GN/m}^2$; for brass, $E = 80\ \text{GN/m}^2$

10. A steel bar 15 mm in diameter is rigidly clamped at its ends at a temperature of 20°C. The bar forms part of the construction of a furnace and in use is subjected to a maximum temperature of 175°C. What is the greatest stress created in the bar?

Coefficient of linear expansion of steel = 11×10^{-6}/°C.
Modulus of elasticity = $200\ \text{GN/m}^2$

11. A steel tube having an external diameter of 36 mm and an internal diameter of 30 mm has a brass rod of 20 mm diameter inside it, the two materials being rigidly joined at their ends when the ambient temperature is 18°C.
Determine the stresses in the two materials:

(*a*) when the temperature is raised to 68°C;
(*b*) when a compressive load of 20 kN is applied at the increased temperature.

For steel: modulus of elasticity = $210\ \text{GN/m}^2$;
Coefficient of expansion = 11×10^{-6}/°C
For brass: modulus of elasticity = $80\ \text{GN/m}^2$;
Coefficient of expansion = 17×10^{-6}/°C

12. A compound bar comprises a gunmetal rod 40 mm diameter which passes through a steel tube having an external diameter of 60 mm. The rod and tube are close fitting and are rigidly fixed together at their ends. Determine the stresses in the two materials when the compound bar is subjected to a temperature rise of 80°C.

A load is now applied to the steel tube such that the stress in the gunmetal rod is removed. What is the resulting stress in the steel tube?

For steel: $\alpha = 11 \times 10^{-6}$/°C; $E = 200\ \text{GN/m}^2$
For gunmetal: $\alpha = 20 \times 10^{-6}$/°C; $E = 90\ \text{GN/m}^2$

13. A compound column is made from steel of square section, 200 mm external side with a wall thickness of 10 mm, filled with concrete. It has an overall length of 6 m.

Calculate:

(*a*) the maximum compressive force that may be supported if the stress in the concrete must not exceed 5 MN/m^2,
(*b*) the reduction in length under this load.

(E for steei = 210 kN/mm^2, E for concrete = 15 kN/mm^2)

(Y.C.F.E.)

14. A brass rod 20 mm diameter is enclosed in a steel tube, 40 mm external diameter and 20 mm internal diameter. The rod and tube are each 1·2 m long at a temperature of 15°C. If the materials are fastened together by two rivets, one at either end find the fixing force supplied by each rivet if the temperature is raised to 135°C.

If the allowable stress for the rivets is 120 MN/m^2 determine the necessary rivet diameter.

(For brass, E = 100 GN/m^2, α = 18·6 x 10^{-6}/°C;
For steel, E = 210 GN/m^2, α = 11·6 x 10^{-6}/°C).

15. A thin cylinder is made of steel having a tensile strength of 490 N/mm^2. The internal diameter of the cylinder is 2 m and the wall thickness is 10 mm. It is subjected to an internal pressure of 700 kN/m^2. Determine from first principles:

(*a*) the hoop stress in the material due to the pressure;
(*b*) the factor of safety used in the design of the cylinder.

(U.L.C.I.)

16. A thin steel plate having a tensile strength of 440 MN/m^2 and a density of 7·8 Mg/m^3 is formed into a circular drum of mean diameter 0·8 m. Determine the greatest speed at which the drum can be revolved if there is to be a safety factor of 8.

Modulus of elasticity = 210 GN/m^2.

17. Derive an expression for the instantaneous acceleration of a particle rotating in a circular path of radius r with an angular speed ω.

Hence deduce an expression for the tensile stress produced in a thin rotating rim due to centrifugal force.

Calculate the maximum speed of rotation of a thin rim of a cast iron flywheel, 1·8 m diameter, if the tensile strength of cast iron is 160 MN/m^2 and a factor of safety of 6 is applied.

(The density of cast iron is 7 Mg/m^3).

(N.C.T.E.C.)

18. Derive expressions for the longitudinal and circumferential stresses in a thin boiler shell of thickness t which is subjected to an internal pressure p.

A cylindrical boiler, of 1·8 m diameter, is constructed from mild steel plate having a tensile strength of 320 MN/m^2. If there is to be a safety factor of 6 when the boiler is working at its maximum pressure of 1·6 MN/m^2 determine the required thickness of the plates.

Assume that efficiency of the longitudinal rivet joint is 80%.

19. A cylindrical shell has a diameter of 1 m and is constructed from 10 mm thick mild steel plates having a tensile strength of 360 MN/m^2. The efficiencies of the longitudinal and circumferential joints are 72% and 42% respectively.

Determine the maximum gas pressure which the shell can withstand if there is to be a safety factor of 4.

20. Define the terms resilience and proof resilience.

A link in a machine mechanism has a diameter of 15 mm and is 0·3 m long. If the stress at the elastic limit is 280 MN/m^2, calculate:

(*a*) the proof resilience;
(*b*) the magnitude of the gradually applied force that would produce the same resilience;
(*c*) the elongation of the link produced by the force obtained in (*b*).

Modulus of elasticity = 210 GN/m^2.

21. A steel bar, 1 m long, is 40 mm diameter for 0·4 m of its length and 60 mm diameter for the remainder. Determine the resilience of the bar when the maximum stress is 200 MN/m^2.

Modulus of elasticity = 210 GN/m^2.

22. A cast iron bar of 30 mm square section is strengthened by the addition of a steel plate, 30 mm wide and 10 mm thick, to each of two opposite sides. The plates are rigidly connected to the bar by fixed pins near the ends. The compound bar has a length of 1·5 m and is subjected to an axial pull of 75 kN which is uniformly distributed over the total area of the end sections. Determine:

(*a*) the stresses in the two materials and the extension of the bar;
(*b*) the strain energy of each material;
(*c*) the force transmitted by the fixing pins.

The moduli of elasticity for steel and cast iron are 200 GN/m^2 and 100 GN/m^2 respectively.

23. A cantilever frame is pinned to a rigid vertical column at A and at B, vertically below A. The top member AC is horizontal, being 2 m long and having a cross-sectional area of 600 mm^2. There are two equal members AD and CD, each 1·2 m long and 1000 mm^2 in cross-section. The remaining member, BD, is 2 m long with a cross-sectional area of 800 mm^2. The frame is pin-jointed throughout.

Determine the forces in the members and the total strain energy of the frame when a load of 50 kN is applied vertically at C.

Modulus of elasticity = 200 GN/m^2.

24. A tie member in a roof truss is required to carry an axial load of 500 kN. The member is a flat bar of constant width and 15 mm thickness.

Design a double cover butt joint for this member. The allowable stresses are: Tension, 150 MN/m^2; Bearing 180 MN/m^2; Shear 90 MN/m^2.

4

Bending of beams

4.1 Pure bending

In Chapter 2 we discussed the internal forces that arise in a beam when it is subjected to an external loading. These opposing internal forces represent the resistance offered by the beam to bending and shearing and as a result bending and shear stresses occur. We know, from Chapter 3, that whenever stresses are set up within a material there is an accompanying distortion of the material.

In general, any section of a beam will be subjected to both bending and shearing actions and therefore bending and shear stresses will occur simultaneously. The resulting deflection will be the cumulative effect of both actions. Thus, to examine the effects of bending only, the beam must be loaded such that the shear force, and hence the shear stress, is zero. This can be achieved by loading a beam as shown in Fig. 4.1(*a*). The shear force and bending moment diagrams are shown in (*b*) and (*c*) respectively.

It will be seen that between B and C the shear force is zero and the bending moment is constant. The beam is then said to be in a state of pure bending between B and C.

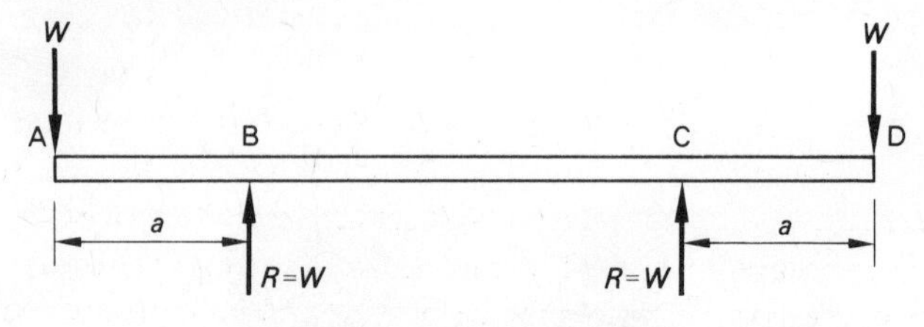

(a) Loading diagram

Fig. 4.1

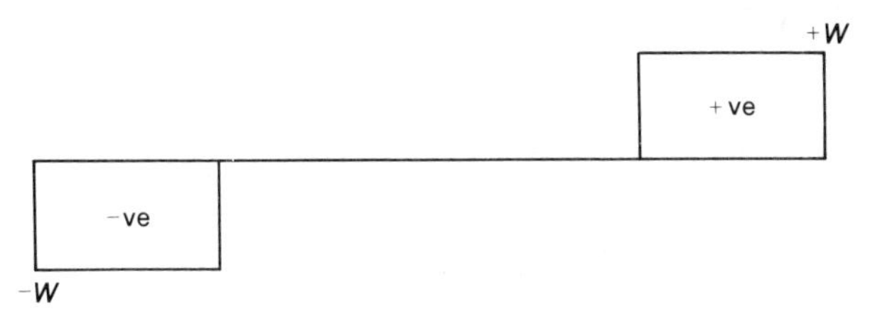

(b) Shear force diagram

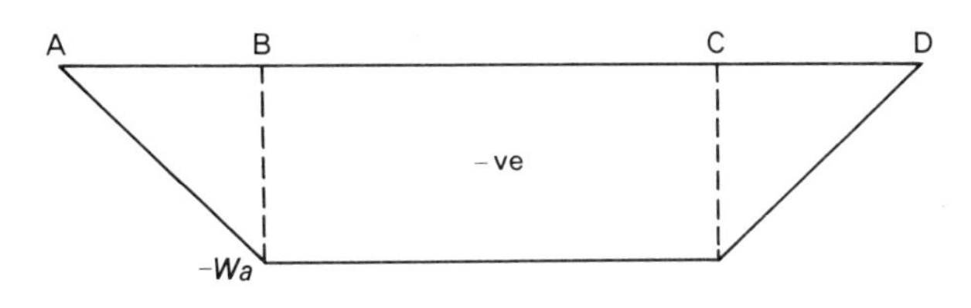

(c) Bending moment diagram

Fig. 4.1 Loading of a beam to produce pure bending

4.2 Bending stresses

Let us now consider the portion BC of the beam shown in Fig. 4.1 which is subjected to a constant bending moment $M = Wa$, as shown in Fig. 4.2.

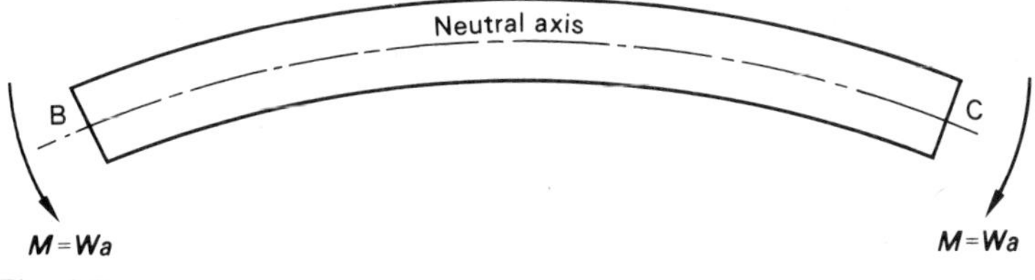

Fig. 4.2

For equilibrium of the beam between B and C the internal stresses must provide an internal moment of resistance which balances the applied moment. It will be apparent that the inside surface of the beam will be compressed while the outside surface is elongated. Thus, at some intermediate position within the beam there will be an unstressed layer. This is known as the neutral axis, as shown on Fig. 4.2. The position at which the neutral axis occurs is derived in Section 4.3.

We therefore need to determine how the stress is distributed over a section in order to provide this internal moment. To do this the following assumptions are made:

(*i*) that the beam is considered to be made up of an infinite number of longitudinal layers which remain in contact during bending but do not exert forces on each other;
(*ii*) that the material is subjected to longitudinal stress only and that there is no resultant longitudinal force on any section of the beam;
(*iii*) that the stress is proportional to the strain and within the limit of proportionality of the material;
(*iv*) that the modulus of elasticity of the material remains constant throughout (i.e. is the same for both tension and compression);
(*v*) that transverse plane sections before bending remain plane after bending (i.e. plane section AC in Fig. 4.3(*a*) remains plane section A′C′ in Fig. 4.3(*c*) after bending);
(*vi*) that the radius of curvature is large in relation to cross-sectional beam dimensions.

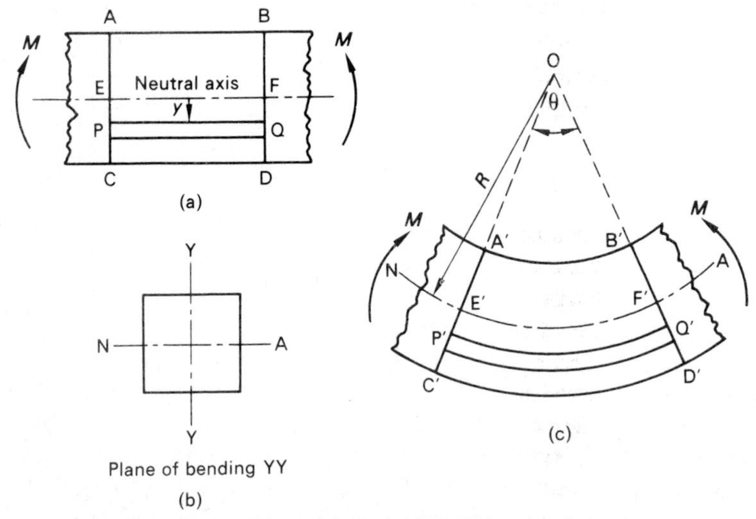

Fig. 4.3 Pure bending of a rectangular beam

Consider a section of a rectangular beam as in Fig. 4.3(*a*). Under the action of pure bending the sections AC and BD rotate to new positions A′C′ and B′D′, as shown in Fig. 4.3(*c*), and subtend an angle θ at the centre of curvature O. The neutral axis EF bends to a new position E′F′, the radius of curvature to this axis being R.

Let us now consider a section PQ at a distance y from the neutral axis which bends to new position P′Q′.

As the neutral axis is, by definition, unstrained during the bending action it follows that

$$EF = E'F' = R\theta$$

Then, initial length of PQ = EF

$$= R\theta$$

and, final length of PQ when bent = P′Q′

$$= (R + y)\theta$$

$$\therefore \quad \text{strain of PQ} = \frac{\text{change in length}}{\text{initial length}}$$

$$= \frac{(R + y)\theta - R\theta}{R\theta}$$

$$= \frac{y}{R}$$

Stress on PQ = E x strain

$$\sigma = \frac{Ey}{R}$$

or

$$\frac{\sigma}{y} = \frac{E}{R} \tag{4.1}$$

But for a given value of applied moment M the radius of curvature is a constant. Thus,

$$\frac{\sigma}{y} = \text{constant}$$

This shows that the stress distribution across the beam is linear and that the stress in any layer of material is proportional to the distance of that layer from the neutral axis. Hence the maximum stresses occur at the outer surfaces of the material which are furthest from the neutral axis. This applies for both symmetrical and unsymmetrical sections.

4.3 Position of neutral axis

Consider an element of area δA which is at a distance y from the N.A. as shown in Fig. 4.4(b). The distance of the N.A. from the outer surfaces are y_t and y_c as shown in Fig. 4.4(a), the stresses at these surfaces being σ_t and σ_c as indicated on the stress diagram Fig. 4.4(c).

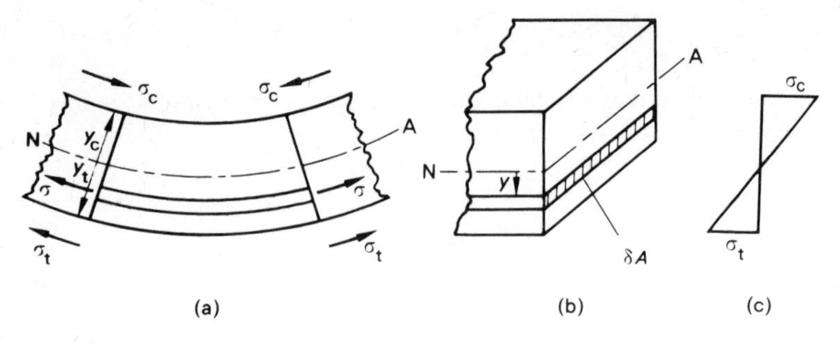

Fig. 4.4

Let the stress on the element be σ. Then, longitudinal force on element of area $\delta A = \sigma \,.\, \delta A$

But from eq. (4.1) $\quad \sigma = \dfrac{E}{R} y$

$\therefore \quad$ Force on element $= \dfrac{E}{R} y \,.\, \delta A$

and total force acting on beam section $= \sum \dfrac{E}{R} y \,.\, \delta A$

As the beam is in equilibrium under the action of a pure moment there cannot be a resultant longitudinal force on the beam section.

Therefore $\qquad \sum \dfrac{E}{R} y \,.\, \delta A = 0$

or, $\qquad \dfrac{E}{R} \sum y \,.\, \delta A = 0$

For this to be zero $\quad \Sigma\, y \,.\, \delta A = 0$

But $\Sigma y \,.\, \delta A$ is the first moment of area of the section about the neutral axis and can be written as $A\bar{y}$,

where $\quad A = \Sigma \delta A$ = total area of section,

and $\qquad \bar{y}$ = distance of centroid from the neutral axis

Therefore $\quad A\bar{y} = 0$

Now as A cannot equal zero then $\bar{y}$ must be zero, so indicating that the neutral axis passes through the centroid of the section. This is the case irrespective of the shape of the beam cross-section.

4.4 Moment of resistance

In the previous section it was shown that the force acting on the element of area δA was $(E/R)y \,.\, \delta A$. Then,

$$\text{Moment of force on element } \delta A \text{ about N.A.} = \frac{E}{R} y \,.\, \delta A \times y$$

$$= \frac{E}{R} y^2 \,.\, \delta A$$

$$\text{Total moment over the whole cross-section} = \sum \frac{E}{R} y^2 \,.\, \delta A$$

$$= \frac{E}{R} \sum y^2 \,.\, \delta A$$

$$\therefore \quad \text{Total moment of resistance} = \frac{E}{R} I$$

where $I = \Sigma y^2 \,.\, \delta A$ = second moment of area of cross-section about neutral axis (frequently erroneously referred to as the moment of inertia).

But for equilibrium the moment of resistance must equal the applied moment M.

Therefore $\quad M = \dfrac{E}{R} I$

or $\quad \dfrac{M}{I} = \dfrac{E}{R}$

Combining this equation with eq. (4.1) gives

$$\frac{M}{I} = \frac{\sigma}{y} = \frac{E}{R} \qquad (4.2)$$

This is the basic equation for bending and is referred to as the 'Engineers Theory of Bending' equation.

From eq. (4.2):

$$\sigma_{max} = \frac{My_{max}}{I}$$

or $\quad M = \sigma_{max} Z \qquad (4.3)$

where $\quad Z = \dfrac{I}{y_{max}}$ = section modulus

Note on simple bending theory

The theory given in the preceding sections provides a fairly accurate means of determining bending stresses in beams that are subjected to pure bending. However, as mentioned in section 4.1, a beam will generally be subjected to both bending and shear stresses. As a result of experiments it is known that the simple theory of bending is not limited to the case of pure bending but will give satisfactory results for varying bending moments, provided that stresses are not required too near the load positions.

Units

M is expressed in newton metres (Nm).

The basic units for E and σ are newtons per square metre (N/m^2). However to avoid very large numbers it is usual to employ multiple units.

R and y have units of metres (m).

I has basic units of $(\text{metres})^4$ but it is quite usual to use $(\text{millimetres})^4$.

4.5 Second moments of area

In section 4.4 it was stated that $\Sigma y^2 . \delta A$ was equal to the second moment of area of the section, I. The value of I will now be obtained for some typical beam sections.

(*a*) RECTANGLE (ABOUT N.A.)

Consider a rectangle of width b and depth d. To obtain the second moment of area about the neutral axis consider an elemental strip of thickness δy at a distance y from the neutral axis as shown in Fig. 4.5.

Area of elemental strip $= b . \delta y$

Second moment of area of elemental strip about N.A. $= by^2 . \delta y$

The total second moment of area of the section is equal to the summation over the whole area of all such elements; i.e.,

$$I = \int_{-d/2}^{d/2} by^2 . dy$$

$$= \left[\frac{by^3}{3}\right]_{-d/2}^{d/2}$$

$$= \frac{bd^3}{12}$$

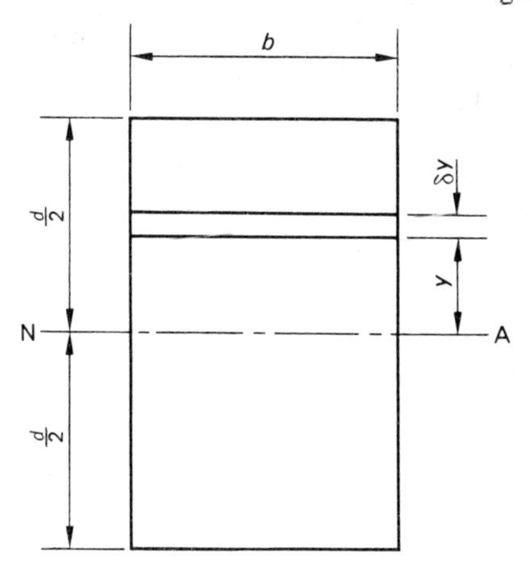

Fig. 4.5

(b) RECTANGLE (ABOUT AXIS XX)

If the second moment of area of the section is required about an axis other than the N.A., e.g. axis XX on Fig. 4.6, then this may be achieved by:

(*i*) integrating about axis XX; or
(*ii*) applying parallel axis theorem.

(*i*) *Integration about axis XX:*
Referring to Fig. 4.6:

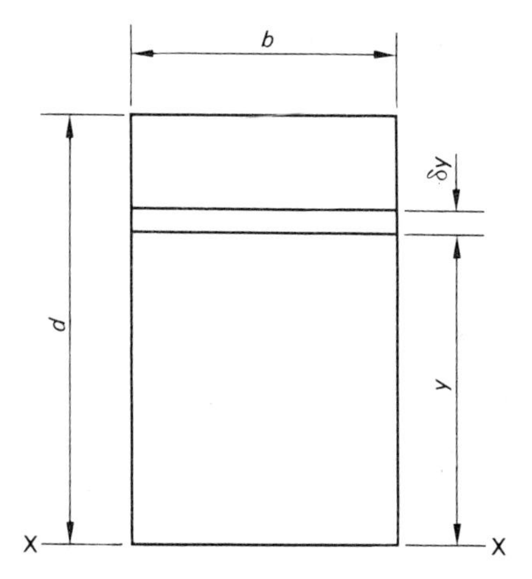

Fig. 4.6

$$\text{Second moment of section about XX} = I_{XX} = \int_0^d by^2 \, . \, dy$$

$$= \left[\frac{by^3}{3}\right]_0^d$$

$$= \frac{bd^3}{3}$$

(*ii*) *Parallel axis theorem:*

When the second moment of area of a section about an axis through its centroid is known, the second moment of area about any parallel axis may be obtained by applying the parallel axis theorem.

Consider a section whose X and Y axes pass through the centroid G, as shown in Fig. 4.7. Suppose the second moment of area is required about axis X′X′ which is parallel to the X axis but at a distance h from it.

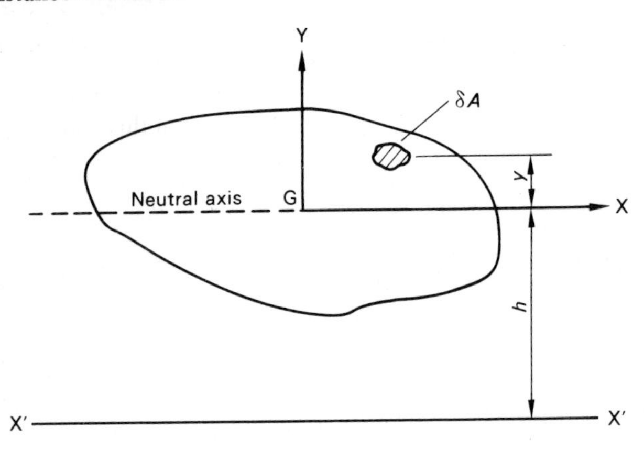

Fig. 4.7

Consider an elemental area δA at a distance y from the X axis.

$$\text{Then, } I_{X'X'} = \int (y+h)^2 \, . \, \delta A = \int (y^2 + 2hy + h^2) \, . \, \delta A$$

$$= \int y^2 \, . \, \delta A + \int 2hy \, . \, \delta A + \int h^2 \, . \, \delta A$$

$$= I_{XX} + 2h \int y \, . \, \delta A + h^2 \int \delta A$$

$$= I_{NA} + 0 + h^2 A$$

The second term is zero since $\int y \, . \, \delta A$ is the first moment of area about the axis X through the centroid G.

Then, $I_{X'X'} = I_{NA} + Ah^2$

Thus the second moment of area of a section about an axis parallel with an axis through its centroid is equal to the second moment of area about the axis through the centroid plus the area of the section multiplied by the square of the distance between the parallel axes.

It follows from the above expression that the second moment of area will always be a minimum for the axis through the centroid.

Applying this to the rectangle shown in Fig. 4.6 gives

$$I_{XX} = \frac{bd^3}{12} + bd\left(\frac{d}{2}\right)^2$$

$$= \frac{bd^3}{3}$$

(c) I SECTION

Referring to Fig. 4.8 it will be seen that the value of I_{NA} for this section is equal to the difference in I values of two rectangles.

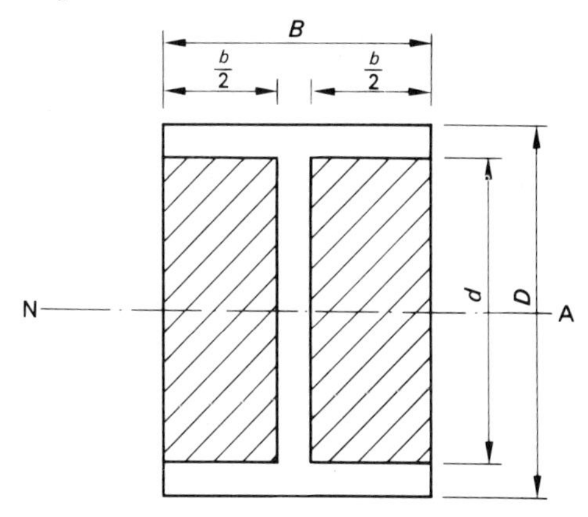

Fig. 4.8

Thus, $I_{NA} = \frac{BD^3}{12} - \frac{bd^3}{12}$

(d) CIRCULAR SECTION

Although the second moment of area can be obtained directly about a diameter such as XX in Fig. 4.9, the integration is not a simple one and it is much easier to determine the polar second moment of area, as indicated in section 5.3 and then apply the perpendicular axis theorem.

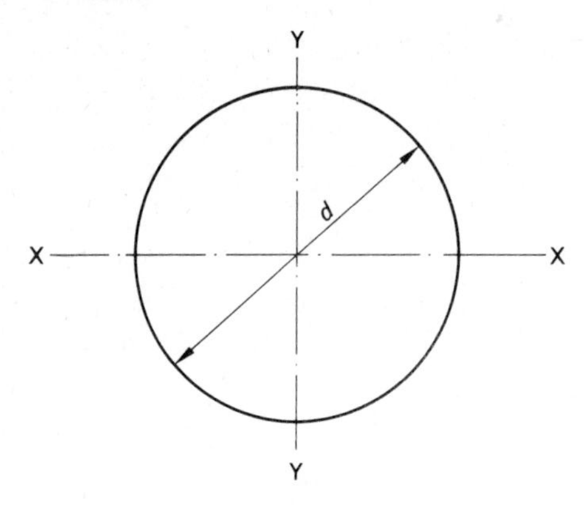

Fig. 4.9

Perpendicular axis theorem:

Consider the section shown in Fig. 4.10 where O is the point of intersection of the three perpendicular axes X, Y, and Z – the Z axis being perpendicular to the plane of the diagram and frequently referred to as the polar axis.

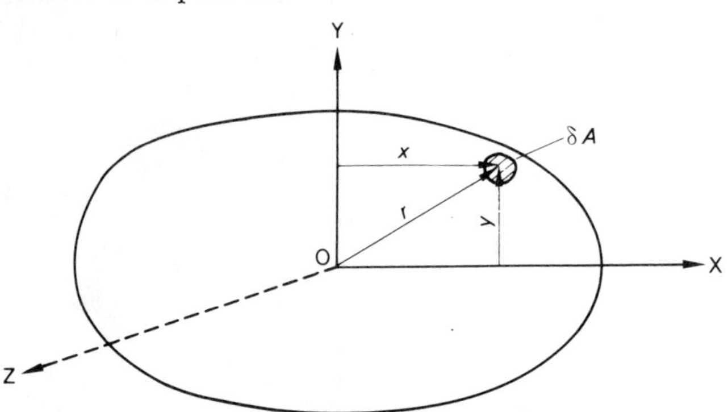

Fig. 4.10

For any element of area δA:

$$I_X = \int y^2 . \delta A$$

$$I_Y = \int x^2 . \delta A$$

and $$I_Z = \int r^2 . \delta A$$

But $$r^2 = x^2 + y^2$$

$\therefore$ $$I_Z = \int (x^2 + y^2) . \delta A$$

$$I_Z = I_X + I_Y$$

Thus the second moment of area with respect to an axis through any point O perpendicular to the section is equal to the sum of the second moments of area with respect to any two mutually perpendicular axes through the same point.

The second moment of area about axis Z is known as the polar second moment of area J and from section 5.3, $J = \pi d^4/32$ for a circular section. Also for a circular section it follows that $I_X = I_Y = I_{DIA}$, where I_{DIA} is the second moment of area about any diameter

Thus $$J = 2I_{DIA}$$

$$I_{DIA} = \frac{\pi d^4}{64}$$

Example 4.1

Determine the second moment of area about the neutral axis for the section shown in Fig. 4.11.

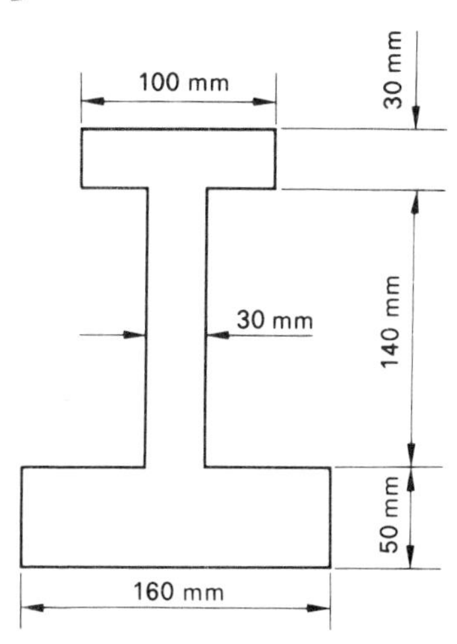

Fig. 4.11 Example 4.1

Solution

The second moment of area about the NA for this, or any other section, can be determined by either of two methods:

(1) by determining the position of the NA and then obtaining the second moment of area about this axis;

(2) by determining the position of the NA and second moment of area relative to a datum axis and then using the parallel axis theorem to evaluate I_{NA}.

The second method lends itself to a tabular layout and is strongly recommended. The two methods will now be illustrated.

METHOD (1)

The position of the neutral axis is obtained by taking moments about axis XX (ref. Fig. 4.12).

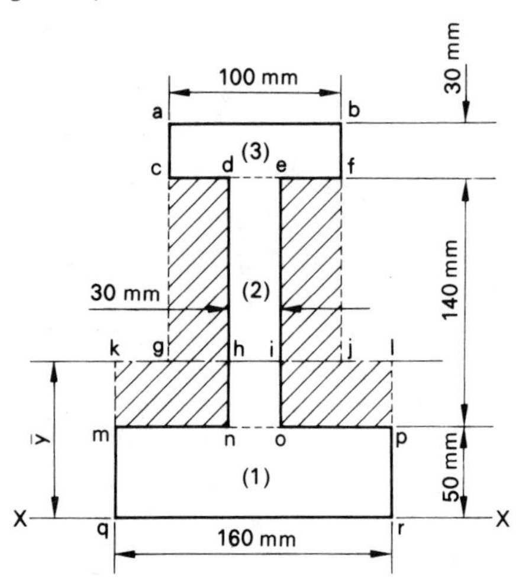

Fig. 4.12 Determination of position of neutral axis–Example 4.1

$$\text{Area (1)} = 160 \times 50 = 8000 \text{ mm}^2$$

Distance of centroid of area (1) from XX = 25 mm

$$\text{Area (2)} = 30 \times 140 = 4200 \text{ mm}^2$$

$$\text{Distance of centroid of area (2) from XX} = 50 + \frac{140}{2} = 120 \text{ mm}$$

$$\text{Area (3)} = 100 \times 30 = 3000 \text{ mm}^2$$

$$\text{Distance of centroid of area (3) from XX} = 50 + 140 + \frac{30}{2} = 205 \text{ mm}$$

Then by moments about XX:

$$(8000 + 4200 + 3000)\,\bar{y} = (8000 \times 25 + 4200 \times 120 + 3000 \times 205)$$

$$\bar{y} = 86{\cdot}772 \text{ mm}$$

I_{NA} can now be obtained for all the individual rectangles having their base coincident with the NA.

Thus $I_{NA} = I_{abjg} - (I_{cdhg} + I_{efji}) + I_{klrq} - (I_{khnm} + I_{ilpo})$

$$= \frac{100 \times 132{\cdot}228^3}{3} - \frac{70 \times 102{\cdot}228^3}{3} + \frac{160 \times 87{\cdot}772^3}{3} - \frac{130 \times 37{\cdot}772^3}{3}$$

$$\approx 80 \times 10^6 \text{ mm}^4$$

METHOD (2)

	1	2	3	4	5
Section	A [mm²]	y [mm]	Ay [mm³]	Ay^2 [mm⁴]	I_G [mm⁴]
1	160 x 50 = 8000	25	200 000	5 000 000	$\frac{160 \times 50^3}{12} = 1\,666\,666$
2	30 x 140 = 4200	120	504 000	60 480 000	$\frac{30 \times 140^3}{12} = 1\,036\,000$
3	100 x 30 = 3000	205	615 000	126 075 000	$\frac{100 \times 30^3}{12} = 225\,000$
Σ	15 200		1 319 000	191 555 000	2 927 666

Column 1 gives the area of the individual section.

Column 2 gives the distance of the centroid of the individual section from XX.

Column 3 gives the first moment of area of the individual section about XX.

Column 5 gives the value of I about the centroid of the individual section from $bd^3/12$.

Columns (5 + 4) then gives the value of I for the individual section about the datum axis XX (c.f. expression in parallel axis theorem, section 4.5).

Then $\bar{y} = \frac{\Sigma Ay}{\Sigma A} = \frac{1\,319\,000}{15\,200} = 86{\cdot}772 \text{ mm}$

and $I_{XX} = \Sigma I_G + \Sigma Ay^2$

$= 2\,927\,666 + 191\,555\,000$

$= 194\,482\,666 \text{ mm}^4$

Using the parallel axis expression gives:

$$I_{NA} = I_{XX} - (\Sigma A)\bar{y}^2$$

$$= 194\,482\,666 - 15\,200 \times 86{\cdot}772^2$$

$$\approx 80 \times 10^6 \text{ mm}^4$$

These two methods should be carefully studied and proficiency obtained in one of them. While the authors recommend method (2) it is largely a matter of individual choice.

Example 4.2

A steel beam, having the cross-section used in example 4.1, is 12 m long. The beam is simply supported at its ends. If the density of steel is 7900 kg/m^3 determine the maximum stress created in the beam under its own weight.

If the maximum allowable working stresses are 150 MN/m^2 in tension and 120 MN/m^2 in compression what is the maximum concentrated load that can be carried at the centre of the beam?

Solution

From example 4.1 the second moment of area about the neutral axis was found to be 80 x 10^6 mm^4 and the area of the section to be 15 200 mm^2.

The weight of the beam is equivalent to a uniformly distributed loading. Thus:

$$\text{Mass of beam} = \frac{15\,200}{10^6}\,[\text{m}^2] \times 7900 \left[\frac{\text{kg}}{\text{m}^3}\right]$$

$$= 120{\cdot}08 \text{ kg/metre length}$$

$$\therefore \quad \text{Weight of beam} = 120{\cdot}08 \times 9{\cdot}81$$

$$= 1178 \text{ N/m}$$

From example 2.6, the maximum bending moment for a beam of length l simply supported at its ends and carrying a uniformly distributed load w is:

$$M_{max} = \frac{wl^2}{8} \quad \text{at the centre span}$$

Thus, for this problem,

$$M_{max} = \frac{1178\ [\text{N/m}] \times 12^2\ [\text{m}^2]}{8}$$

$$= 18\,848\ \text{N/m}$$

As the beam is subjected to sagging bending the upper flange will be in compression and the lower flange in tension.

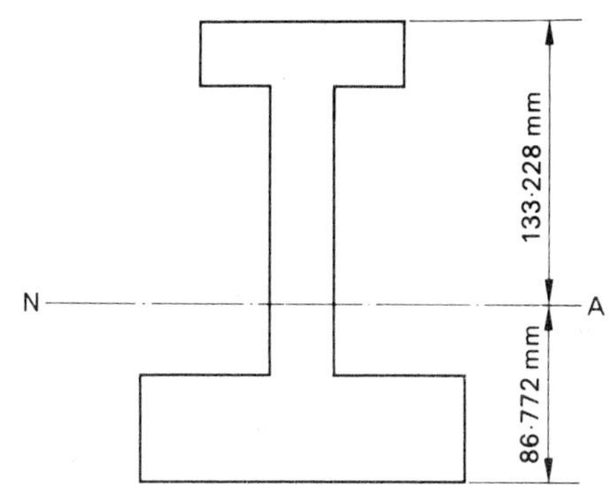

Fig. 4.13 Example 4.2

Then, referring to Fig. 4.13:

$$\text{Maximum tensile stress} = \frac{M_{max}\, y_t}{I}$$

$$= \frac{18\,848\ [\text{N m}] \times 86{\cdot}772/10^3\ [\text{m}]}{80/10^6\ [\text{m}^4]}$$

$$= 20{\cdot}44 \times 10^6\ \text{N/m}^2$$

$$= 20{\cdot}44\ \text{MN/m}^2$$

$$\text{Maximum compressive stress} = \frac{M_{max}\, y_c}{I}$$

$$= \frac{18\,848 \times 133{\cdot}228/10^3}{80/10^6}$$

$$= 31{\cdot}39\ \text{MN/m}^2$$

The maximum stress created in the beam due to its own weight is $31{\cdot}39\ \text{MN/m}^2$.

It is evident that as the allowable stress in compression is less than that in tension then the compression flange will be the critical one since the stress produced is greater for this flange.

The maximum allowable bending moment is therefore given by

$$M_{all} = \frac{\sigma_c I}{y_c}$$

where σ_c = allowable compressive stress

$= 120\ MN/m^2$

Thus, $$M_{all} = \frac{120 \times 10^6 \times 80/10^6}{133{\cdot}228/10^3}$$

$= 72\,000\ N\,m$

∴ Allowable bending moment due to concentrated load

$= 72\,000 - 18\,848$

$= 53\,152\ N\,m$

But the maximum bending moment due to a concentrated load W applied at the centre span is $Wl/4$ (see section 2.1) and occurs at the centre span.

Thus, $$53\,152 = \frac{W \times 12}{4}$$

$W = 17\,717\ N$

$= 17{\cdot}717\ kN$

The maximum concentrated load that can be applied at the centre of the beam is 17·717 kN.

4.6 Effect of web of an I-section

In the case of a thin web I-section beam the contribution of the web to the resisting moment at the section is very small and for initial design purposes it is customary to consider the bending moment to be resisted by the flanges only. This assumption may be justified by considering the stress distribution due to the weight of the beam in example 4.2.

Figure 4.14 shows the section and the calculated stress distribution.

Stress at inside surface of compression flange

$$= \frac{103{\cdot}228}{133{\cdot}228} \times 31{\cdot}39$$

$= 24{\cdot}32\ MN/m^2$

∴ Average stress on compression flange $= \left(\frac{31{\cdot}39 + 24{\cdot}32}{2}\right)$

$= 27{\cdot}855\ MN/m^2$

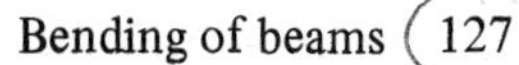

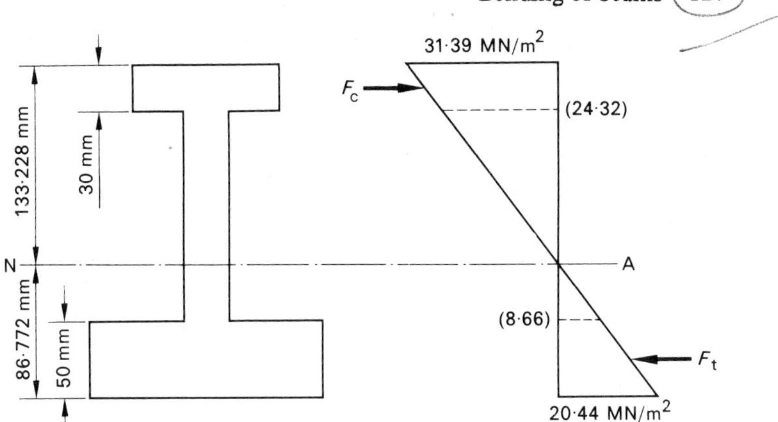

Fig. 4.14 Stress distribution for *I*-beam

Force on compression flange, F_c = average stress x area

$$= 27{\cdot}855 \times 10^6 \left[\frac{\text{N}}{\text{m}^2}\right] \times \frac{3000}{10^6} [\text{m}^2]$$

$$= 83\,565 \text{ N}$$

$$\text{Moment of } F_c \text{ about neutral axis} = 83\,565 \text{ [N]} \times \frac{118{\cdot}228}{10^3} \text{ [m]}$$

$$= 9880 \text{ N m}$$

Similarly, stress at inside surface of tension flange

$$= \frac{36{\cdot}772}{86{\cdot}772} \times 20{\cdot}44$$

$$= 8{\cdot}662 \text{ MN/m}^2$$

$$\text{Average stress on tension flange} = \left(\frac{20{\cdot}44 + 8{\cdot}662}{2}\right)$$

$$= 14{\cdot}551 \text{ MN/m}^2$$

$\therefore$ Force on tension flange, F_t $= 14{\cdot}551 \times 8000$

$$= 116\,408 \text{ N}$$

$$\text{Moment of } F_t \text{ about neutral axis} = 116\,408 \times \frac{61{\cdot}772}{10^3}$$

$$= 7191 \text{ N m}$$

∴ Contribution of flanges to total resisting moment $= 9880 + 7191$

$= 17\,071$ N m

But the total resisting moment is 18 848 N m (see example 4.1).

Thus, contribution of flanges $= \dfrac{17\,071}{18\,848} \times 100$

$= 90{\cdot}55\%$

This indicates that for thin web sections it is satisfactory as a first approximation to assume that the bending moment is resisted only by the flanges.

Example 4.3

A cantilever having a length of 1·25 m carries a concentrated load of 10 kN at its free end. The section of the cantilever is shown in Fig. 4.15. Determine the maximum stresses in the material and show the stress distribution across the section.

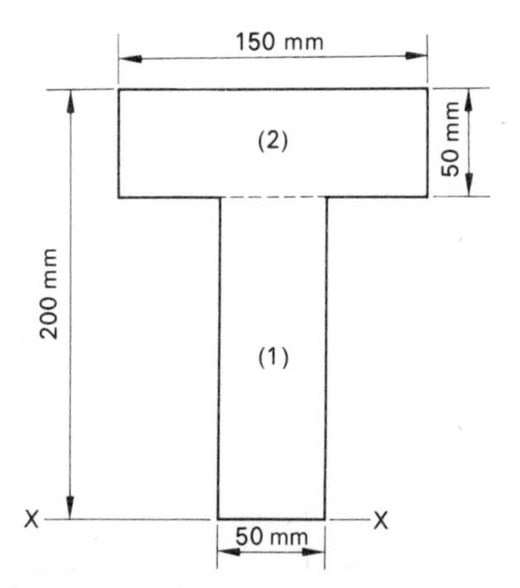

Fig. 4.15 Example 4.3

Solution

The position of the neutral axis and the value of the second moment of area about this axis are obtained using the tabular method set out in Example 4.1.

Section	A [mm^2]	y [mm]	Ay [mm^3]	Ay^2 [mm^4]	I_G [mm^4]
1	50 x 150 = 7500	75	562 500	42 187 500	$\frac{50 \times 150^3}{12} = 14\,062\,500$
2	150 x 50 = 7500	175	1 312 500	229 687 500	$\frac{150 \times 50^3}{12} = 1\,562\,500$
Σ	15 000		1 875 000	271 875 000	15 625 000

$$\bar{y} = \frac{\Sigma Ay}{\Sigma A} = \frac{1\,875\,000}{15\,000} = 125 \text{ mm}$$

$$I_{XX} = \Sigma I_G + \Sigma Ay^2$$
$$= 15\,625\,000 + 271\,875\,000$$
$$= 287\,500\,000 \text{ mm}^4$$

$$I_{NA} = I_{XX} - (\Sigma A)\,\bar{y}^2$$
$$= 287\,500\,000 - 15\,000 \times 125^2$$
$$= 53\,125\,000 \text{ mm}^4$$

The maximum moment will occur at the built-in end, i.e.,

$$M_{max} = 10 \text{ [kN]} \times 1{\cdot}25 \text{ [m]}$$
$$= 12{\cdot}5 \text{ kN m}$$

Referring to Fig. 4.16(*a*) and applying

$$\sigma = \frac{My}{I} \qquad \text{[from eq. (4.4)]}$$

gives $$\sigma_c = \frac{12\,500 \text{ [N m]} \times 125 \times 10^{-3} \text{ [m]}}{53{\cdot}125 \times 10^{-6} \text{ [m}^4\text{]}}$$
$$= 29{\cdot}41 \text{ MN/m}^2$$

and $$\sigma_t = \frac{12\,500 \times 75 \times 10^{-3}}{53{\cdot}125 \times 10^{-6}}$$
$$= 17{\cdot}65 \text{ MN/m}^2$$

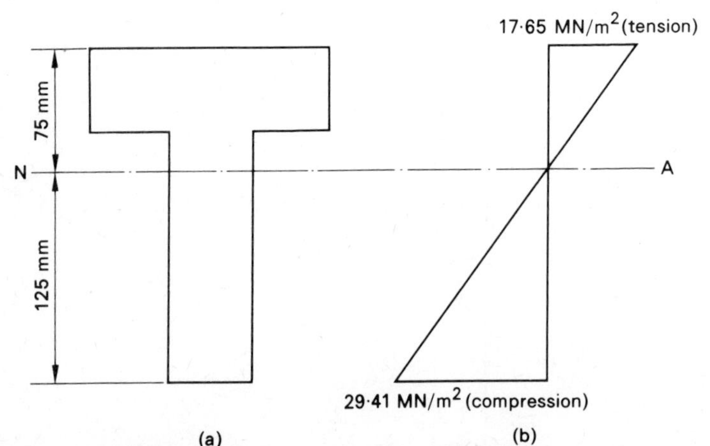

Fig. 4.16 Position of neutral axis and stress distribution–Example 4.3

The maximum compressive stress is 29·41 MN/m^2, the maximum tensile stress is 17·65 MN/m^2 and the stress distribution is as shown in Fig. 4.16(*b*).

Example 4.4

A steel tube having an external diameter of 20 mm and a length of 4 m is simply supported, as shown in Fig. 4.17, and carries a load of 26 N at each end. Under these conditions the deflection at the centre span is 15 mm.

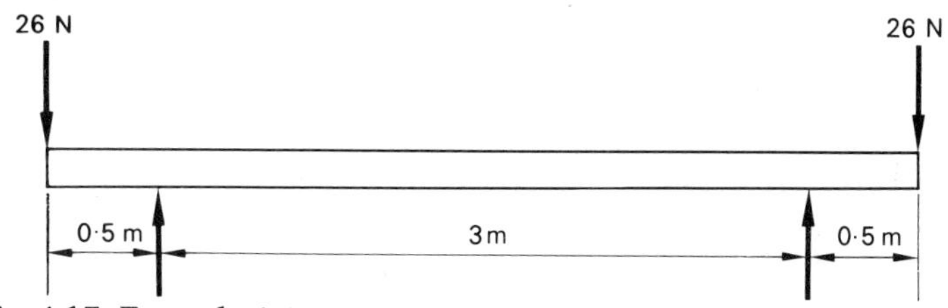

Fig. 4.17 Example 4.4

Determine the internal diameter of the tube and the maximum stress created. The weight of the tube is to be neglected.
Modulus of elasticity = 210 x 10^9 N/m^2.

Solution

Between the two supports the beam is subjected to pure bending and will therefore bend into a circular arc. If the central deflection is δ and the radius of curvature R then from Fig. 4.18, by the property of intersecting chords

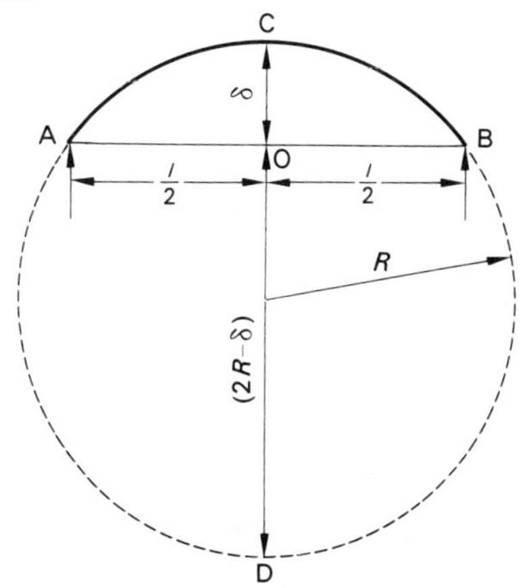

Fig. 4.18 Circular arc bending–Example 4.4

$$AO \times BO = CO \times DO$$

$$(l/2)^2 = (2R - \delta)\delta$$

$$= 2R\delta - \delta^2$$

Now if the stresses are within the elastic limit of the material then δ will be small in comparison with R. Hence δ^2 can be neglected in comparison with $2R\delta$.

Thus $\frac{l^2}{4} \approx 2R\delta$

or $R \approx \frac{l^2}{8\delta}$

From eq. (4.2):

$$\frac{M}{I} = \frac{E}{R}$$

$$\therefore \quad I = \frac{MR}{E}$$

$$= \frac{Ml^2}{8E\delta}$$

Substituting $M = 26 \times 0{\cdot}5 = 13$ N m, $l = 3$ m, $E = 210 \times 10^9$ N/m^2 and $\delta = 0{\cdot}015$ m, gives

$$I = \frac{13 \times 3^2}{8 \times 210 \times 10^9 \times 0{\cdot}015}$$

$$= \frac{4{\cdot}64}{10^9} \text{ m}^4$$

If the internal diameter of the tube is d [mm] then

$$I = \frac{\pi}{64}\left(\frac{20^4 - d^4}{10^{12}}\right) \text{ m}^4$$

Equating expressions for I

$$\frac{4{\cdot}64}{10^9} = \frac{\pi}{64}\left(\frac{20^4 - d^4}{10^{12}}\right)$$

$$(20^4 - d^4) = 94\,526$$

$$d = 16 \text{ mm}$$

The inside diameter of the tube is 16 mm.

$$\text{Maximum bending stress, } \sigma = \frac{My}{I}$$

$$= \frac{13 \times 0{:}010}{4{\cdot}64/10^9}$$

$$= 28 \text{ MN/m}^2$$

As the section is symmetrical this stress applies to both tension and compression surfaces.

Example 4.5

A uniform beam of symmetrical section is 6 m long and is freely supported at its ends. It carries a distributed load of 40 kN/m over the complete span and concentrated loads of 30 kN at 1·5 m from each end. The beam is built up of a standard **I**-section, 0·3 m deep and having a second moment of area of 200×10^6 mm^4 with a 0·2 m wide by 20 mm thick plate riveted to each flange. Calculate the maximum bending stress in the beam.

What is the percentage increase in the strength of the beam due to the addition of the plates?

Solution

A diagram is given in Fig. 4.19.

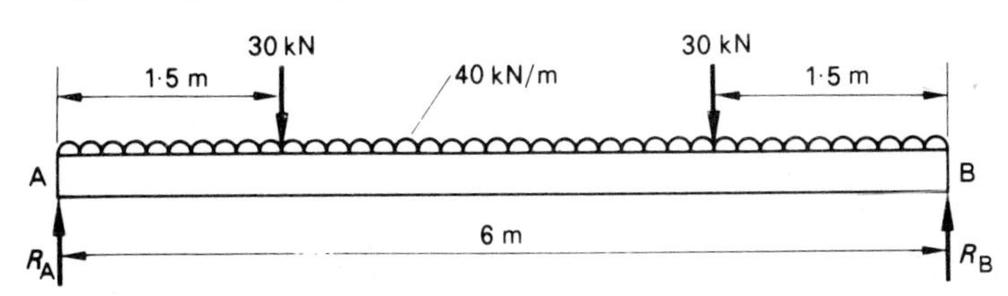

Fig. 4.19 Example 4.5

For vertical equilibrium of the beam

$$R_A + R_B = 30 + 30 + 6 \times 40$$

$$= 300 \text{ kN}$$

Thus, by symmetry $R_A = R_B = 150$ kN

The maximum moment will occur at the centre of the beam, i.e.,

$$M_{max} = 150 \times 3 - 30 \times 1{\cdot}5 - (40 \times 3) \times 1{\cdot}5$$

$$= 225 \text{ kN m}$$

Second moment of area of I-section $= 200 \times 10^6$ mm^4

Second moment of area of plates $= \dfrac{0{\cdot}2 \times 0{\cdot}34^3}{12} - \dfrac{0{\cdot}2 \times 0{\cdot}30^3}{12}$ [m^4]

$= 205 \times 10^6$ mm^4

$\therefore$ Second moment of area of beam $= 200 \times 10^6 + 205 \times 10^6$

$= 405 \times 10^6$ mm^4

$= 405 \times 10^{-6}$ m^4

Then $\sigma = \dfrac{My}{I}$

$= \dfrac{225 \times 10^3 \times 0{\cdot}17}{405 \times 10^{-6}}$

$= 94{\cdot}44$ MN/m^2

The maximum bending stress in the beam is 94·44 MN/m^2 for both tension and compression.

From eq. (4.3)

$$M = \sigma_{max} Z$$

where $Z = \dfrac{I}{y_{max}}$ = section modulus

Thus, for a given maximum stress σ_{max}, the moment that can be carried by the beam is proportional to its section modulus.

Thus, for I-section only, $Z = \dfrac{200 \times 10^6}{150} = 1{\cdot}333 \times 10^6$ mm^3

and for I-section plus plates, $Z = \dfrac{405 \times 10^6}{170} = 2{\cdot}382 \times 10^6$ mm^3

Therefore, increase in strength due to addition of plates

$= \left(\dfrac{2{\cdot}382 - 1{\cdot}333}{1{\cdot}333}\right) \times 100$

$= 78{\cdot}7\%$

Problems

1. Determine the radius of curvature to which a steel strip 5 mm thick and 40 mm wide may be bent if the maximum stress is not to exceed 100 MN/m^2.
Modulus of elasticity = 210 GN/m^2.

2. A mild steel has an elastic limit stress of 300 MN/m^2 and a modulus of elasticity of 200 GN/m^2. What is the smallest diameter drum around which a 4 mm thick strip can be wound without the elastic limit being exceeded?

3. Derive the equations of the simple bending theory and state clearly the assumptions made.

Determine the maximum stress set up in a steel beam 100 mm wide and 6 mm deep bent into a circular arc of radius 6 m and calculate the value of the bending moment required to produce this curvature. Modulus of elasticity = 200 GN/m^2.

(E.M.E.U.)

4. An axle having a diameter of 150 mm and a length of 2·5 m is supported symmetrically in two bearings that are 2 m apart. A vertical load of 150 kN is applied at each end of the axle. Determine the maximum bending stress in the axle and its radius of curvature. Modulus of elasticity = 210 GN/m^2.

5. A steel bar of rectangular cross-section is bent into a circular arc of radius 15 m. The width of the section is twice its depth and the depth is radial. The maximum stress created due to bending is 60 MN/m^2. Determine the dimensions of the section and the magnitude of the applied bending moment.
Modulus of elasticity = 200 GN/m^2.

6. State three assumptions made in the theory of simple bending. Assuming the expression $\sigma/y = E/R$ for simple bending, derive the expression $M/I = E/R$, stating the meaning of all the symbols.

A horizontal wooden joist of rectangular section is 200 mm deep and 75 mm wide. It is simply supported at its ends over a span of 6 m. Calulate the maximum intensity of the uniformly distributed load it can carry over the whole span if the maximum tensile stress in the joist is not to exceed 6·8 MN/m^2.

(N.C.T.E.C.)

7. A steel pipe having an external diameter of 100 mm and an internal diameter of 80 mm is used as a beam over a span of 3 m.

Determine the maximum central load that can be applied if the bending stress is not to exceed 125 MN/m^2. The weight of the pipe is to be ignored.

How does the strength of this beam compare with that of a solid circular bar having the same cross-sectional area?

8. A rolled steel joist (R.S.J.) has top and bottom flanges each 150 mm x 25 mm and a web 250 mm x 15 mm. The joist is simply supported over a span of 4 m and carries a uniformly distributed load, which includes its own weight, of 30 kN/m. If the maximum bending stress is not to exceed 80 MN/m^2 what is the maximum central load that can be applied in addition to the uniformly distributed load?

9. Derive the expression $\sigma/y = E/R$ for a beam subjected to simple bending stating the meaning of the symbols.

A uniform horizontal beam, 10 m long, is simply supported at its ends and carries a uniformly distributed load, including its own weight, of 2·5 kN/m over the whole span, together with a central load of 30 kN. The section of the beam has an overall depth of 250 mm and is symmetrical about the neutral axis for bending. The second moment of area of the section about this axis is 270×10^{-6} m^4 and the modulus of elasticity for the material is 206 GN/m^2. Calculate:

(*a*) the maximum bending stress in the material;
(*b*) the radius of curvature at the mid-span. (N.C.T.E.C.)

10. A steel girder, 8 m long, is simply supported at its ends and concentrated loads of 60 kN and 90 kN are applied at distances of 2 m and 5 m respectively from the left-hand end. The girder is of **I**-section, 140 mm deep, with flanges 100 mm wide by 12 mm thick and a web thickness of 8 mm.

Determine:

(*a*) the magnitude of the maximum bending moment and the position at which it occurs;
(*b*) the maximum bending stresses;
(*c*) the radius of curvature at the section where the moment is a maximum.

Modulus of elasticity = 200 GN/m^2.

11. A cantilever has a free length of 3 m. It is of **T**-section with a flange 100 mm x 20 mm and a web 200 mm x 12 mm, the flange being at the top. If the allowable tensile stress is 40 MN/m^2 what uniformly distributed load can be applied over the whole length of the cantilever? What is then the maximum compressive stress?

12. A 2-m length of channel having the section shown in Fig. 4.20 is arranged as a horizontal cantilever. If the maximum bending stress is not to exceed 120 N/mm^2, determine the magnitude of the maximum concentrated load that can be supported at the free end. (U.L.C.I.)

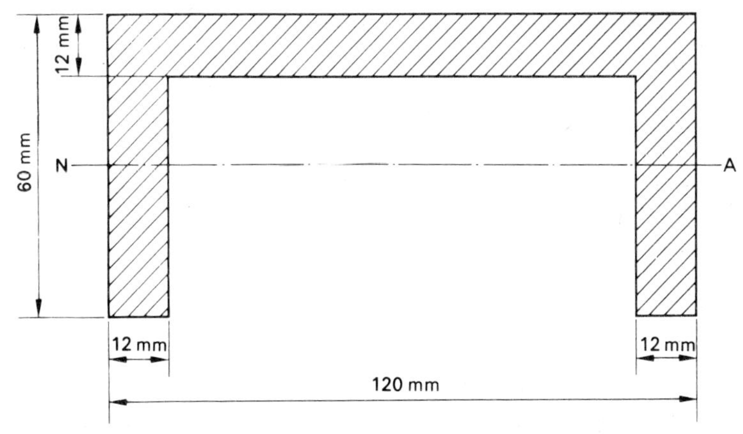

Fig. 4.20

13. A horizontal cantilever supports loads as shown in Fig. 4.21 and has the uniform T-section shown in Fig. 4.22. The flange is uppermost and horizontal.

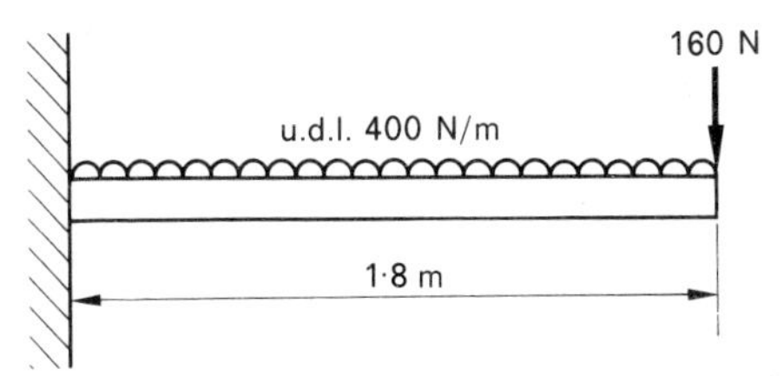

Fig. 4.21

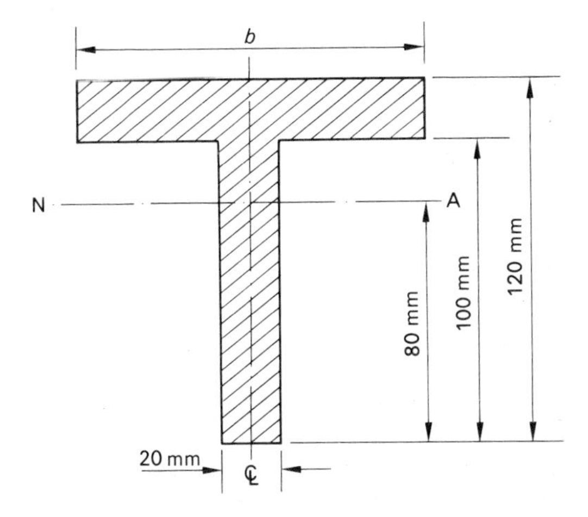

Fig. 4.22

Calculate:

(*a*) the width b of the flange;
(*b*) the second moment of area about NA;
(*c*) the maximum tensile and compressive stresses set up.

(N.C.T.E.C.)

14. A cast iron beam, 8 m long, is simply supported at its ends and has the cross-section shown in Fig. 4.23. If the density of cast iron is 7·9 Mg/m^3 determine the maximum tensile and compressive stresses set up due to the weight of the beam.

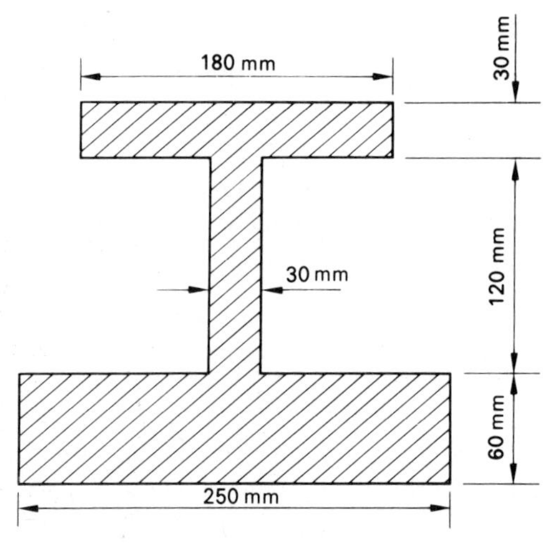

Fig. 4.23

15. A floor is carried by timber beams pitched at 0·4 m and simply supported over a 5 m span. The floor loading is 4 kN/m^2 and there is also a concentrated load of 1·5 kN per beam which acts at a distance of 1 m from one support. The beam depth is to be three times its width and the maximum stresses are not to exceed 6 MN/m^2 in tension and 8 MN/m^2 in compression. Determine the position and magnitude of the maximum bending moment and evaluate suitable beam dimensions.

16. A steel beam 10 m long has a cross-section as shown in Fig. 4.24. The beam is simply supported at its ends and carries point loads of 20 kN each at a distance of 2 m from each end.

Calculate the maximum bending stress and radius of curvature for that part of the beam between the loads.
Modulus of elasticity of steel = 200 GN/m^2.

(E.M.E.U.)

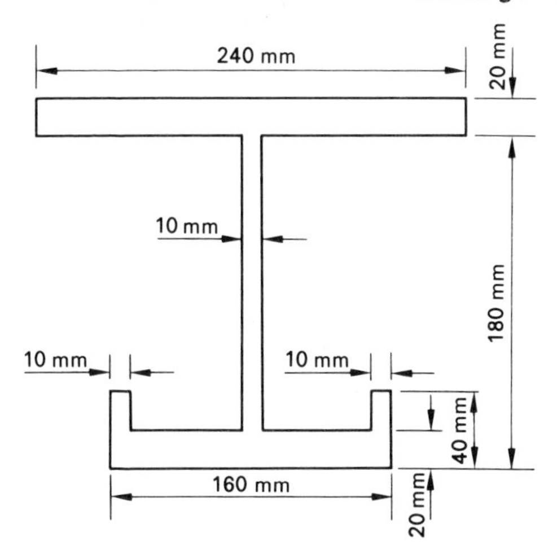

Fig. 4.24

17. A beam of length 8 m is simply supported at its ends and carries a uniformly distributed load of 18 kN/m over the complete span and a central concentrated load of 80 kN. The beam is formed from a standard steel **I**-section, 350 mm deep with 150 mm wide flanges and a steel plate 200 mm wide by 15 mm thick is riveted to each flange. The second moment of area of the **I**-section alone is 230×10^{-6} m^4.

Determine the maximum bending stress in the plate and flanges and the proportion of the bending moment carried by the plates. Neglect the effect of the rivet holes.

18. A steel **T**-section is simply supported over a span of 4 m. The flange is 120 mm wide, the total depth of section is 160 mm and the material has a thickness of 20 mm throughout.

The beam is to carry three equal concentrated loads at equal distances across the span and the maximum tensile stress is not to exceed 120 MN/m^2.

Determine whether the flange should be at the top or bottom and the magnitude of the allowable loads.

5

Torsion of circular shafts

5.1 Torsional stress

When a shaft of circular cross-section is acted upon by a torque about its longitudinal axis it can be confirmed experimentally that provided the angle of twist is small then:

(*i*) all circular sections of the shaft remain sensibly circular during twisting and their diameters remain unchanged;
(*ii*) plane cross-sections remain plane (for other cross-sections this is not generally the case).

Consider the shaft shown in Fig. 5.1 to be made up of a series of short cylinders. When a torque is applied to the shaft resisting shear stresses will be set up throughout the complete length of the shaft and therefore a relative movement will occur between the cylinders. These

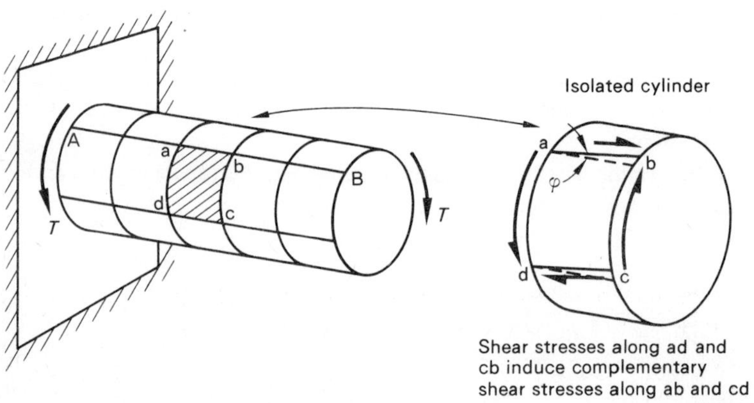

Fig. 5.1 Torsion of a circular shaft

shearing stresses act tangentially and will introduce complementary shear stresses as shown on the isolated cylindrical element in Fig. 5.1.

As the end faces of this short cylindrical element remain plane and parallel, any rectangular element abcd on the surface of the shaft will be in a state of pure shear. The effect of this is to cause longitudinal distortion of the material so that the lengths ab and dc are each sheared through angle ϕ. This twisting effect will be the same for each individual cylinder so that an original line AB drawn on the surface of the shaft parallel with the axis will be twisted through angle ϕ to AB′ when the torque is applied, as shown in Fig. 5.2. (AB will actually form a helix when the shaft is twisted.)

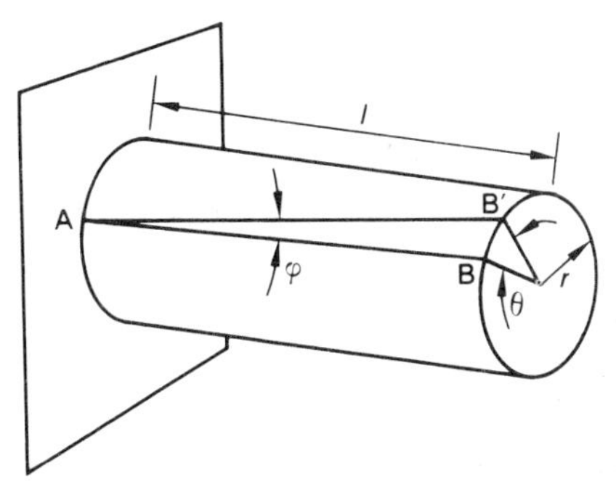

Fig. 5.2 Strain of a circular shaft due to torsion

Now, for pure shear the angle ϕ represents the shear strain, and if τ is the shear stress on the surface of the shaft then

$$\frac{\text{Shear stress}}{\text{Shear strain}} = \text{modulus of rigidity}$$

$$\frac{\tau}{\phi} = G \qquad (5.1)$$

From Fig. 5.2

$$\text{Arc BB}' = l\phi \quad \text{if } \phi \text{ is small}$$

and also $\quad \text{Arc BB}' = r\theta$

where θ is the angle of twist over the length of shaft l.

$$\therefore \qquad l\phi = r\theta$$

$$\phi = \frac{r\theta}{l}$$

Substituting for ϕ into eq. (5.1) gives

$$\frac{\tau}{r\theta/l} = G$$

or
$$\frac{\tau}{r} = \frac{G\theta}{l} \qquad (5.2)$$

From Fig. 5.2 it will be seen that θ is constant for all radii at a particular cross-section of the shaft, and as G and l are also constant then eq. (5.2) may be written

$$\frac{\tau}{r} = \text{constant}$$

Now, if τ_1 is the shear stress at any radius r_1 then τ_1/r_1 is also a constant.

$$\therefore \quad \frac{\tau}{r} = \frac{\tau_1}{r_1} = \frac{\tau_2}{r_2} = - - - - - = \frac{\tau_n}{r_n} = \text{constant} \qquad (5.3)$$

Thus, the shear stress at any point within the cross-section of the shaft is proportional to the radius. Hence, the stress varies uniformly from zero at the centre of the shaft to a maximum at the outside radius (see Fig. 5.5 – solid shaft).

5.2 Moment of resistance

Consider a cross-section of the shaft, as shown in Fig. 5.3, and let τ_1 be the shear stress acting over an elemental ring of radius r_1 and thickness δr_1.

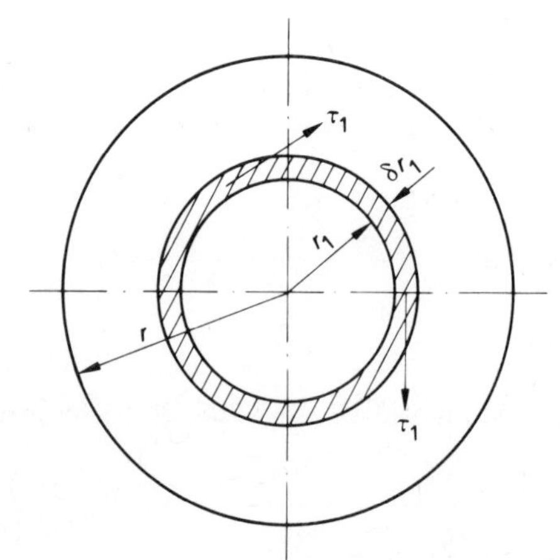

Fig. 5.3

Shearing force on elemental ring = shear stress x area

$$= \tau_1 \times 2\pi r_1 . \delta r_1$$

Moment of this shearing force about shaft axis

$$= \tau_1 . 2\pi r_1 . \delta r_1 \times r_1$$

But from eq. (5.3)

$$\frac{\tau_1}{r_1} = \frac{\tau}{r}$$

Substituting for τ_1 into the previous equation gives:

Moment of elemental shearing force about shaft axis

$$= \frac{\tau}{r} . 2\pi r_1^3 . \delta r_1$$

The total internal resisting moment offered by the shaft will be the sum of all such moments which, for equilibrium, must balance the applied torque T.

$$\text{Hence} \quad T = \int_0^r \frac{\tau}{r} . 2\pi r_1^3 . dr_1$$

$$= \frac{\tau}{r} \int_0^r 2\pi r_1^3 . dr_1$$

$$= \frac{\tau}{r} \left[\frac{2\pi r_1^4}{4} \right]_0^r$$

$$= \frac{\tau}{r} . \frac{\pi r^4}{2}$$

Now $(\pi r^4/2)$ is equal to the *polar second moment of area*, J, of the cross section about the shaft axis (see section 5.3) and so the above equation becomes

$$T = \frac{\tau J}{r}$$

$$\therefore \quad \frac{T}{J} = \frac{\tau}{r} \tag{5.4}$$

Combining eqs. (5.2) and (5.4) gives the general torsion equation.

$$\frac{T}{J} = \frac{\tau}{r} = \frac{G\theta}{l} \quad * \tag{5.5}$$

* *Note:* Equation (5.5) only applies to shafts of circular cross-section.

Equation (5.4) may be written in the form

$$T = \tau Z_p$$

where $Z_p = \frac{J}{r}$ = polar modulus of section

This should be compared with the bending equation $M = \sigma_{max} Z$ where Z = section modulus = I/y_{max} [eq. (4.3)].

5.3 Polar second moment of area

For a solid shaft the polar second moment of area is obtained by summing (i.e., integrating) all the individual second moments of area from $r_1 = 0$ at the shaft centre to $r_1 = r$ at the circumference, where r = the shaft radius.

Thus,
$$J = \int_0^r r_1^2 . \mathrm{d}A$$
$$= \int_0^r 2\pi r_1^3 . \mathrm{d}r_1$$
$$= \left[\frac{2\pi r_1^4}{4}\right]_0^r = \frac{\pi r^4}{2}$$

Substituting $r = \frac{d}{2}$ (d = shaft diameter)

gives $J = \frac{\pi d^4}{32}$ $\left(\text{c.f. } I = \frac{\pi d^4}{64} \text{ for bending}\right)$

For a hollow shaft having an external diameter D and an internal diameter d the integration must be between the limits $r_1 = d/2$ to $r_1 = D/2$, i.e.,

$$J = \int_{d/2}^{D/2} 2\pi r_1^3 . \mathrm{d}r_1$$

$$J = \frac{\pi}{32}(D^4 - d^4)$$

5.4 Transmission of power

When a shaft transmits power, and hence a torque, it twists until the internal resisting torque balances the torque to be transmitted. If the torque is constant then:

Power transmitted = ωT watts

where ω = angular velocity of shaft (rad/s)

$= 2\pi n$ rad/s

where n = frequency of rotation (rev/s)

and T = torque transmitted (Nm)

Thus, Power transmitted = $2\pi n T$ watts

In many instances of power transmission the torque does not remain constant with time. This is illustrated in Fig. 5.4 which represents the torque diagram for one cycle of a single acting four-stage gas engine drawn on an angle base.

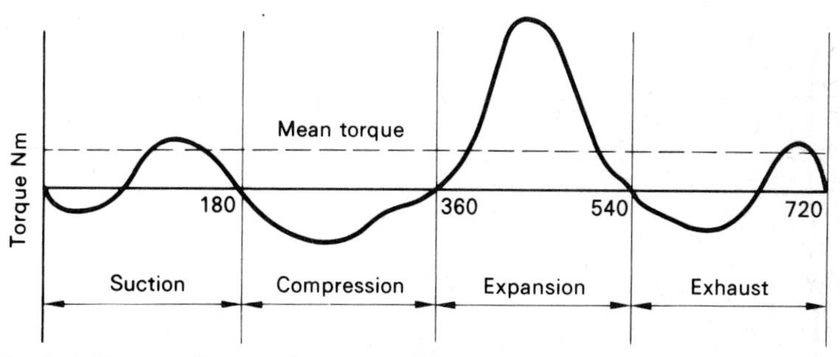

Fig. 5.4 Torque diagram for gas engine

It will be apparent that to determine the power which is transmitted in this case the mean torque should be used. However, for shaft design calculations the maximum torque must be used since this gives the maximum stress condition. Hence when dealing with problems involving a varying torque the relevant equations should be used in the form

$$\frac{T_{max}}{J} = \frac{\tau}{r} = \frac{G\theta}{l} \quad \text{and} \quad \text{Power} = 2\pi n T_{mean} \text{ watts}$$

Example 5.1

Calculate the diameter of a solid steel shaft which is to transmit 80 kW at a speed of 300 rev/min if the angle of twist is not to exceed 0·5° over a length of 4 m. What will be the maximum shear stress in the material under these conditions?

$G = 70 \text{ GN/m}^2$

Solution

The torque being transmitted by the shaft can be calculated from

$$\text{Power} = 2\pi n T$$

where $$n = \frac{300}{60} = 5 \text{ rev/s}$$

Then $$80 \times 10^3 = 2\pi \times 5 \times T$$

$$T = \frac{8000}{\pi} = 2546 \text{ Nm}$$

From eq. (5.5)

$$J = \frac{Tl}{G\theta}$$

$$= \frac{2546 \text{ [Nm]} \times 4 \text{ [m]}}{70 \times 10^9 \text{ [N/m}^2\text{]} \times 0{\cdot}5 \times \pi/180 \text{ [rad]}}$$

$$= \frac{1667}{10^8} \text{ m}^4$$

But $$J = \frac{\pi d^4}{32}$$

$$\therefore \quad d^4 = \frac{32}{\pi} \times \frac{1667}{10^8} = \frac{16\,979}{10^8} \text{ m}^4$$

$$d = \frac{11{\cdot}41}{10^2} \text{ m}$$

$$= 114{\cdot}1 \text{ mm}$$

To determine the maximum stress we again use eq. (5.5).

$$\frac{\tau}{r} = \frac{G\theta}{l}$$

$$\therefore \quad \tau = \frac{G\theta r}{l} = \frac{70 \times 10^9 \text{ [N/m}^2\text{]} \times (0{\cdot}5\pi/180) \text{ [rad]} \times (57{\cdot}05/10^3) \text{ [m]}}{4 \text{ [m]}}$$

$$= 87{\cdot}38 \times 10^6 \text{ N/m}^2$$

$$= 87{\cdot}38 \text{ MN/m}^2$$

The required shaft diameter is 114·1 mm and the resulting stress is 87·38 MN/m².

5.5 Comparison of solid and hollow shafts

We have seen in section 5.1, eq. (5.3), that the shear stress created in a shaft is proportional to the radius from the axis of the shaft. This statement is valid for both solid and hollow circular shafts. From eq. (5.4), $\tau = Tr/J$ and therefore the stress in a solid shaft of diameter D_s varies from zero at the shaft centre to $(TD_s/2J_s)$ at the outside radius. This means that when the limiting shear stress of the material is reached at the outside surface the central region is still very lowly stressed. However, with a hollow shaft having an external diameter D and an internal diameter d the stress will vary from $(Td/2J_h)$ at the inside diameter to $(TD/2J_h)$ at the outside diameter. These two conditions are illustrated in Fig. 5.5.

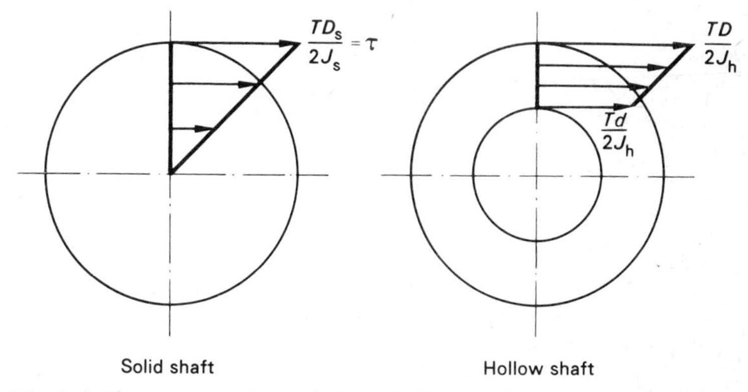

Fig. 5.5 Shear stresses in a solid and hollow shaft

Hence, for the same limiting shear stress it is apparent that the average intensity of stress is considerably greater in the hollow shaft than in the solid shaft. Consequently, for a specified cross-sectional area a hollow shaft can transmit a greater torque, and hence more power. This can be shown as follows:

For equal shaft areas $\frac{\pi}{4} D_s^2 = \frac{\pi}{4}(D^2 - d^2)$

i.e. $D_s^2 = D^2 - d^2$

If $d = \mathrm{k}D$, where k is a constant (less than one)

Then $D_s^2 = D^2 - \mathrm{k}^2 D^2$

$$D_s = D\sqrt{(1 - \mathrm{k}^2)}$$

From the torque equation it follows that for equal maximum stresses in the two shafts then

$$\frac{T_s D_s}{2J_s} = \frac{T_h D}{2J_h}$$

where subscripts s and h refer to the solid and hollow shafts respectively.

$$\therefore \quad \frac{T_s D_s}{(\pi D_s^4/32)} = \frac{T_h D}{\{(\pi/32)(D^4 - k^4 D^4)\}}$$

$$\frac{T_h}{T_s} = \frac{D^3(1-k^4)}{D_s^3}$$

Substituting for D_s from above gives

$$\frac{T_h}{T_s} = \frac{(1-k^4)}{(1-k^2)^{3/2}} = \frac{(1+k^2)(1-k^2)}{(1-k^2)^{3/2}}$$

$$= \frac{1+k^2}{\sqrt{(1-k^2)}}$$

A graph is plotted in Fig. 5.6 showing the variation of T_h/T_s with k.

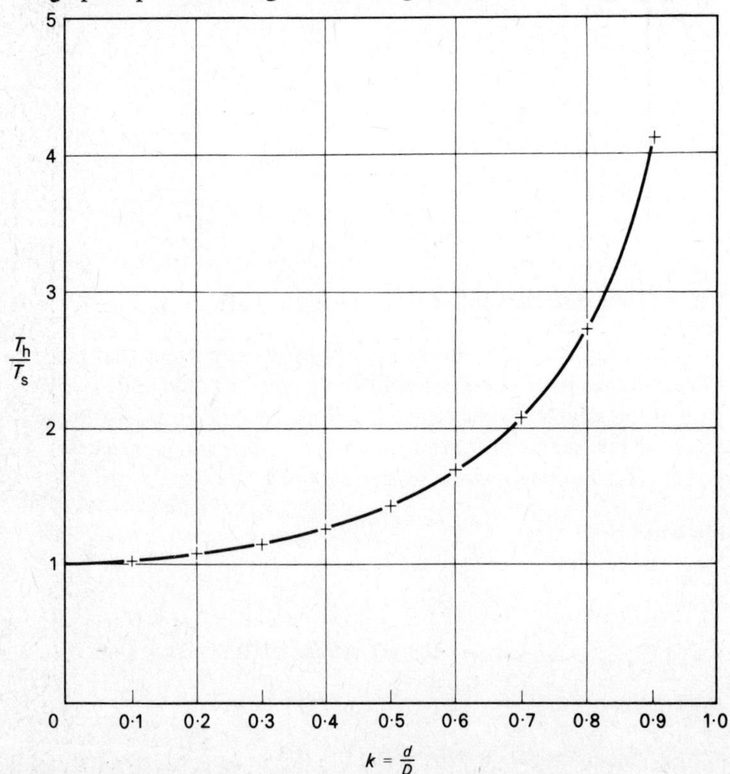

Fig. 5.6 Comparison of torque carrying capability for hollow and solid shafts with equal area and equal maximum shear stress

Thus, it will be seen that when the ratio of internal to external diameters of the hollow shaft is 0·6 the hollow shaft can transmit 1·7 times as much power as a solid one of the same cross-sectional area.

Example 5.2

Determine the maximum torque that can be transmitted by a hollow shaft having an external diameter of 250 mm and an internal diameter of 150 mm if the maximum shear stress is not to exceed 70 MN/m^2. What will then be the power that can be transmitted by the shaft when rotating at 120 rev/min if the maximum torque exceeds the mean torque by 25%?

Solution

The maximum shear stress will occur at the outside of the shaft when the torque is also a maximum. Hence, using

$$\frac{T_{max}}{J} = \frac{\tau}{r}$$

where $J = \frac{\pi}{32}(250^4 - 150^4) = 3{\cdot}38 \times 10^8 \text{ mm}^4$

$\tau = 70 \times 10^6 \text{ N/m}^2$

and $r = 125 \text{ mm} = 0{\cdot}125 \text{ m}$

gives $$T_{max} = \frac{3{\cdot}38 \times 10^8 \text{ [m}^4\text{]}}{10^{12}} \times \frac{70 \times 10^6 \text{ [N/m}^2\text{]}}{0{\cdot}125 \text{ [m]}}$$

$$= 186{\cdot}9 \times 10^3 \text{ Nm}$$

Then $$T_{mean} = \frac{T_{max}}{1{\cdot}25} = \frac{186{\cdot}9 \times 10^3}{1{\cdot}25} = 149{\cdot}5 \times 10^3 \text{ Nm}$$

$$\text{Power transmitted} = 2\pi n T_{mean}$$

$$= 2\pi \,.\, \frac{120}{60} \,.\, 149{\cdot}5 \times 10^3$$

$$= 1879 \times 10^3 \text{ W}$$

$$= 1{\cdot}879 \text{ MW}$$

The allowable torque is 186·9 x 10^3 Nm and the power transmitted at 120 rev/min is 1·879 MW.

Example 5.3

A hollow shaft, whose internal diameter is 0·6 times the external diameter, is required to transmit 800 kW at 150 rev/min under the action of a constant torque.

The maximum shearing stress in the shaft must not exceed 60 MN/m^2 and the twist over a length of 3 m must not be greater than 1°. Determine the minimum external diameter of the shaft that will satisfy these conditions.
Modulus of rigidity = 70 GN/m^2.

Solution

In this problem two separate conditions are specified and a suitable shaft diameter must be calculated to satisfy them both.

The torque being transmitted must first be obtained from

$$\text{Power} = 2\pi n T$$

$$800 \times 10^3 = 2\pi \cdot \frac{150}{60} T$$

$$T = 50\,830 \text{ Nm}$$

SHEAR STRESS CONDITION

From the torsion equation

$$\frac{T}{J} = \frac{\tau}{r}$$

Now $$J = \frac{\pi}{32}(D^4 - d^4)$$

But $$d = 0{\cdot}6\,D$$

$\therefore$ $$J = \frac{\pi}{32} D^4 (1 - 0{\cdot}6^4) = 0{\cdot}0845\,D^4$$

Thus $$\frac{50\,830}{0{\cdot}0845\,D^4} = \frac{2 \times 60 \times 10^6}{D}$$

$$D^3 = \frac{50\,830}{0{\cdot}0845 \times 120 \times 10^6} = 4957 \times 10^{-6} \text{ m}^3$$

$$D = 17{\cdot}05 \times 10^{-2} \text{ m}$$

$$= 170{\cdot}5 \text{ mm}$$

ANGLE OF TWIST CONDITION

To determine the necessary diameter to satisfy this condition we must use

$$\frac{T}{J} = \frac{G\theta}{l}$$

Then $$\frac{50\,830}{0{\cdot}0845\,D^4} = \frac{70 \times 10^9 \times 1 \times \pi/180}{3}$$

$$D^4 = \frac{3 \times 50\,830 \times 180}{0{\cdot}0845 \times 70 \times 10^9 \times \pi} = 14{\cdot}77 \times 10^{-4}\ \text{m}^4$$

$$D = 1{\cdot}96 \times 10^{-1}\ \text{m}$$

$$= 196\ \text{mm}$$

Thus, although a diameter of 170·5 mm is sufficient to satisfy the stress condition such a diameter would result in an angle of twist of greater than 1° over a 3 m length. Similarly, a diameter of 196 mm, which is necessary to satisfy the twist condition, will result in a maximum stress well below the allowable value of 60 MN/m^2.

Hence, in order to satisfy both conditions the minimum diameter of the shaft must be 196 mm.

Note: A frequent error with this type of problem is to use the relationship

$$\frac{\tau}{r} = \frac{G\theta}{l}$$

This *cannot* be used because the question does not state that the twist is 1° over a 3 m length when the stress is 60 MN/m^2.

Example 5.4

A hollow steel shaft, whose external diameter to internal diameter are to be in the ratio 4 : 3 is to be designed to transmit 100 kW at 400 rev/min. The maximum torque is known to be 1·36 times the mean torque. If the maximum shear stress that is permissible is 80 MN/m^2 determine suitable shaft diameters.

Two lengths of the shaft are to be connected together by means of a flanged coupling which is to have 6 bolts on a pitch circle diameter of 120 mm. Determine the necessary bolt diameter if their maximum working stress is to be 50 MN/m^2. Assume that bearing failure of the flange does not occur.

Solution

$$\text{Power} = 2\pi n T_{\text{mean}}$$

$$100 \times 10^3 = 2\pi \,.\, \frac{400}{60} \,.\, T_{\text{mean}}$$

$$T_{\text{mean}} = \frac{6000 \times 10^3}{800\pi} = 2387\ \text{Nm}$$

Then $T_{max} = 1{\cdot}36 \times 2387$

$= 3247$ Nm

Using $\dfrac{T}{J} = \dfrac{\tau}{r}$

where $J = \dfrac{\pi}{32}(D^4 - d^4)$

$= \dfrac{\pi}{32}(D^4 - (\tfrac{3}{4})^4 D^4)$ since $\dfrac{d}{D} = \dfrac{3}{4}$

$= 0{\cdot}0671\, D^4$

gives $\dfrac{3247}{0{\cdot}0671\, D^4} = \dfrac{80 \times 10^6}{D/2}$

$$D^3 = \frac{3247}{0{\cdot}0671 \times 160 \times 10^6} = 302{\cdot}4 \times 10^{-6}\ \text{m}^3$$

$$D = 6{\cdot}712 \times 10^{-2}\ \text{m}$$

$$= 67{\cdot}12\ \text{mm}$$

Then, if the external diameter is rounded up to 68 mm the inside diameter becomes $0{\cdot}75 \times 68 = 51$ mm

Let the force on each bolt be P newtons.
Then, torque that can be carried by the coupling

$= 6P \times$ pitch circle radius

$= 6P \times \dfrac{60}{1000}$ [Nm]

$= 0{\cdot}36P$ [Nm]

But this must equal the maximum torque carried by the shaft; i.e.,

$$0{\cdot}36P = 3247$$

$$P = 9020\ \text{N}$$

Shear stress in bolt $= \dfrac{\text{force}}{\text{cross-sectional area}}$

$$50 \times 10^6 = \frac{9020}{(\pi d^2/4)}$$

where d = effective bolt diameter

$\therefore$ $d^2 = \dfrac{9020 \times 4}{\pi \times 50 \times 10^6} = 229{\cdot}7 \times 10^{-6}\ \text{m}^2$

$$d = 15{\cdot}16 \times 10^{-3}\ \text{m}$$

$$= 15{\cdot}16\ \text{mm}$$

As the effective diameter of a bolt does not include the thread, i.e., it is the core diameter, a suitable diameter would be 20 mm.

5.6 Compound shafts

Problems involving shafts made up of two or more materials are solved by applying the principles already presented. However, it should be appreciated that when the compound shaft consists of separate shafts connected in series (Fig. 5.7(*a*)) then the applied torque will be carried by each shaft, whereas if the shafts are rigidly connected in parallel (Fig. 5.7(*b*)) then each shaft carries a proportion of the applied torque such that their angles of twist are equal.

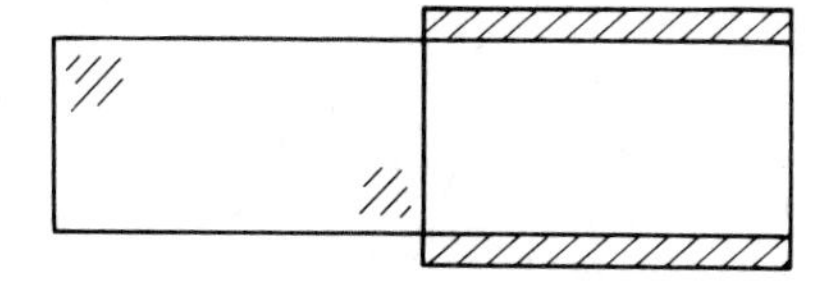

(a) Solid and hollow shafts in series

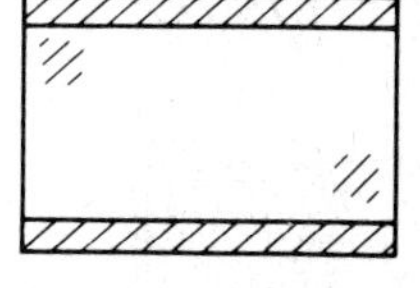

(b) Solid and hollow shafts in parallel

Fig. 5.7 Compound shafts

Example 5.5

A solid alloy shaft having a diameter of 60 mm is to be coupled in series with a hollow steel shaft of the same external diameter. What should be the internal diameter of the steel shaft if its angle of twist per unit length is to be the same as that of the alloy shaft?

Such a shaft is required to transmit 200 kW, the maximum shearing stresses in the alloy and the steel being limited to 50 MN/m^2 and 80 MN/m^2 respectively. Determine the speed at which the compound shaft should be driven.

For the steel, $G = 80$ GN/m^2

For the alloy, $G = 30$ GN/m^2

Solution

As the shafts are coupled in series they must transmit the same torque. Thus, equating expressions for T, from eq. (5.5), gives:

$$\frac{G_A \theta_A J_A}{l_A} = \frac{G_S \theta_S J_S}{l_S}$$

But as $\dfrac{\theta_A}{l_A} = \dfrac{\theta_S}{l_S}$ is a specified condition

then $G_A J_A = G_S J_S$

$$30 \times 10^9 \times \frac{\pi \times 60^4}{32 \times 10^{12}} = 80 \times 10^9 \times \frac{\pi}{32} \frac{(60^4 - d^4)}{10^{12}}$$

where d is the required internal diameter of the steel shaft.

$$\frac{30 \times 60^4}{80} = 60^4 - d^4$$

$$d^4 = 60^4 \left(1 - \frac{30}{80}\right)$$

$$d = 60 \times \sqrt[4]{\frac{50}{80}}$$

$$= 53{\cdot}32 \text{ mm}$$

The required internal diameter of the steel shaft is 53·32 mm.

Now $T = \dfrac{\tau J}{r}$ [eq. (5.5)]

and the maximum torque that can be carried by the alloy shaft is given by

$$T_A = \frac{\tau_A J_A}{r_A}$$

$$= \frac{50 \times 10^6 \text{ [N/m}^2\text{]} \times \dfrac{\pi}{4} \times \dfrac{60^4}{10^{12}} \text{ [m}^4\text{]}}{30/10^3 \text{ [m]}}$$

$$= 16\,964 \text{ Nm}$$

Similarly, the maximum torque that can be carried by the steel shaft is

$$T_S = \frac{\tau_S J_S}{r_S}$$

$$= \frac{80 \times 10^6 \times \dfrac{\pi}{4} \times \left(\dfrac{60^4 - 53{\cdot}32^4}{10^{12}}\right)}{30/10^3}$$

$$= 10\,340 \text{ Nm}$$

Hence, the maximum torque that can be transmitted by the compound shaft is 10 340 Nm as any torque in excess of this would cause the allowable stress in the steel to be exceeded. The alloy shaft will not be working at its maximum allowable stress.

Then Power $= \omega T$ watts

$$200 \times 10^3 = \omega \times 10\,340$$

$$\omega = 19{\cdot}34 \text{ rad/s}$$

$$= \frac{19{\cdot}34}{2\pi} \times 60 = 184{\cdot}7 \text{ rev/min}$$

The speed of rotation to transmit 200 kW is 184·7 rev/min.

Example 5.6

A round steel bar is surrounded by a close fitting duralumin tube, the two being rigidly fixed together to form a compound shaft. The shaft is to withstand a torque of 800 Nm and the maximum shearing stresses in the steel and duralumin are not to exceed 100 MN/m^2 and 70 MN/m^2 respectively. Determine the necessary diameter of the steel bar and the external diameter of the duralumin tube. What will be the twist per metre length of such a shaft?

For steel: Modulus of rigidity = 80 GN/m^2

For duralumin: Modulus of rigidity = 28 GN/m^2

Solution

The two shafts are connected in parallel and therefore must have the same twist per unit length.

Thus, from $\dfrac{G\theta}{l} = \dfrac{\tau}{r}$ [eq. (5.5)]

we get

$$\frac{\tau_S}{r_S G_S} = \frac{\tau_D}{r_D G_D}$$

(suffixes S and D referring to the steel and duralumin respectively)

If d = diameter of steel bar and D = external diameter of duralumin tube then

$$\frac{100 \times 10^6}{d/2 \times 80 \times 10^9} = \frac{70 \times 10^6}{D/2 \times 28 \times 10^9}$$

$$\therefore \quad D = 2d$$

For shafts in parallel

Total torque = torque carried by steel + torque carried by duralumin

Thus:
$$800 = \frac{\tau_S J_S}{r_S} + \frac{\tau_D J_D}{r_D}$$

$$= \frac{100 \times 10^6}{d/2} \times \frac{\pi d^4}{32} + \frac{70 \times 10^6}{D/2} \times \frac{\pi (D^4 - d^4)}{32}$$

$$= \frac{10^6 \times \pi}{16} \left(100\, d^3 + \frac{70\,((2d)^4 - d^4)}{2d}\right)$$

$$= \frac{10^6 \pi}{16} (100\, d^3 + 525\, d^3)$$

$$d^3 = \frac{800 \times 16}{10^6 \pi \times 625} \text{m}^3$$

$$d = 0{\cdot}018\,68 \text{ m}$$

$$= 18{\cdot}68 \text{ mm}$$

$\therefore$
$$D = 37{\cdot}36 \text{ mm}$$

The diameter of the steel bar must be 18·68 mm and the external diameter of the duralumin tube 37·36 mm.

The angle of twist per metre length is given by

$$\frac{\theta}{l} = \frac{\tau}{rG}$$

This can be applied to either the steel or duralumin.

Hence, for steel

$$\frac{\theta}{l} = \frac{100 \times 10^6 \text{ [N/m}^2\text{]}}{(9{\cdot}34/10^3) \text{ [m]} \times 80 \times 10^9 \text{ [N/m}^2\text{]}}$$

$$= 0{\cdot}134 \text{ rad/m}$$

$$= 7{\cdot}768^\circ\text{/m}$$

The rate of twist of the compound bar is 7·768° per metre length.

Problems

1. A brass test specimen to be used in a torsion test has a diameter of 10 mm and a gauge length of 0·2 m. It was found from the test that the limit of proportionality of the material was reached at a torque of of 35 Nm.

At this torque the angle of twist over the gauge length was 10·75°. Calculate the modulus of rigidity of the material.

2. A solid steel shaft, having a diameter of 80 mm, transmits a steady torque at a speed of 140 rev/min. The angle of twist must not exceed 1° over a 3 m length. Determine the maximum shear stress in the shaft and the power which is transmitted.
Modulus of rigidity = 75 GN/m^2.

3. Determine the power that can be transmitted by a hollow steel shaft if the allowable shear stress is 65 MN/m^2. The external and internal diameters are 250 mm and 125 mm respectively and the speed of rotation is 300 rev/min.

What will be the twist over a 4 m length under these conditions?
Modulus of rigidity = 80 GN/m^2.

4. A shaft is to be designed to transmit 1200 kW at a speed of rotation of 200 rev/min. The following conditions are specified:

(*a*) the maximum shear stress must not exceed 55 MN/m^2;
(*b*) the twist must not exceed 1° over a length of 18 diameters.

Determine the required shaft diameter and the actual working stress.
Modulus of rigidity = 75 GN/m^2.

5. The propeller shaft of a car is a hollow steel tube 50 mm outside diameter and 3 mm thick. Determine the shear stress in the shaft when it is transmitting 75 kW at a speed of 4800 rev/min.

6. A hollow steel shaft has an external diameter of 150 mm and an internal diameter of 100 mm. What is the maximum power that can be transmitted by the shaft if the allowable shear stress is 90 MN/m^2 and the speed of rotation is 360 rev/min. Assume that the maximum torque is 20% greater than the mean value.

7. The ratio of internal diameter to external diameter of a shaft is to be 0·6. Determine suitable diameters of a shaft that is required to transmit 1·8 MW at 250 rev/min if the allowable shear stress is 50 MN/m^2.

Assume $T_{max} = 1{\cdot}12\ T_{mean}$.

8. A hollow shaft 5 m long, rotating at 50 rev/s is found to twist by 11° when transmitting 502 kW. If the maximum shear stress is 80 N/mm^2, determine the outside and inside diameters of the shaft. Take G for the shaft material to be 83 kN/mm^2.

(U.L.C.I.)

9. A steel bar, 20 mm diameter, reached the elastic limit shear stress when subjected to an applied torque of 185 Nm. At this stress the angle of twist, measured over a length of 0·15 m, was found to be 1·3°. Determine the shear stress at the elastic limit and the modulus of rigidity of the steel.

What would be the minimum diameter of a solid shaft of this steel if it were required to transmit 400 kW at 500 rev/min with a safety factor of 4 based upon this elastic limit shear stress?

10. A hollow steel shaft has an internal diameter of 200 mm and an external diameter of 250 mm. Calculate the maximum shear stress in the shaft when it is transmitting 5 MW at 360 rev/min.

11. A hollow steel shaft is to transmit 1·8 MW at a speed of 5·2 rev/s. The internal diameter is 0·6 times the external diameter and the maximum shear stress is not to exceed 60 MN/m^2. Calculate:

(*a*) the least dimensions of the shaft;
(*b*) the angle of twist in degrees over a length of 2·5 m when transmitting the above power if the modulus of rigidity for steel is 80 GN/m^2.

(N.C.T.E.C.)

12. Determine the power that can be transmitted safely by a hollow shaft of external diameter 300 mm and internal diameter 150 mm at a speed of 6 rev/s. The shaft may be assumed to be subject to pure torsion and the angle of twist is not to exceed 3° in 10 m of shaft length. Calculate the maximum intensity of shear stress induced in the shaft. Take the modulus of rigidity of the shaft material as 80 GN/m^2.

(U.E.I.)

13. The power consumption of an electric motor running at 1450 rev/min is 60 kW when transmitting torque through a steel shaft 40 mm diameter and 500 mm long. If the efficiency of the motor is 85% determine the maximum shear stress and the corresponding angle of twist between the ends of the shaft.
Modulus of rigidity of steel = 80 GN/m^2.

(E.M.E.U.)

14. The solid steel propeller shaft on a ship has a diameter of 0·25 m and transmits 900 kW at 75 rev/min. It is decided to replace the solid shaft by a hollow steel shaft whose internal diameter is to be half its external diameter but which is to transmit 1500 kW at 75 rev/min with the maximum shear stress being the same as that for the solid shaft.

Determine the required dimension of the hollow shaft and the ratio of the mass of the hollow shaft to that of the solid shaft, assuming them to be of equal length.

15. The propeller shaft on a car is required to transmit 75 kW at a speed of 5000 rev/min. From tests it is known that the maximum torque is 17% greater than the mean value. If the external diameter of the shaft is limited to 50 mm and the maximum shear stress is not to exceed 38 MN/m^2 what thickness should the shaft wall be?

Compare the mass and angle of twist of this shaft with a solid shaft to satisfy the same conditions, assuming the shafts to be of the same length and of similar material.

16. A shaft is 60 mm diameter for part of its length and 30 mm diameter for the remainder. If the maximum shear stress must not exceed 50 MN/m^2 what is the maximum power that can be transmitted by the shaft at 500 rev/min? Calculate the angle of twist per metre length in the 60 mm diameter section when the shaft is transmitting this power. Modulus of rigidity = 80 GN/m^2.

17. Two solid shafts of 60 mm diameter are to be connected in series by a close fitting tube of similar material. If the twist per unit length of the shaft is twice that of the tube determine the external diameter of the tube.

18. A steel shaft AB, 50 mm diameter, is screwed into the end of a steel tube BC, 45 mm internal diameter and 55 mm external diameter. The effective lengths of the shaft and tube after assembly are AB = 1 m and BC = 0·6 m. The open end C of the tube is firmly held and a torque applied to end A of the shaft.

If the maximum allowable shear stress is 60 MN/m^2 determine:

(*a*) the maximum torque which can be transmitted;
(*b*) the actual shear stresses in the shaft and the tube;
(*c*) the angle of twist of end A.

Modulus of rigidity = 80 GN/m^2.

19. A shaft, ABC, consists of two parts, AB and BC, with a coupling at B. The shaft rotates at 300 rev/min and power is supplied at B. Shaft BA is solid, 2 m in length, and transmits 250 kW while shaft BC is hollow, with an internal diameter of 60 mm, 3 m in length and transmits 300 kW. Determine the necessary shaft diameters if the shear stress in BA is not to exceed 40 MN/m^2 and the angles of twist of BA and BC are to be equal.

Modulus of rigidity = 80 GN/m^2.

20. A composite shaft having a circular cross-section is 0·7 m long. It is rigidly built in at each end and consists of a 0·4 m length of 30 mm diameter steel joined to a 0·3 m length of 50 mm diameter bronze. If the limiting shear stress in the bronze is 35 MN/m^2 determine the maximum torque that can be applied at the joint. What is the maximum shear stress in the steel?

For steel: $G = 80$ GN/m^2.

For bronze: $G = 40$ GN/m^2.

6

Materials testing

6.1 Purpose of material testing

The determination of a theoretical stress analysis is a major part of the design of many engineering products but to carry out such an analysis the engineer must possess information on the physical properties of the materials which are to be used. In addition to knowing the moduli of elasticity and rigidity, the tensile strength, yield stress or proof stress, etc. it is also important to possess information on the ductility, hardness, and toughness of a material.

Various standard test procedures will be discussed in this chapter but other properties, such as fatigue and creep, must be left for future studies. Briefly, fatigue failure can occur when a relatively small alternating stress is applied over a prolonged period while creep refers to the slow but gradual extension of a material that is subjected to a small stress, although for most materials this only occurs at high temperatures.

6.2 The tensile test

The results obtained from a tensile test have been discussed in section 3.5, but no mention was made of the testing machine or conditions of the test.

The tensile test, which is probably the most common test carried out on a material, must be performed on a machine that is capable of applying a true axial load to the test specimen. The machine must incorporate a means of measuring the applied load, and to determine the extension of the test specimen an extensometer is attached to it.

A typical machine, which is suitable for both tensile and compressive tests, is the Denison T42 B3, shown in Fig. 6.1. With this machine the

Fig. 6.1 Denison T42 B3 testing machine

force is applied to the test specimen via hydraulic rams. A series of knife edges enables the operator to select the load range required for the test and a fine straining control, which operates a relief valve in the hydraulic system, enables the load to be applied slowly, held almost constant or reduced slowly. The straining head is connected to a multiple lever system at the top of the machine and the load recorded on a dial.

The main disadvantage of a machine such as that outlined above is that a constant rate of loading or constant rate of straining cannot be applied to the test specimen. To do this a servo-controlled machine such as the Dartec shown in Fig. 6.2 is required. An automatic graph plotter is incorporated and both load and deflection are displayed numerically to great accuracy. Furthermore, the load can be cycled between two prescribed limits at a constant change rate.

Fig. 6.2 Dartec Servo-controlled testing machine

It is important that a tensile test specimen conforms to the dimensions specified by The British Standards Institution. This is because the total extension consists of a uniform extension and a local extension, as mentioned in section 3.6. The specimen dimensions and test procedure should be in accordance with BS18 for metals and BS2782 for plastics, the usual section being circular for metals and rectangular for plastics.

In the case of plastic materials the stress-strain relationship is not linear and the modulus of elasticity is defined as the stress/strain ratio at a strain of 0·2%. Plastics do not behave as do metals when subjected to a stress, as the strain is dependent on both the magnitude of the load and the time for which it is applied. Hence no definite strain can be associated with a given stress and a value for the modulus of elasticity does not have much meaning. Furthermore, a plastic material only returns to its original length over a period of time following the removal of the load and as such is said to be viscoelastic.

6.3 Extensometers

When the determination of the modulus of elasticity of a material is required an accurate stress-strain or load-extension graph must be obtained. This necessitates the measurement of very small extensions and the use of an extensometer.

Fig. 6.3 Lindly extensometer

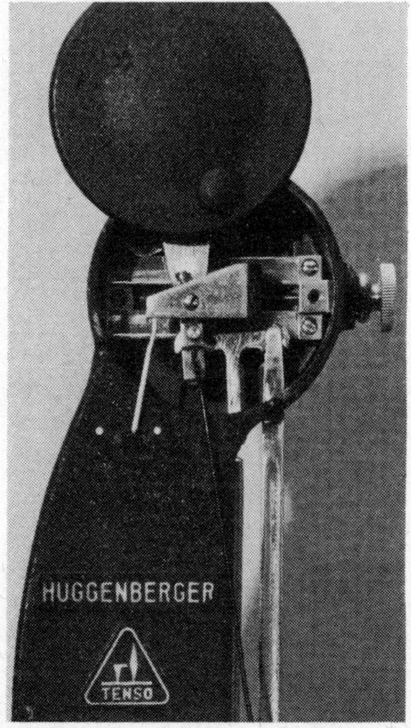

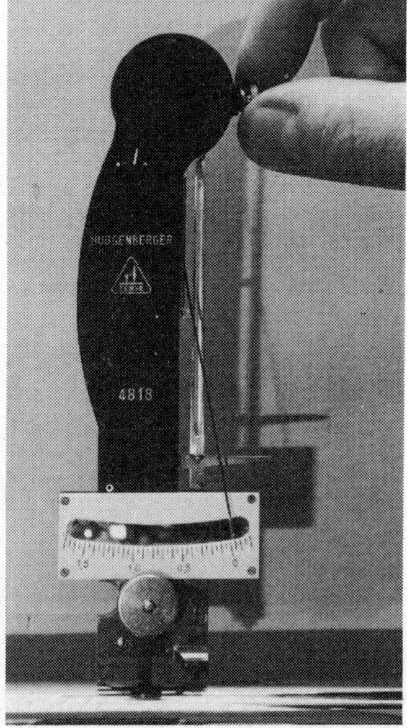

Fig. 6.4 Huggenburger extensometer

The Lindley extensometer shown in Fig. 6.3 consists of two arms that are hinged by a spring steel strip. The extensometer is clamped to the test specimen by two pairs of screws which are 50 mm apart and the extension is recorded on a dial gauge. The distances of the points and dial gauge from the hinge are such that a magnificiation of 2 is produced and each division represents an extension of 0·001 mm.

The Huggenburger extensometer shown in Fig. 6.4 is clamped to the test specimen by two knife-edges which are either 10 mm or 20 mm apart and an arrangement of levers produces a magnification of up to 2000. This extensometer enables extensions to be recorded to an accuracy of 8×10^{-4} mm.

6.4 The compression test

A compression test can often be performed on the same machine as a tensile test, but for the testing of building materials special compression testing machines are usually used.

For metals, the test specimen is usually cylindrical and in the case of a ductile material such as mild steel the stress-strain diagram is of the form shown in Fig. 6.5. The diagram is initially linear and within this

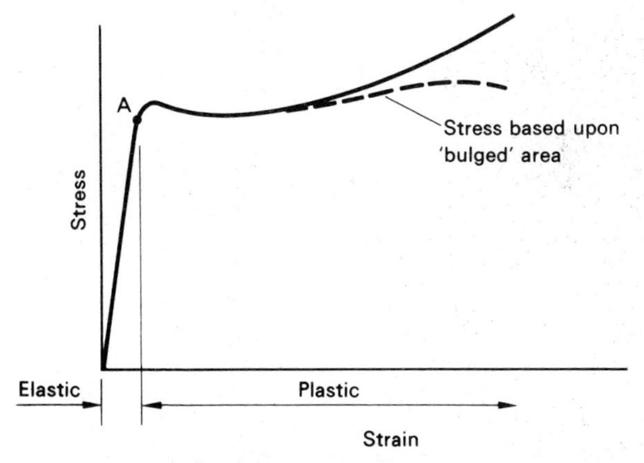

Fig. 6.5 Typical stress-strain graph for mild steel in compression

range the stress is proportional to the strain so satisfying Hooke's law. The strain is elastic within this range but at A yielding occurs and the material enters the plastic region. As for the tensile test on a similar material, the yielding stage produces considerable deformation at fairly constant load and an increase in load is then required to produce further compression. The test specimen becomes barrel shaped at this stage, mainly due to the friction between the specimen and the compression faces of the testing machine, which restricts lateral expansion at the ends. Surface cracks tend to occur but disintegration of the specimen does not take place and no maximum load can be determined. The true stress-strain diagram for such a test is shown dotted on Fig. 6.5 and various stages during such a test are shown in Fig. 6.6.

Fig. 6.6 Stages during the compression of a short cylindrical copper test specimen

The behaviour of brittle materials in compression is completely different however and following an initially elastic stage they fail by cracking or splitting, as shown in Fig. 6.7. This failure is due to shear stresses created on the plane of cracking, which is at an angle of approximately 45° to the axis of the specimen.

Fig. 6.7 Compressive failure of brass test specimen

6.5 The torsion test

A typical torsion testing machine, such as the Avery machine shown in Fig. 6.8, consists of two straining heads to hold the specimen. One of these is attached to the measuring system and a torque is applied to the other by means of an electric motor drive through a reduction gearing. The applied torque is measured via a system of levers (or by a torque arm acting against a spring) and recorded on a dial. The rotation of the specimen over a certain gauge length is obtained by means of a torsion-meter which is clamped to the specimen.

A torsion test enables the modulus of rigidity (ref. section 3.15) of a material to be determined since pure shear stresses are produced within

Fig. 6.8 Avery torsion testing machine

the material. A typical torque-twist diagram for a ductile material such as mild steel is shown in Fig. 6.9 and the initial straight portion of the graph, where the material is within its limit of proportionality, is used to determine the modulus of rigidity. Following yielding of the material at A the angle of twist increases rapidly with increasing torque. At A only the outer surface of the material is at the yield stress but as the

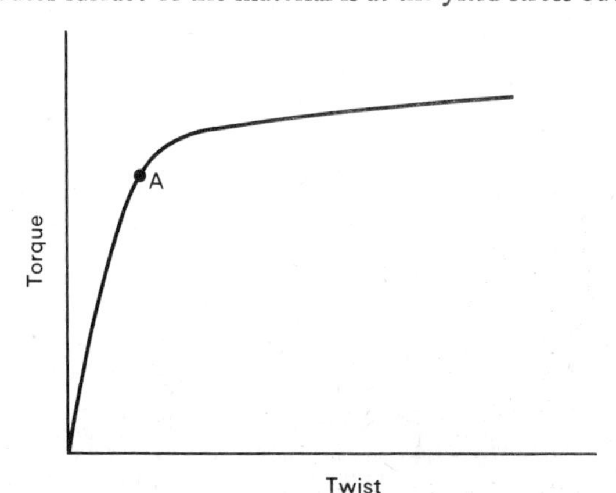

Fig. 6.9 Typical torque-twist diagram for a ductile material

torque is increased so more of the material reaches the yield stress until just prior to fracture the test specimen is almost wholly plastic. The torque-twist graph does not represent the shear stress-shear strain relationship because once the outer surface of the material has yielded the stress is no longer proportional to the radius and hence not proportional to the torque.

6.6 Hardness tests

The hardness of a material is determined by its ability to withstand indentation. Although one tends to associate resistance to abrasion as being a measure of the hardness of a material the wear properties are not dependent upon hardness alone. There are four hardness tests which may be used.

1. THE BRINELL HARDNESS TEST (Fig. 6.10)
A hardened steel ball, of diameter D, is pressed against the flat surface of the specimen with a certain force, which depends upon the

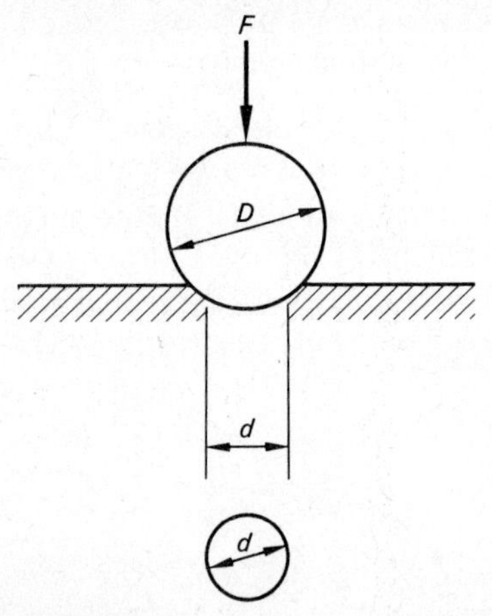

Fig. 6.10 Brinell hardness test

size of the ball being used. The standard ball sizes are 1, 2, 5, and 10 mm diameter and the associated forces are specified in BS240. The diameter d of the resulting indentation is measured with a microscope. Then,

$$\text{Brinell hardness number} = \frac{\text{applied force } F}{\text{curved area of indentation } A}$$

where $$A = \tfrac{1}{2}\pi D\{D - \sqrt{(D^2 - d^2)}\}$$

The Brinell hardness number is obtained from a set of tables in BS240 which are based on the above formula.

2. THE VICKERS HARDNESS TEST (Fig. 6.11)

This test is similar to the Brinell test except that the indentor is a diamond, whose shape is a square based pyramid with an apex angle of 136°. The diamond is pressed against the test specimen with a certain force, as contained in BS427, and for a specified time.

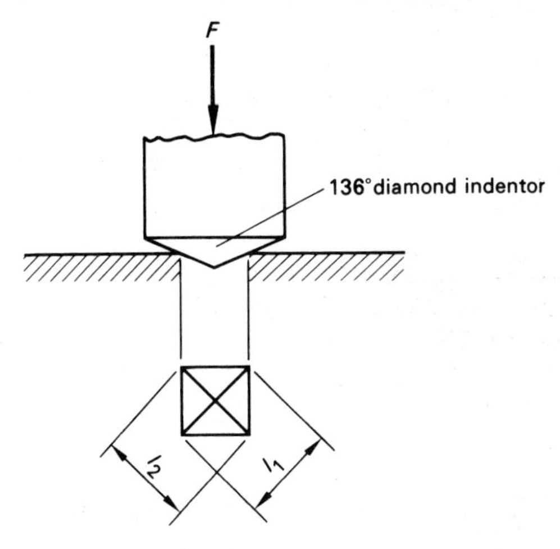

Fig. 6.11 Vickers hardness test

The lengths l_1 and l_2 of the resulting indentation are measured with a microscope. Then,

$$\text{Vickers pyramid number} = \frac{\text{applied force } F}{\text{surface area of indentation } A}$$

where $$A = l^2/2 \sin 68° = l^2/1{\cdot}854$$

l being the average length of the diagonals, i.e., $\frac{1}{2}(l_1 + l_2)$.

Although the hardness number can be calculated from the above equation it is usually obtained from tables contained in BS427.

3. THE ROCKWELL HARDNESS TEST

This test uses an indentor having a 120° conical diamond with a rounded apex for hard materials, or a steel ball for softer materials. However, the hardness is determined by the depth of penetration of the indentor instead of by the surface area of the indentation as in the case of the two previous tests. A minor load F is applied to cause a small indentation as indicated in Fig. 6.12(*a*). The major load F_M is then applied and removed after a specified time to leave the minor load F still acting. These two stages are shown in Figs. 6.12(*b*) and (*c*). Thus, the permanent increase in the depth of penetration caused by the major load is d[mm] and the Rockwell hardness number H_R is then given by

$$H_R = K - 500d$$

where K is a constant having a value of 100 for the diamond indentor and 130 for the steel ball indentor.

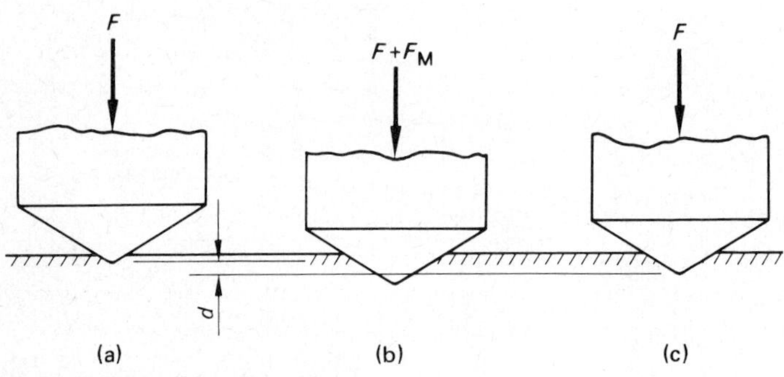

Fig. 6.12 Rockwell hardness test

The hardness number is usually read off a dial incorporated in the testing machine.

Various combinations of major load and indentor can be used and these are denoted by letters, e.g., A, B, C, and F for metals and K, E, M, L, and R for plastics. Thus, the scale used must always be indicated when quoting a Rockwell hardness number and the details of this test are contained in BS891.

4. THE SHORE SCLEROSCOPE HARDNESS TEST

This is a dynamic test in which a small indentor, whose tip is a steel ball or a rounded diamond, is dropped from a height of 250 mm onto the material whose hardness is required. The indentor is dropped down a glass tube whose length is calibrated from 0 to 140. On impact some of the energy possessed by the indentor will be absorbed by the material, depending on its hardness, and the height to which the indentor rebounds is a measure of the hardness of the material. This is read off from the calibrated tube as the 'Scleroscope number'. The harder the material, so the smaller the indentation produced and hence less energy is absorbed and the rebound height is greater.

The Brinell, Vickers, and Rockwell tests can be carried out on a single machine, known as a Universal Hardness Tester. A magnified image of the indentation produced is reflected onto a ground glass screen so enabling the Brinell and Vickers Pyramid hardness numbers to be quickly obtained. The Rockwell hardness number is recorded immediately from a dial. Such a testing machine avoids the necessity of separate machines for each test.

The tests so far described can only be carried out on small test specimens or components. For larger components a portable machine, which is clamped to the component, or a hardness comparator is used.

Fig. 6.13 Avery impact testing machine

The latter works on the Brinell principle and compares the indentation produced in the component by a steel ball with that produced on a surface of known hardness. Both indentations are produced simultaneously by a single hammer blow.

6.7 Impact testing

The toughness of a material is defined as its ability to withstand a shock loading without fracture. For certain structures, the ability to withstand an impact from a moving body or to absorb a sudden impact load may be very important and will require a material possessing both strength and ductility. The impact test is designed to test this property of a material.

The test specimen is a notched bar, the purpose of the notch being to produce a stress concentration and simulate brittle fracture conditions. In practice, notches and surface irregularities produce stress concentrations and are therefore a possible source of cracks. These tend to spread rapidly so appearing to be a brittle fracture although the material itself may be quite ductile.

The two principal impact tests are the Izod and the Charpy. In both instances the test specimen is rigidly supported and is impacted by a striker attached to a pendulum which is released from a certain position. The difference between the height from which the pendulum is released and the height to which it rises after impact gives a measure of the energy absorbed by the specimen and this is recorded on a dial mounted on the tester, as shown in Fig. 6.13. In order to reduce friction the bearing should be well lubricated.

1. THE IZOD TEST

The test specimen, which is either square or circular, is clamped vertically as a cantilever with the notch facing the striker, as shown in Fig. 6.14. The conditions for the test are specified in BS131: Part 1 and the energy absorbed is termed the 'Izod impact energy'.

2. THE CHARPY TEST

The test specimen, which is of square section, is mounted horizontally with the notch central and opposite the point of impact of the striker as shown in Fig. 6.15.

The types of notch are shown in Fig. 6.16 and the conditions governing the test are specified in BS131: Parts 2 and 3. The test results are quoted as 'impact energy' or 'impact energy per minimum cross-sectional area of specimen'.

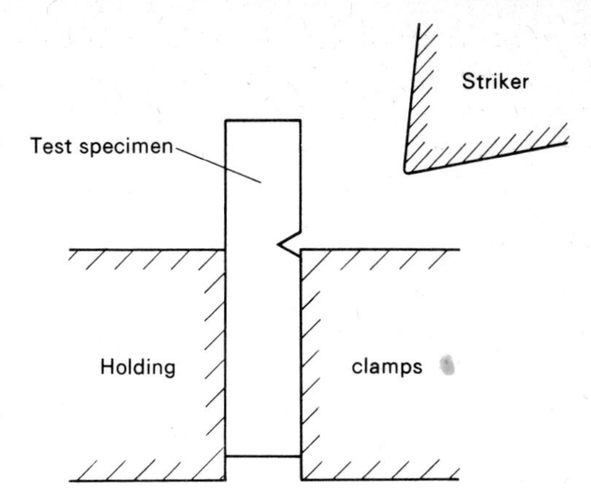

Fig. 6.14 The Izod test

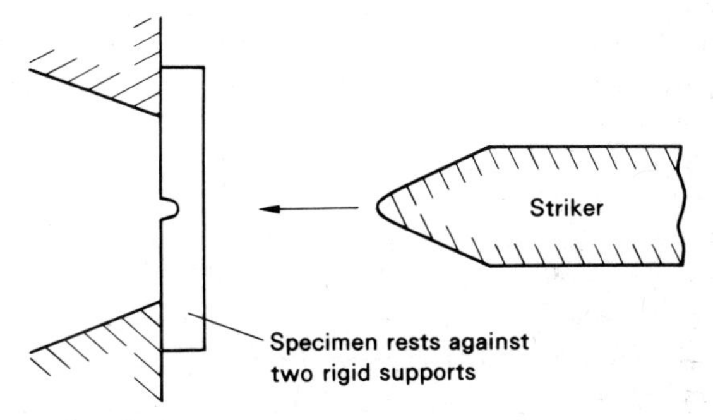

Fig. 6.15 The Charpy test

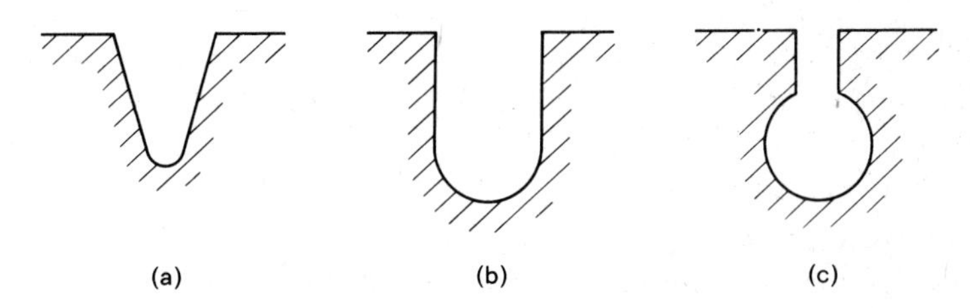

Fig. 6.16 Notch shapes for the Charpy test

7

Kinematics

7.1 Kinematics

Kinematics is the section of dynamics which deals with the motion of bodies without considering the forces which are necessary to produce the motion.

7.2 Revision of terms

The velocity of a body is a measure of its rate of change of displacement with respect to time.

As velocity has both magnitude and direction it is a vector quantity.

The acceleration of a body is the rate of change of its velocity with respect to time.

As velocity is specified by both magnitude (speed) and direction it follows that if a change occurs in either the speed or direction of motion of the body then the body is subject to an acceleration. This is readily illustrated by considering the motion of a car which is travelling around a circular track at a constant speed. Because the direction of motion is continually changing it follows that both the velocity and acceleration are also continually changing – although the speed of the car remains constant.

7.3 Derivation of equations of motion

The general equations of motion may be derived from the displacement-time graph.

Consider the motion of a body in a straight line (i.e., the velocity is numerically equal to the speed) and let the displacement s at any time

t be as shown on Fig. 7.1. At any given instant the velocity is defined as being the rate of change of displacement with respect to time; i.e.,

$$v = \frac{ds}{dt} = \text{slope of displacement-time graph} \tag{7.1}$$

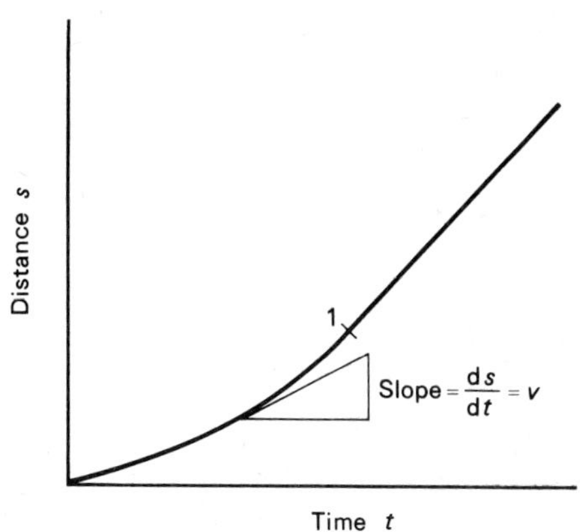

Fig. 7.1 Distance-time graph for varying velocity

If this slope is determined at several points on the graph then the velocity-time graph indicated in Fig. 7.2 is obtained. Note that as the slope of the distance-time graph is shown to be constant beyond point 1 then the velocity is constant beyond this point.

The acceleration of a body at any instant is defined as the rate of change of velocity with respect to time; i.e.,

$$a = \frac{dv}{dt} = \text{slope of velocity-time graph} \tag{7.2}$$

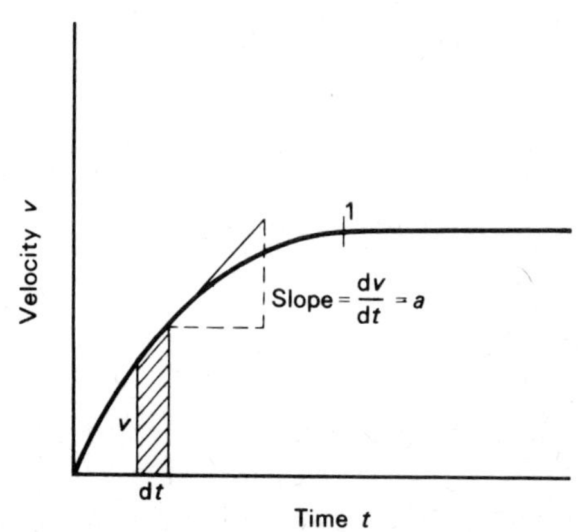

Fig. 7.2 Velocity-time graph for varying velocity

Again, by determining the slope at various points on the graph the acceleration-time graph indicated in Fig. 7.3 is obtained. Note that as the velocity beyond point 1 is constant then the acceleration beyond this point is zero.

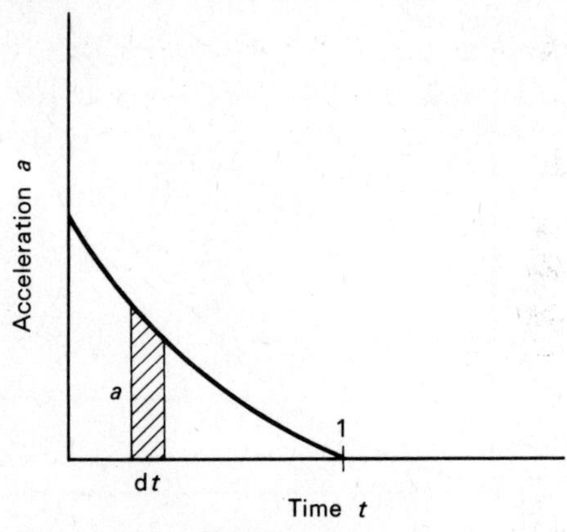

Fig. 7.3 Acceleration-time graph for varying velocity

Differentiating eq. (7.1) with respect to t gives

$$\frac{dv}{dt}=\frac{d^2s}{dt^2}$$

where d^2s/dt^2 is the second differential coefficient of s with respect to t.

Then, from eq. (7.2)

$$a=\frac{dv}{dt}=\frac{d^2s}{dt^2}$$

Thus if s = displacement

$$\frac{ds}{dt} = \text{rate of change of displacement with time} = \text{velocity}$$

$$\frac{d^2s}{dt^2} = \text{rate of change of velocity with time} = \text{acceleration}$$

Referring to Fig. 7.2 it will be seen that the area of an element of the velocity-time graph is $v\,dt$.

But from eq. (7.1),

$$v=\frac{ds}{dt}$$

Integrating: $\int ds = \int v\,dt$

gives displacement = area under velocity-time graph (7.3)

Similarly, from eq. (7.2),

$$\int dv = \int a\,dt$$

$$\text{velocity} = \text{area under acceleration-time graph} \tag{7.4}$$

Having established the general relationships (7.3) and (7.4) connecting the three variables – displacement, velocity and acceleration – with time, we can now apply them to the particular case of uniform acceleration, as depicted by the graphs in Fig. 7.4.

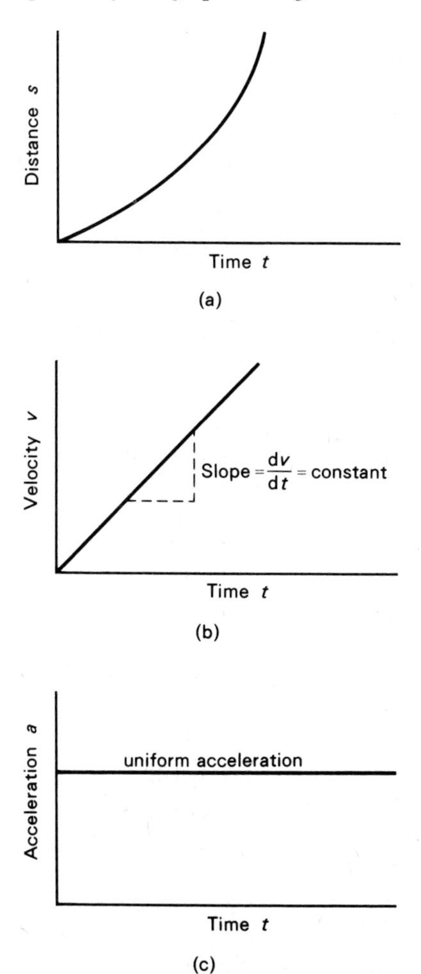

Fig. 7.4 Graphs for uniform acceleration

For uniformly accelerated motion the magnitude of a is constant as shown in Fig. 7.4(c). Hence, eq. (7.2) may be rewritten, when integrating, in the form

$$\int \mathrm{d}v = a \int \mathrm{d}t$$

If it is assumed that the displacement $s = 0$ at $t = 0$, that the initial velocity is v_1 at $t = 0$ and that v_2 is the velocity at time t the above equation becomes

$$\int_{v_1}^{v_2} \mathrm{d}v = a \int_0^t \mathrm{d}t$$

$$\Big[v\Big]_{v_1}^{v_2} = a\Big[t\Big]_0^t$$

$$v_2 - v_1 = at$$

$$v_2 = v_1 + at \qquad (7.5)$$

This equation gives a straight line plot (c.f. $y = mx + c$) as shown in Fig. 7.4(b).

From eqs. (7.1) and (7.2)

$$a = \frac{\mathrm{d}v}{\mathrm{d}t} = \frac{\mathrm{d}v}{\mathrm{d}s} \cdot \frac{\mathrm{d}s}{\mathrm{d}t} = v \cdot \frac{\mathrm{d}v}{\mathrm{d}s}$$

Then $\quad a = v\dfrac{\mathrm{d}v}{\mathrm{d}s}$

and integrating, with acceleration a constant, gives

$$\int v\mathrm{d}v = a \int \mathrm{d}s$$

$$\int_{v_1}^{v_2} v\mathrm{d}v = a\int_0^s \mathrm{d}s$$

$$\left[\frac{v^2}{2}\right]_{v_1}^{v_2} = a\Big[s\Big]_0^s$$

$$v_2^2 - v_1^2 = 2as$$

$$v_2^2 = v_1^2 + 2as \qquad (7.6)$$

Further equations may now be obtained by eliminating variables in turn. Thus, substituting for v_2 from eq. (7.5) in eq. (7.6) gives

$$(v_1 + at)^2 = v_1^2 + 2as$$

$$v_1^2 + 2atv_1 + a^2t^2 = v_1^2 + 2as$$

$$v_1t + \tfrac{1}{2}at^2 = s \qquad (7.7)$$

Similarly, substituting for v_1 from eq. (7.5) into eq. (7.6) gives

$$s = v_2 t - \tfrac{1}{2}at^2 \qquad (7.8)$$

Adding eq. (7.7) and (7.8) gives

$$2s = (v_1 + v_2)t$$

$$s = \frac{(v_1 + v_2)t}{2} \qquad (7.9)$$

The eqs. (7.5) to (7.9) are the equations for uniformly accelerated linear motion. The equations for uniformly accelerated angular motion can be derived in a similar manner using the first and second differential coefficients of θ for angular velocity and acceleration, i.e., angular displacement = θ, angular velocity $\omega = (\mathrm{d}\theta/\mathrm{d}t)$ and angular acceleration

$$\alpha = \frac{\mathrm{d}\omega}{\mathrm{d}t} = \frac{\mathrm{d}^2\theta}{\mathrm{d}t^2}$$

Thus, for uniformly accelerated motion we have the following equations:

	Linear motion	*Angular motion*
	$s = \dfrac{(v_1 + v_2)t}{2}$	$\theta = \dfrac{(\omega_1 + \omega_2)t}{2}$
	$v_2 = v_1 + at$	$\omega_2 = \omega_1 + \alpha t$
	$s = v_1 t + \tfrac{1}{2}at^2$	$\theta = \omega_1 t + \tfrac{1}{2}\alpha t^2$
	$v_2^2 = v_1^2 + 2as$	$\omega_2^2 = \omega_1^2 + 2\alpha\theta$
	$s = v_2 t - \tfrac{1}{2}at^2$	$\theta = \omega_2 t - \tfrac{1}{2}\alpha t^2$
where the notation is:		
Displacement	s	θ
Initial velocity (at $t = 0$)	v_1	ω_1
Final velocity (at time t)	v_2	ω_2
Acceleration	a	α
Time	t	t

Example 7.1

A body travels a distance of 5 m during the first second of its motion, 21 m in the third and 45 m in the sixth second. Show that these distances are consistent with the body having a uniform acceleration.

Solution

By using the information given up to the end of the third second and assuming a uniform acceleration we can calculate this acceleration. This value can then be used to determine the distance travelled in the sixth second and this calculated distance compared with the distance given.

Let v_0 = initial velocity at $t = 0$

v_1 = velocity when $t = 1$ second

v_2 = velocity when $t = 2$ seconds, etc.

and s_1 = distance travelled at end of 1 second ($t = 1$ s)

s_2 = distance travelled at end of 2 seconds ($t = 2$ s), etc.

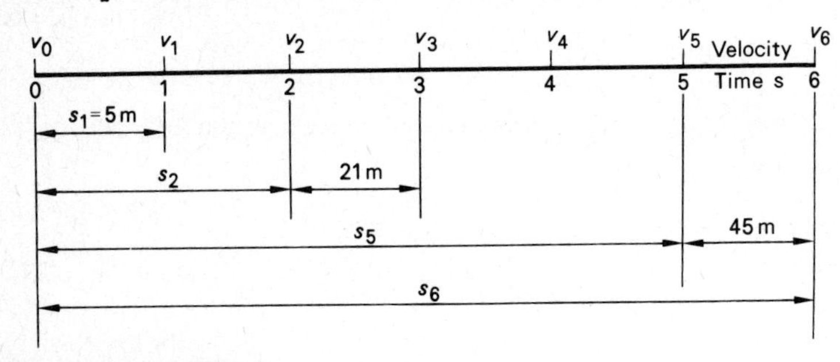

Fig. 7.5 Example 7.1

Referring to Fig. 7.5 the distance travelled during the first second can be calculated by applying eq. (7.7) in the form

$$s = v_0 t + \tfrac{1}{2}at^2$$

Then, for $t = 1$ $\quad 5 = v_0 + \tfrac{1}{2}a$ (1)

Similarly for $t = 2$ $\quad s_2 = 2v_0 + \tfrac{1}{2}a \, . \, 4$ (2)

and for $t = 3$ $\quad s_3 = 3v_0 + \tfrac{1}{2}a \, . \, 9$ (3)

Substituting for v_0 from (1) into eqs. (2) and (3) gives

$$s_2 = 2\,(5 - \tfrac{1}{2}a) + 2a$$

$$s_2 = 10 + a \qquad (4)$$

and $s_3 = 3\,(5 - \tfrac{1}{2}a) + \tfrac{9}{2}a$

$$s_3 = 15 + 3a$$

But $s_3 = S_2 + 21$

Substituting for s_2 from eq. (4) and equating values of s_3 gives

$$10 + a + 21 = 15 + 3a$$
$$2a = 16$$
$$a = 8 \text{ m/s}^2$$

Then from eq. (1) $v_0 = 1$ m/s

The distance travelled up to the end of the fifth second is then given by

$$s_5 = 5\,v_0 + \tfrac{1}{2}a.\,25$$
$$= 5 \times 1 + \frac{25 \times 8}{2}$$
$$= 105 \text{ m}$$

Similarly, the distance travelled up to the end of the sixth second is given by

$$s_6 = 6\,v_0 + \tfrac{1}{2}a\,.\,36$$
$$= 6 + 18 \times 8$$
$$= 150 \text{ m}$$

Distance travelled in sixth second $= s_6 - s_5$

$$= 150 - 105$$
$$= 45 \text{ m}$$

This is the same as the distance given and is therefore consistent with the body having a uniform acceleration.

Example 7.2

A body is projected upwards from a position on a level ground with a velocity V at an angle θ to the horizontal. Assuming that the body undergoes uniformly accelerated motion, determine general expressions for:

(*a*) the time of flight;
(*b*) the distance from the point of projection to the point where the body strikes the ground;
(*c*) the maximum height reached.

Hence, determine the angle at which a shell having a projection velocity of 480 m/s should be fired if it is to have a range of 15 km with a minimum time of flight. What will be the maximum height reached and the time of flight of the shell? Assume standard gravitational acceleration.

Solution

Referring to Fig. 7.6 the initial projection velocity V can be resolved into two rectangular components:

Horizontal component of projection velocity = $V \cos \theta$

Vertical component of projection velocity = $V \sin \theta$

If air resistance is neglected the only force acting on the body is its weight (i.e., the gravitational force on its mass), and the body will therefore be subjected to the local gravitational acceleration g downwards.

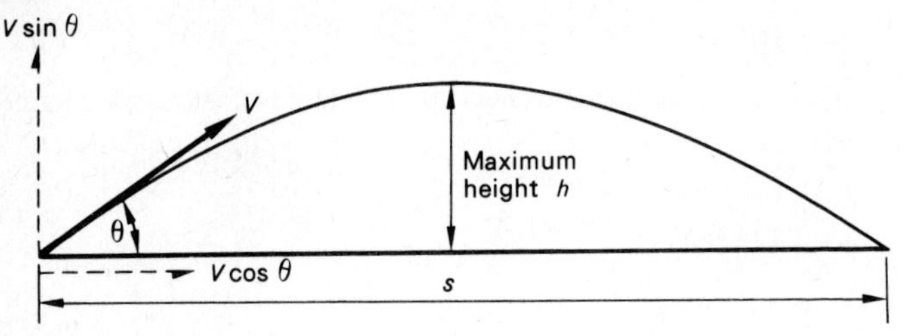

Fig. 7.6 Example 7.2

As there is no force acting in the horizontal direction the horizontal component of the projection velocity remains constant throughout the flight.

(*a*) The total time of flight can be obtained by applying $s = v_1 t + \frac{1}{2}at^2$ to the vertical flight.

When the body has returned to the ground the net vertical distance travelled is zero. Therefore $s = 0$, $v_1 = V \sin \theta$ and $a = -g$.

Then $0 = (V \sin \theta)\, t - \frac{1}{2}gt^2$

$$t = \frac{2\, V \sin \theta}{g}$$

(*b*) The horizontal distance travelled during the flight is called the range and as the horizontal component of the velocity remains constant we have:

$$\text{Range} = \text{horizontal component of velocity} \times \text{time of flight}$$

$$= V \cos \theta \times \frac{2\, V \sin \theta}{g}$$

$$= \frac{V^2 \sin 2\theta}{g} \quad (\text{since } \sin 2\theta = 2 \sin \theta \cos \theta)$$

(*c*) The maximum height reached during the flight can be obtained from applying

$$v_2^2 = v_1^2 + 2as$$

where $v_2 = 0$ since at the maximum height the vertical component of velocity is instantaneously zero.

$$v_1 = V \sin\theta = \text{initial vertical velocity}$$

$$a = -g$$

and $s = h$ (maximum height)

Thus $0 = (V \sin\theta)^2 - 2gh$

$$h = \frac{V^2 \sin^2\theta}{2g}$$

For the shell problem, using

$$\text{Range} = \frac{V \sin 2\theta}{g}$$

then $15 \times 10^3 = \dfrac{480^2 \sin 2\theta}{9{\cdot}81}$

$$\sin 2\theta = 0{\cdot}6388$$

$$\theta = 19^\circ\ 51' \quad \text{or} \quad 70^\circ\ 9'$$ (greater time of flight – and greater height)

$$\text{Time of flight} = \frac{2\,V \sin\theta}{g}$$

$$= \frac{2 \times 480 \times \sin 19^\circ\ 51'}{9{\cdot}81}$$

$$= 33{\cdot}23 \text{ s}$$

$$\text{Maximum height reached} = \frac{V^2 \sin^2\theta}{2g}$$

$$= \frac{480^2 \times \sin^2 19^\circ\ 51'}{2 \times 9{\cdot}81}$$

$$= 1354 \text{ m}$$

$$= 1{\cdot}354 \text{ km}$$

The shell should be fired at an angle of $19^\circ\ 51'$ to the horizontal, when its time of flight will be 33·23 s and the maximum height reached 1·354 km.

Example 7.3

A ball is thrown with a velocity of 20 m/s and at an angle of 30° to the horizontal from a position on the side of a hill which has an inclination of 15°. The direction is such that the ball travels down the slope of the hill. Determine the time of flight and the range measured down the hill.

Solution

This problem is solved by considering the motion perpendicular to the ground.

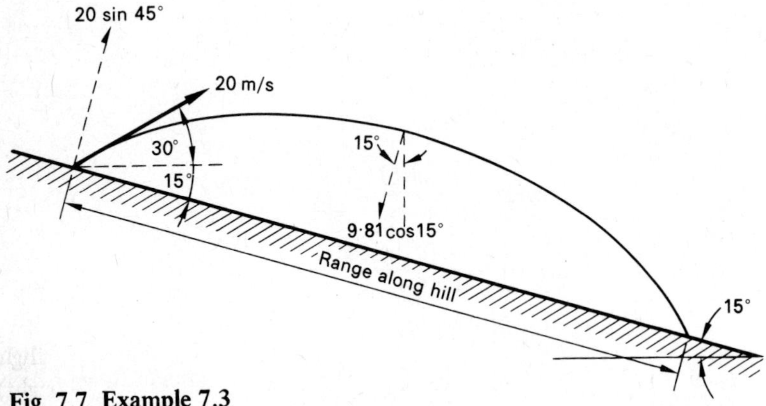

Fig. 7.7 Example 7.3

Referring to Fig. 7.7, the initial velocity component perpendicular to the ground is 20 sin 45° m/s while the gravitational acceleration component perpendicular to the ground is 9·81 cos 15° m/s^2.

Then, applying $s = v_1t + \frac{1}{2}at^2$ to the motion perpendicular to the ground gives, since the net distance travelled perpendicular to the ground is zero,

$$0 = (20 \sin 45°)t - \tfrac{1}{2}(9{\cdot}81 \cos 15°)t^2$$

$$t = \frac{40 \sin 45°}{9{\cdot}81 \cos 15°} = 2{\cdot}985 \text{ s}$$

The horizontal component of the projection velocity is 20 cos 30° = 17·32 m/s. This remains unchanged during flight irrespective of the inclination of the ground. Thus,

$$\text{Horizontal range} = 17{\cdot}32 \times 2{\cdot}985$$
$$= 51{\cdot}7 \text{ m}$$

From Fig. 7.7,

$$\text{Range along hill} = \frac{\text{horizontal range}}{\cos 15^\circ}$$

$$= \frac{51{\cdot}7}{\cos 15^\circ}$$

$$= 53{\cdot}52 \text{ m}$$

The time of flight is 2·985 s and the range along the hill is 53·52 m.

Example 7.4

Determine the time required for a wheel to reach a speed of 180 rev/min from rest if it is subjected to a uniform angular acceleration of 0·3 rad/s^2. How many revolutions will the wheel make in reaching this speed?

Solution

Applying $\omega_2 = \omega_1 + \alpha t$

where ω_1 = initial angular velocity = 0

$$\omega_2 = \text{final angular velocity} = \frac{180 \times 2\pi}{60} = 6\pi \text{ rad/s}$$

$$\alpha = 0{\cdot}3 \text{ rad/s}^2$$

and t = time taken

gives $6\pi = 0{\cdot}3t$

$$t = 20\pi$$

$$= 62{\cdot}84 \text{ s}$$

The angular distance θ travelled in this time is given by

$$\theta = \frac{(\omega_1 + \omega_2)t}{2}$$

$$= \frac{6\pi}{2} \,.\, 20\pi$$

$$= 60\pi^2 \text{ rad}$$

$$\therefore \quad \text{Number of revolutions} = \frac{60\pi^2}{2\pi} = 30\pi$$

$$= 94{\cdot}26$$

The time to accelerate to 180 rev/min is 62·84 s and the number of revolutions in this time is 94·26.

Example 7.5

The distance s travelled by a body moving in a straight line with a variable acceleration is given by the relationship

$$s = \frac{4}{3}t^3 + t^2 + 6,$$

where t is the time in seconds. What time will elapse whilst the velocity changes from 6 m/s to 30 m/s and what distance will the body travel in this time? What will be the acceleration of the body when its velocity is 30 m/s?

Solution

The given relationship for distance travelled is

$$s = \frac{4}{3}t^3 + t^2 + 6$$

Differentiating this with respect to time gives

$$\frac{ds}{dt} = v = 4t^2 + 2t$$

Thus, the time t_1 taken to reach a velocity of 6 m/s is given by

$$6 = 4t_1^2 + 2t_1$$

i.e., $$4t_1^2 + 2t_1 - 6 = 0$$

$$(t_1 - 1)(4t_1 + 6) = 0$$

$\therefore$ $$t_1 = 1 \text{ s is only positive solution}$$

Similarly the time t_2 taken to reach a velocity of 30 m/s is given by

$$30 = 4t_2^2 + 2t_2$$

$$4t_2^2 + 2t_2 - 30 = 0$$

$$(2t_2 - 5)(2t_2 + 6) = 0$$

$$t_2 = 2{\cdot}5 \text{ s is only positive solution}$$

Hence, the time for the velocity to change from 6 m/s to 30 m/s is 1·5 s.

The distance travelled s_1 in 1 s is given by:

$$s_1 = \frac{4}{3} . 1^3 + 1^2 + 6$$

$$= 8{\cdot}33 \text{ m}$$

and the distance s_2 travelled in 2·5 s is given by

$$s_2 = \frac{4}{3} \times (2{\cdot}5)^3 + (2{\cdot}5)^2 + 6$$

$$= 33{\cdot}08 \text{ m}$$

Therefore the distance covered while the velocity changes from 6 m/s to 30 m/s = 33·08 − 8·33 = 24·75 m.

To obtain the acceleration at any instant the expression for velocity must be differentiated with respect to time.

i.e., $v = 4t^2 + 2t$

$$\frac{dv}{dt} = a = 8t + 2$$

Thus, after 2·5 s, when the velocity is 30 m/s, the acceleration is given by

$$a = 8 \times 2{\cdot}5 + 2$$

$$= 22 \text{ m/s}^2$$

Example 7.6

A body moving in a straight line with a uniform acceleration of 0·8 m/s^2 passes a fixed point A with a velocity of 4 m/s. What will be the distance of the body from A, 6 seconds later, and what is its velocity at this time?

Solution

From eq. (7.2):

$$a = \frac{dv}{dt} = 0{\cdot}8$$

Integrating $v = 0{\cdot}8t + C$ (1)

where C is a constant of integration

Equation (1) gives the velocity at any time t.

If t is taken as zero when the body is at A, then $v = 4$ m/s when $t = 0$.

Thus, in (1) $4 = 0 + C$

$\therefore$ $C = 4$ m/s

The equation for the velocity of the body is therefore

$$v = 0{\cdot}8t + 4$$

But from eq. (7.1):

$$v = \frac{ds}{dt} = 0{\cdot}8t + 4$$

Again integrating with respect to t gives

$$s = \frac{0{\cdot}8t^2}{2} + 4t + D$$

where D is a constant of integration.

Now when $t = 0$ the body is at A, i.e.,

$$s = 0$$

at $t = 0$

$\therefore$ $D = 0$

Then $s = 0{\cdot}4t^2 + 4t$ (2)

Equation (2) gives the distance of the body from A at any time t.

When $t = 6$ s

$$s = 0{\cdot}4 \times 6^2 + 4 \times 6 = 38{\cdot}4 \text{ m}$$

and $v = 0{\cdot}8 \times 6 + 4 = 8{\cdot}8$ m/s

After 6 seconds the body will be 38·4 m from A and have a velocity of 8·8 m/s.

7.4 Relationship between linear and angular motion

Consider a particle which is rotating about a fixed centre O with a varying angular acceleration. Suppose that in a small interval of time δt the body moves through a small peripheral distance δs as indicated in Fig. 7.8.

Then, the instantaneous linear velocity v is given by

$$v = \frac{\delta s}{\delta t}$$

But, $\delta s = r\,.\,\delta\theta$

where $\delta\theta$ is the angle subtended at O by the peripheral distance δs.

Then $v = \dfrac{r\delta\theta}{\delta t} = r\dfrac{d\theta}{dt}$ (in the limit as $t \to 0$)

$$v = r\omega \tag{7.11}$$

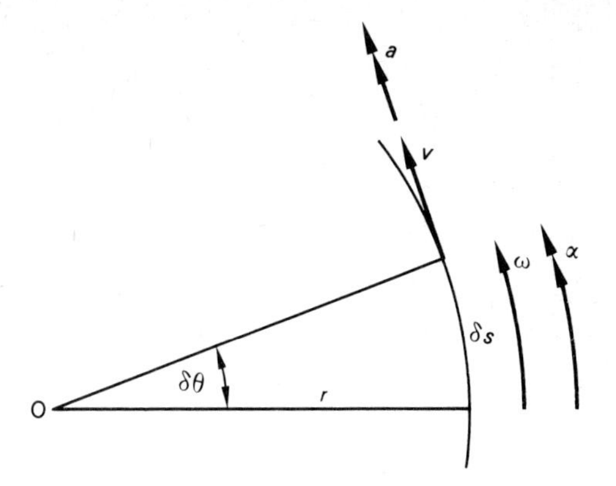

Fig. 7.8

where $$\omega = \frac{d\theta}{dt} = \text{instantaneous angular velocity}$$

Thus, at any instant

Linear velocity = radius x angular velocity

Also $$a = \frac{dv}{dt}$$

Therefore, differentiating the relationship for v [eq. (7.11)] gives

$$\frac{dv}{dt} = a = r\frac{d\omega}{dt}$$

$$a = r\alpha \tag{7.12}$$

where $$\alpha = \frac{d\omega}{dt} = \frac{d^2\theta}{dt^2}$$
is the instantaneous angular acceleration of the particle.

Thus, at any instant,

Linear acceleration = radius x angular acceleration

7.5 Centripetal acceleration

Consider a particle to be moving on a circular path of radius r, as shown in Fig. 7.9(a), with a constant angular velocity ω.

When the body is at A it possesses an instantaneous linear velocity v, tangential to the circle. Suppose that during the small interval of time δt the body moves from A to B, the arc AB subtending the small angle

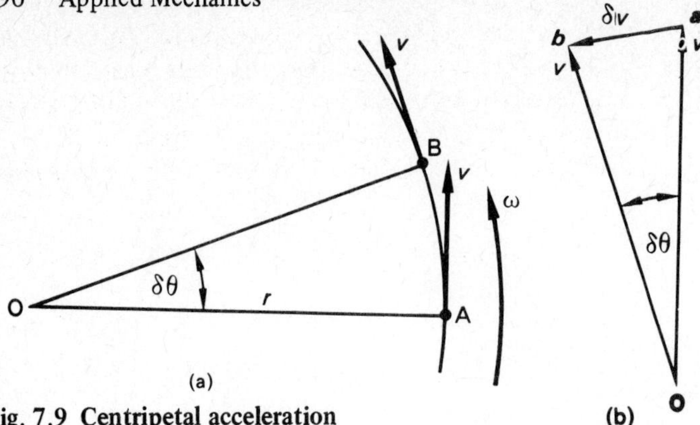

Fig. 7.9 Centripetal acceleration

$\delta\theta$ at the centre of the circle O. Now although the body is moving with a uniform speed in a circular path the direction of the instantaneous linear velocity is continually changing and the body is therefore subject to an acceleration in the direction of the instantaneous change in linear velocity.

Referring to the velocity vector diagram (Fig. 7.9(*b*)), vector **O*a*** represents the velocity at A while vector **O*b*** represents the velocity at B. It will be seen that there is a vector change in velocity δv during the motion from A to B of magnitude ***ab***, acting in a radial direction.

$$\text{Vector change in velocity} = \boldsymbol{ab}$$

$$\delta v = v \,.\, \delta\theta$$

Now this change in velocity occurs in a time δt. Therefore,

$$\text{Acceleration} = \frac{\delta v}{\delta t} = v\,\frac{\delta\theta}{\delta t}$$

$$= v\,\frac{\mathrm{d}\theta}{\mathrm{d}t} \quad \text{(in the limit as } t \to 0\text{)}$$

But $$\frac{\mathrm{d}\theta}{\mathrm{d}t} = \omega = \frac{v}{r}$$

$$\therefore \quad \text{Acceleration} = \frac{v^2}{r}$$

$$= \omega^2 r \qquad (7.13)$$

The direction of this instantaneous acceleration $\omega^2 r$ is towards the centre of the circle O and is given the particular name *centripetal* which means tending towards the centre. For uniform speed of rotation the centripetal acceleration is constant in magnitude.

It must be appreciated that the centripetal acceleration is in addition to any angular acceleration that occurs. For example, consider a particle that is instantaneously at point A in Fig. 7.10(*a*) to be rotating about O with a variable speed. If, at the particular instant considered, the angular acceleration is α then the particle at A will have two acceleration components:

1. An instantaneous angular acceleration at right angles to OA of magnitude $r\alpha$.
2. A centripetal acceleration along radius AO of magnitude $\omega^2 r$ where ω is the instantaneous angular velocity.

Hence, at the given instant, the acceleration of the body will be the resultant of the two separate components as indicated in Fig. 7.10(*b*).

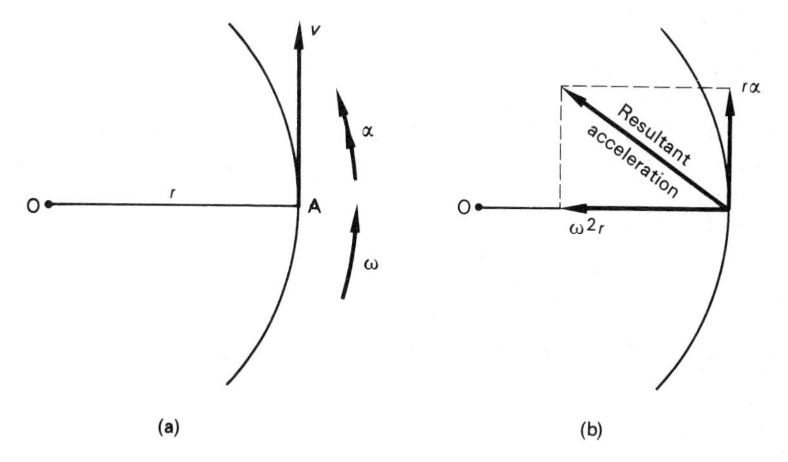

Fig. 7.10 Resultant acceleration of a body subjected to an angular acceleration

Example 7.7

The peripheral speed of a wheel 0·32 m diameter increases uniformly from 8 m/s to 16 m/s during 20 revolutions of the wheel. What is the angular acceleration? If the wheel continues to gain in speed at the same rate, what will be the centripetal acceleration of a point on the circumference after the wheel has turned through a further 10 revolutions?

Solution

When the linear speed is 8 m/s, the angular speed ω_1 can be obtained from eq. (7.11).

i.e., $$\omega_1 = \frac{8}{0\cdot16} = 50 \text{ rad/s}$$

Similarly, when the linear speed is 16 m/s, $\omega_2 = \dfrac{16}{0\cdot16} = 100$ rad/s.

This change in angular speed takes place during 20 revolutions.

i.e., $\theta = 20 \times 2\pi = 40\pi$ rad

The angular acceleration can be obtained by using

$$\omega_2^2 = \omega_1^2 + 2\alpha\theta$$

$$100^2 = 50^2 + 2\alpha \,.\, 40\pi$$

$$\alpha = \frac{7500}{80\pi} = 29\cdot9 \text{ rad/s}^2$$

After a further 10 revolutions the total angle turned through will be $30 \times 2\pi = 60\pi$ rad, and the angular velocity ω_3 is given by

$$\omega_3^2 = \omega_1^2 + 2\alpha\theta$$

$$= 50^2 + 2 \times \frac{7500}{80\pi} \times 60\pi$$

$$= 2500 + 11\,250$$

$$= 13\,750 \text{ (rad/s)}^2$$

Therefore Centripetal acceleration $= \omega_3^2 r$

$$= 13\,750 \text{ [rad/s]}^2 \times 0\cdot16 \text{ [m]}$$

$$= 2200 \text{ m/s}^2$$

The centripetal acceleration after a further 10 revolutions will be 2200 m/s^2.

Problems

1. A train passes three consecutive kilometre posts A, B, and C. The time taken to travel from A to B is 80 s and from B to C is 56 s. Assuming that the acceleration of the train is uniform between A and C determine its speed at each of the posts.
2. A body starts from rest with a constant acceleration of 1·5 m/s^2 which ceases after a certain time. It then moves at a constant speed of 6 m/s for the next 12 s after which it is uniformly retarded and brought to rest. If the complete motion takes 18 s determine the retardation and the total distance travelled.

3. A train takes 12 min to travel between two stations that are 16 km apart. It undergoes a uniform acceleration during the first 45 s and a uniform retardation during the last 60 s of its journey. For the remainder of the time it travels at a constant speed. Draw the speed-time graph and determine the uniform speed, the distance travelled at this speed and the rate of acceleration.

4. A mechanical drive is used to move wagons along a horizontal track at a mine. The wagons are accelerated uniformly to a speed of 40 km/h over a distance of 60 m. The speed is then maintained constant for 2 min and the wagons are then brought to rest with a uniform deceleration. If the rate of acceleration is three times the rate of deceleration determine:

(*a*) the distance travelled,
(*b*) the total time to complete the journey.

5. A shell is to be fired so that the maximum height reached is 0·8 km and its range is 16 km. Determine the velocity at which the shell should be fired and its inclination to the horizontal.

6. A shell is fired from the side of a hill, that is inclined at 30° to the horizontal, with a horizontal velocity of 400 m/s.
Determine:

(*a*) the range of the shell along the hillside;
(*b*) the magnitude and direction of the impact velocity.

7. A particle moves along a straight horizontal line such that its distance from a certain point is given by

$$s = 3t - t^2$$

where s is in metres and t in seconds.
Determine:

(*a*) the velocity of the particle after 1 s and 6 s;
(*b*) the distance of the particle from the fixed point after 1 s and 6 s;
(*c*) the total distance travelled from a time of 1 s to a time of 6 s.

8. The position of a body moving along a horizontal line is given by

$$s = t^3 - 8t^2 + 12t + 2$$

where s is the distance from the origin in metres and t is the time in seconds.

Determine:

(*a*) the times when the velocity of the body is zero;
(*b*) the acceleration when $t = 1{\cdot}5$ s.

9. A body starts from a point A and moves in a straight line such that its velocity at any instant is given by

$$v = 0{\cdot}6t^2 + 0{\cdot}4t$$

Determine:

(*a*) the distance from A and the acceleration of the body after 2 s;
(*b*) the time when the body will be at a distance of 10 m from A;
(*c*) the time when the acceleration of the body will be 4 m/s^2.

10. A vehicle starts from rest and accelerates for 10 s. The acceleration a [m/s^2] is given by the equation

$$a = -0{\cdot}22t + 4{\cdot}4$$

where t is the time in seconds.
Determine:

(*a*) the velocity after 5 s and the distance travelled in this time;
(*b*) the distance travelled during the acceleration time.

11. A ball is thrown vertically upwards with a speed of 30 m/s and 3 s later a second ball is thrown upwards in the same way with a speed of 20 m/s.
Determine:

(*a*) the height when the two balls are the same distance above the ground;
(*b*) the time when this occurs measured from the instant of projection of the first ball;
(*c*) the time interval in the return of the two balls to the ground.

12. A wheel of 0·25 m diameter is accelerated uniformly so that its peripheral speed is 12 m/s after 40 revolutions. Determine the angular acceleration of the wheel.

If the wheel continues to accelerate at the same rate what will be the centripetal acceleration of a point on the periphery after a further 20 revolutions?

13. A wheel which is rotating at 30 rev/min is uniformly accelerated for 90 s during which time it makes 75 revolutions. What is the angular velocity at the end of this time? For how much longer must the wheel be accelerated to reach an angular velocity of 12 rad/s?

14. A flywheel, 1 m in diameter, is accelerated from rest at 0·4 rad/s^2 to a speed of 200 rev/min. It runs for 45 s at this speed and is then brought to rest with a uniform deceleration in 60 revolutions. Determine:

(*a*) the angular deceleration;
(*b*) the total time of motion;
(*c*) the total number of revolutions made;
(*d*) the linear deceleration of a point on the rim;
(*e*) the maximum centripetal acceleration of a point on the rim during the motion of the flywheel.

8

Kinetics — translatory motion

8.1 Kinetics

Kinetics deals with the action of the forces which cause motion as distinct from the kinematics in the previous chapter which dealt only with the geometry of motion. In this chapter we shall cover the relationships between force and motion which result from the application of Newton's *laws of motion* and his statement of the *law of universal gravitation.*

8.2 Momentum

The linear momentum of a body at any instant is defined as the product of its mass m and its velocity at that instant, v; i.e.,

$$\text{Momentum} = mv$$

The basic unit of momentum is kg m/s or N s.

Linear momentum is a vector quantity having the same direction as the velocity of the body.

8.3 Newton's laws of motion

1. A body remains at rest or continues to move with a uniform velocity only if there is no resultant force acting on it.
2. The rate of change of momentum of a body is proportional to the resultant force acting on it and occurs in the direction of the resultant force.
3. The action and reaction forces between bodies in contact are equal and opposite to each other.

The first law of motion is only an extension of the second, since if a body is at rest or is moving with a constant velocity it obviously has no acceleration and, therefore, cannot have a resultant force acting upon it.

The second law provides the basic kinetic equation $F = ma$, where F is the resultant force required to give a body of mass m an acceleration a. The direction of the acceleration a is the same as that of the force F.

The third law, which is fundamental to our whole study of mechanics, tells us that forces always act in pairs of equal and opposite forces. It is most important to appreciate that this law applies to contacting bodies in motion as well as to bodies that are at rest (see section 8.15).

8.4 The kinetic equation of motion

Newton's second law of motion states that the rate of change of momentum of a body is proportional to the resultant force acting on the body. We can therefore express this law in the form:

$$F \propto \frac{\mathrm{d}}{\mathrm{d}t}(mv) \qquad \frac{\mathrm{d}}{\mathrm{d}t} \text{ denoting a rate of change}$$

For a body having a constant mass m it is evident that only the velocity can change with respect to time. Thus

$$F \propto m\frac{\mathrm{d}v}{\mathrm{d}t}$$

But $\frac{\mathrm{d}v}{\mathrm{d}t}$ = linear acceleration a

Therefore, $F \propto ma$

$$F = \mathrm{k}ma$$

where k is a constant.

Now the unit of force, the newton (N), is the force required to give a mass of 1 kilogramme an acceleration of 1 metre/second^2.

$$\therefore \quad 1\ [\mathrm{N}] = \mathrm{k} \times 1\ [\mathrm{kg}] \times 1\ [\mathrm{m/s^2}]$$

The constant k therefore has a value of unity and

$$F = ma \tag{8.1}$$

The mass of a body is a measure of the inertia, or resistance to change in velocity, of that body since the greater the mass of the body the greater is the force required to produce a given acceleration.

8.5 The law of universal gravitation

Sir Isaac Newton was also responsible for stating the law governing the mutual attraction between bodies. This law states that every body in the universe is attracted to every other body by a force that is directly proportional to the product of their masses and inversely proportional to the square of their distance apart. The law may therefore be stated in the form:

$$F = \frac{Gm_1 m_2}{d^2} \tag{8.2}$$

where F = mutual force of attraction between masses m_1 and m_2

d = distance between the centres of masses

G = gravitational constant (independent of the masses involved)

For any two masses which are on the surface of the earth, or any other planet, the force of attraction between them is negligible. However, if one of the masses is the earth, or any other planet, then the force of attraction becomes appreciable and it is this case that will now be examined.

Since the earth exerts a force of attraction on all other bodies it follows from Newton's second law that this force will produce an acceleration.* This acceleration, due to the earth's force of attraction, is given the symbol g, and if air resistance is neglected then g will be the gravitational acceleration of the earth for bodies falling in a vacuum.

If M_0 is the mass of the earth (assumed concentrated at its centre of mass) and M the mass of a body on the surface of the earth, then applying eqs. (8.1) and (8.2) gives:

$$\frac{GM_0 M}{r^2} = F = Mg$$

Where r = radius of the earth

$$\therefore \quad g = \frac{GM_0}{r^2}$$

* This still applies in the case of the moon where the earth's force produces an acceleration of the moon causing its direction of motion to change continually, its speed remaining approximately constant.

By experiment the following values are known:

$$\text{Mass of the earth, } M_0 = 5{\cdot}98 \times 10^{24} \text{ kg}$$

$$\text{Mean radius of earth, } r = 6375 \text{ km} = 6{\cdot}375 \times 10^6 \text{ m}$$

$$G = 6{\cdot}668 \times 10^{-11} \text{ m}^3/\text{kg s}^2$$

Thus

$$g = \frac{6{\cdot}668 \times 10^{-11} \,[\text{m}^3/\text{kg s}^2] \times 5{\cdot}98 \times 10^{24} \,[\text{kg}]}{(6{\cdot}375 \times 10^6)^2 \,[\text{m}^2]}$$

$$\approx 9{\cdot}81 \text{ m/s}^2$$

Thus, a body which is allowed to fall in a vacuum at the surface of the earth will have an acceleration of approximately 9·81 m/s^2. As the earth is not a perfect sphere, being flattened at the poles and bulging at the equator, this acceleration varies over the surface of the earth. The standard gravitational acceleration of the earth is 9·806 65 m/s^2, being the acceleration produced at a latitude of 45°N. However, for most engineering problems a value of 9·81 m/s^2 is sufficiently accurate and has been used throughout this text.

8.6 Mass and weight

The mass of any body is a fundamental property of that body and is a measure of the quantity of matter within the body. This quantity of matter will be the same irrespective of where the body is positioned. Now the particular force which attracts a body to the surface of the earth, or any other planet, is known as the weight of that body. Thus the weight of a body provides its gravitational acceleration. However, it was stated in section 8.5 that the gravitational acceleration varies over the surface of the earth. Furthermore, the gravitational acceleration at the surface of the moon is approximately one sixth of that at the surface of the earth and at the planet Jupiter the gravitational acceleration is 2·65 times that at the earth. Thus, it follows, that the weight of a body is not a fixed quantity and to talk of weight as a fundamental property of a body is strictly incorrect.

Whenever the term weight is used in this text it is understood to imply 'the gravitational force acting upon the mass', and in all instances the gravitational acceleration has been taken as that at the surface of the earth, i.e., 9·81 m/s^2 (for calculations). Thus, whenever a force arises from the gravitational action on a mass the dead load in kilogrammes must be converted into a force in newtons. Thus,

$$\underset{\text{(weight)}}{\text{Gravitational force}} = 9{\cdot}81m \text{ newtons}$$

where m = mass of body in kilogrammes

8.7 Linear translation

If a body is acted upon by a number of external forces such that the resultant of these forces passes through the centre of mass of the body the resulting motion will be translatory (see Fig. 8.1).

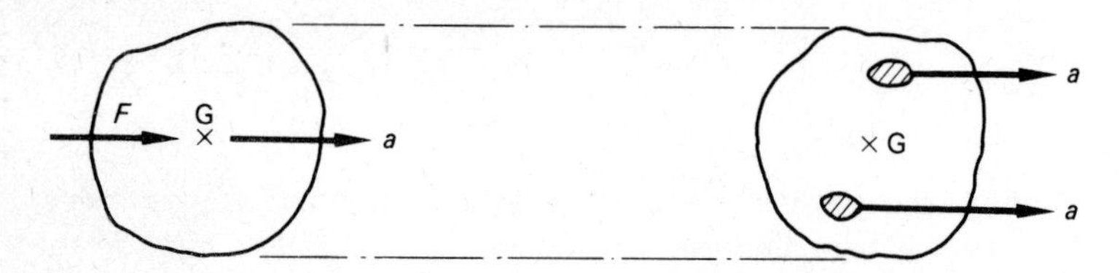

Force F has no moment about the centre of mass G. Therefore the body does not rotate and has translatory motion only.

All parts of the body have the same acceleration.

Fig. 8.1 Linear translation

As the resultant force passes through the centre of mass the moment of the forces about any axis through the centre of mass will always be zero for this type of motion.

Example 8.1

A train having a mass of 300 Mg is travelling at a steady speed of 60 km/h when it begins to climb an incline of 1 in 100. During the climb the power unit exerts a constant tractive force of 25 kN. If the resistances to motion remain constant at 60 N/Mg determine how far the train will travel up the incline before coming to rest.

Solution

$$\text{Gravitational force on train (i.e., its weight)} = 300 \times 10^3 \times 9{\cdot}81$$
$$= 2943 \times 10^3 \text{ N}$$

The forces opposing the motion of the train are:

(*i*) the component of the weight of the train acting down the incline, i.e.,

$$\frac{2943 \times 10^3}{100} = 29\,430 \text{ N}$$

(*ii*) the resistances to motion of 60 N/Mg, i.e.,

$$60\left[\frac{\text{N}}{\text{Mg}}\right] \times 300\ [\text{Mg}] = 18\,000\ \text{N}$$

The forces acting on the train are shown in Fig. 8.2.

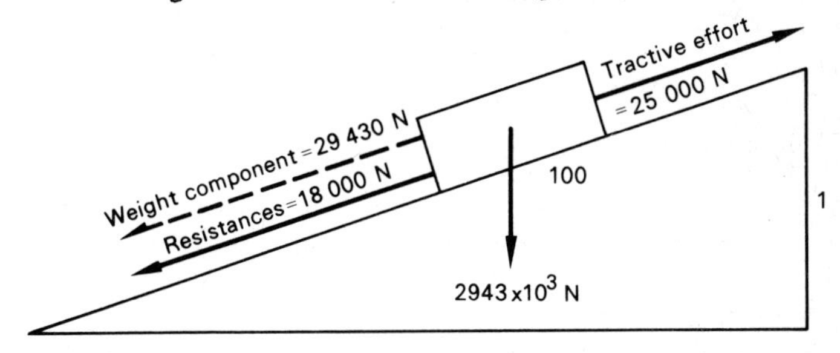

Fig. 8.2 Example 8.1

Thus, resultant force F acting on the train in the direction of its motion

$$= 25\,000 - 29\,430 - 18\,000$$
$$= -22\,430\ \text{N}$$

Applying $\quad F = ma \quad$ [eq. (8.1)]

gives $\quad -22\,430\ [\text{N}] = 300 \times 10^3\ [\text{kg}] \times a$

$$a = -0{\cdot}0748\ \text{m/s}^2 \quad \text{(retardation)}$$

The distance travelled up the incline can now be obtained by applying

$$v_2^2 = v_1^2 + 2as$$

where $\quad v_2 = 0, v_1 = 60\ \text{km/h} = 16{\cdot}67\ \text{m/s}, a = -0{\cdot}0748\ \text{m/s}^2$

$$0^2 = 16{\cdot}67^2 + 2\,(-0{\cdot}0748)\,s$$
$$s = 1856\ \text{m}$$

The distance travelled by the train up the incline before coming to rest is 1856 m.

Example 8.2

The block shown in Fig. 8.3 rests on a horizontal plane and the coefficient of friction between the block and the plane is 0·3. Determine the magnitude of the force P that will cause:

(*a*) the block just to slide across the plane;
(*b*) the block to tip about the right-hand corner A.

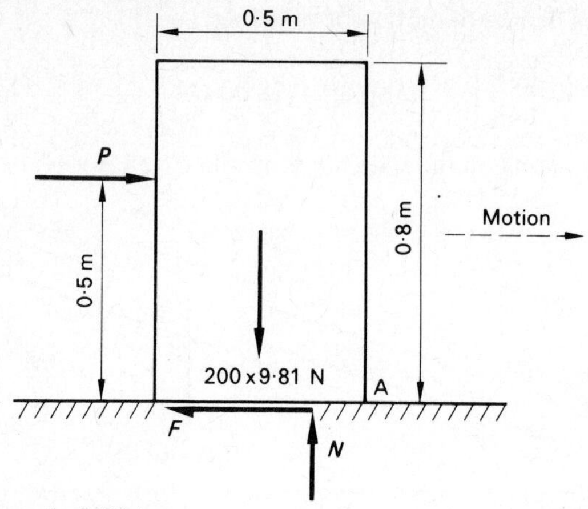

Fig. 8.3 Example 8.2(*a*)

Determine also the maximum acceleration, due to P, that the block can undergo without tipping.

Solution

(*a*) Let the normal reaction between the block and the plane be N and the frictional resistance be F.

Then, for vertical equilibrium

$$N = 200 \times 9{\cdot}81$$
$$= 1962 \text{ N}$$

When the block is on the point of slipping

$$F = \mu N$$
$$= 0{\cdot}3 \times 1962$$
$$= 588{\cdot}6 \text{ N}$$

But for horizontal equilibrium

$$P = F$$
$$= 588{\cdot}6 \text{ N}$$

The force required to cause the block just to slide across the plane is 588·6 N.

(*b*) The force required to cause the block to tip about point A can be obtained by taking moments about A (ref. Fig. 8.4). At the instant of tipping the reaction N will act through A, as shown.

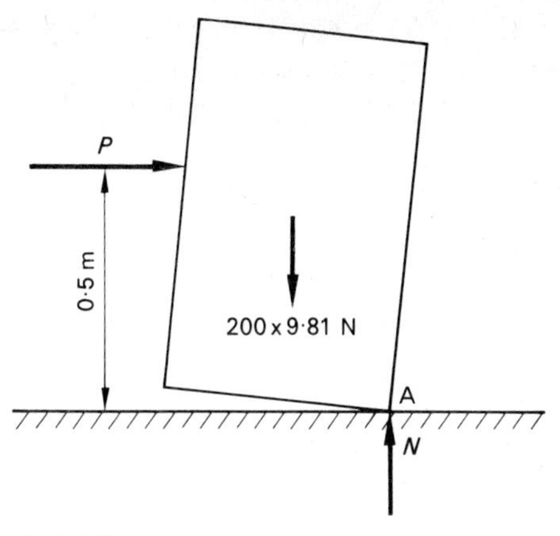

Fig. 8.4 Example 8.2(*b*)

Then,

$$\text{Clockwise moment about A} = \text{anticlockwise moment about A}$$

$$P \times 0{\cdot}5 = 200 \times 9{\cdot}81 \times 0{\cdot}25$$

$$P = 981 \text{ N}$$

The force required to cause the block to tip is 981 N.

As the maximum force P that can be applied without tipping is 981 N and the limiting frictional resistance is 588·6 N it follows that

$$\text{Maximum accelerating force } F = 981 - 588{\cdot}6$$

$$= 392{\cdot}4 \text{ N}$$

Applying

$$F = ma$$

gives

$$392{\cdot}4 = 200\,a$$

$$a = 1{\cdot}962 \text{ m/s}^2$$

The maximum possible acceleration without tipping occurring is 1·962 m/s^2.

8.8 Translation in a curved path

It was shown in section 7.5 that when a body moves in a curved path, having a radius of curvature r, with a uniform speed v, it is subject to an inward radial acceleration, known as centripetal acceleration, of v^2/r.

From Newton's laws of motion it follows that a body cannot be subjected to an acceleration unless a force is acting to cause that acceleration. Thus, referring to Fig. 8.5 and applying Newton's second law:

$$\begin{aligned}\text{Centripetal force} &= \text{mass} \times \text{centripetal acceleration}\\ &= m\frac{v^2}{r}\\ &= m\omega^2 r \quad (\text{since } v = r\omega \text{ for uniform circular motion})\end{aligned}$$

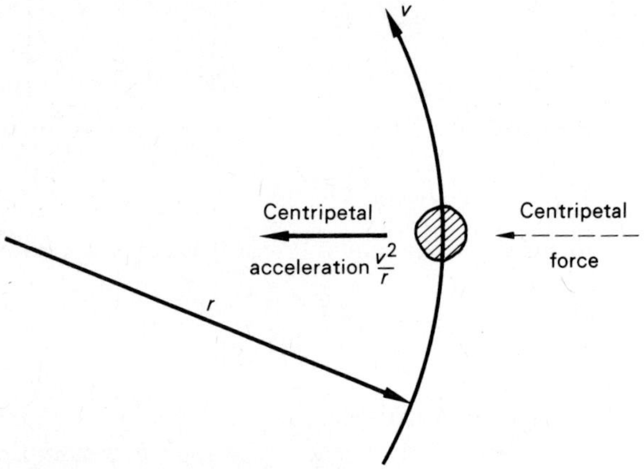

Fig. 8.5 Translation in a curved path

The centripetal force, which must be applied externally to the body, can be created in a variety of ways. In the case of a mass on the end of a string or carried on a rotating arm the tension in the string or in the arm provides the force. In the case of a train rounding a curve, however, the force is provided by the side-thrust of the rails on the flanges of the wheels. In other cases the centripetal force may arise from frictional effects. This is the case when a car negotiates a bend and if the road conditions are such that insufficient frictional force can be created at the tyres then a skid will result.

Now, by Newton's third law, there must be an opposing force of equal magnitude to the centripetal force. This force therefore acts radially outwards and is known as the centrifugal force. It must be appreciated that the centrifugal force is NOT applied to the body, e.g., in the case of a train on a curve, the train pushes outwards on the rails (CENTRIFUGAL) and the rails exert a force of equal magnitude on the train (CENTRIPETAL).

8.9 Vehicle on a curved horizontal track

Consider a vehicle of mass m which is travelling around a curved horizontal track of radius r at a constant speed v. The centripetal force on such a vehicle will be mv^2/r.

Let the normal reactions at the wheels be N_1 and N_2 and the frictional forces between the wheels and the road surface be F_1 and F_2 respectively, as shown in Fig. 8.6(a).

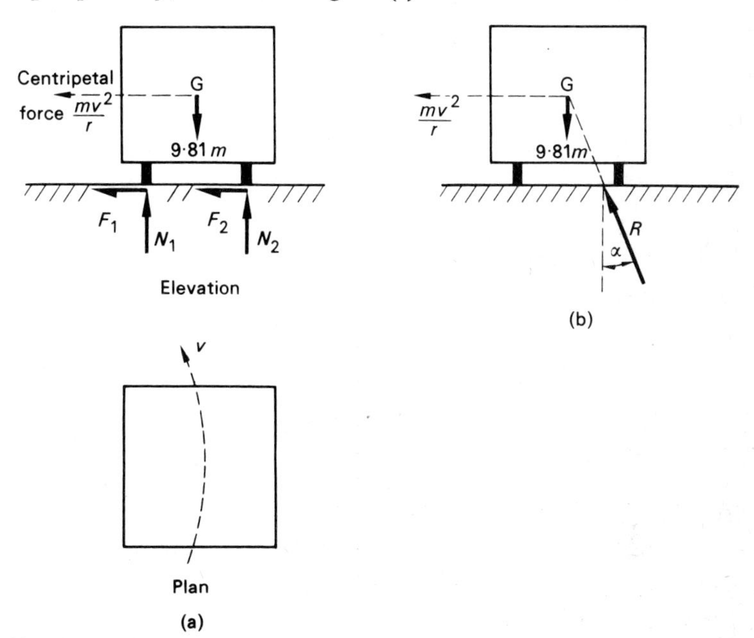

Fig. 8.6 Vehicle on a curved horizontal track

These four forces may be replaced by the resultant R, which will be inclined to the normal to the road surface at some friction angle α.

If there is to be no rotation about the centre of mass G then this resultant force R must pass through G, as indicated in Fig. 8.6(b). Hence the vehicle is acted upon by two forces only, namely its weight and the resultant R. Of these, only R can provide the centripetal force necessary to keep the vehicle moving in a circular path.

Then from Fig. 8.6(b):

$$R \sin \alpha = \frac{mv^2}{r}$$

and $$R \cos \alpha = 9{\cdot}81\, m$$

Thus $$\tan \alpha = \frac{v^2}{9{\cdot}81\, r} \tag{8.3}$$

From Fig. 8.6(*a*) and (*b*) it may be seen that the horizontal component of R is equal to the sum of the frictional forces developed between the wheels and the ground, i.e., $(F_1 + F_2)$. When limiting friction conditions are reached the angle α will attain its maximum value of ϕ (the limiting friction angle) and the centripetal force will also attain its maximum possible value.

Hence, from eq. (8.3), when $\alpha = \phi$

$$\text{Tan}\ \phi = \mu = \frac{v^2}{9{\cdot}81\, r}$$

where μ is the coefficient of friction between the wheels and the road surface.

$$\therefore \quad v^2 = 9{\cdot}81\, \mu r$$

and $$v = \sqrt{9{\cdot}81\, \mu r} \tag{8.4}$$

This is the maximum velocity at which a vehicle can negotiate a bend of given radius r, and skidding will occur if this value is exceeded.

However it is possible for a vehicle to overturn about its outer wheels before the speed is reached at which skidding occurs. At the position when overturning of the vehicle is about to occur the normal reaction N_1, and consequently F_1, will be zero. Therefore, the whole of the weight will be transferred to the outer wheels as indicated in Fig. 8.7.

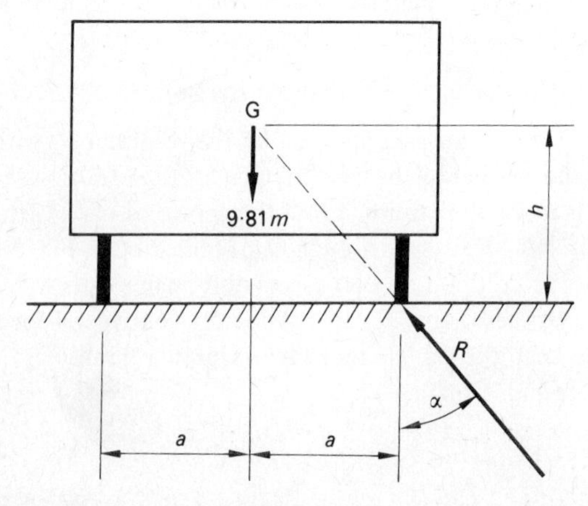

Fig. 8.7 Vehicle about to overturn on curved horizontal track

If the track width of the vehicle is $2a$ and the height of the centre of mass of the vehicle above the road surface is h, then

$$\tan \alpha = \frac{a}{h}$$

Substituting into eq. (8.3) gives:

$$\frac{a}{h} = \frac{v^2}{9{\cdot}81\, r}$$

$$\therefore \quad v^2 = \frac{9{\cdot}81\, ar}{h}$$

$$\text{and} \quad v = \sqrt{\frac{9{\cdot}81\, ar}{h}} \qquad (8.5)$$

Equation (8.5) shows that the tendency of a vehicle to overturn is related to its dimensions since increasing the wheelbase or lowering the centre of mass will enable a vehicle to negotiate a given bend at an increased speed.

Example 8.3

A motor car has a track width of 1·5 m and in a certain loaded condition its centre of mass is 0·7 m above the ground in a plane mid-way between the wheels. If the maximum coefficient of friction between the tyres and the road is 0·48 determine whether the car will slip or overturn when cornering at speed on a level road.

What will be the greatest speed at which a bend of 20 m radius can be negotiated without slipping or overturning occurring?

Solution

The maximum speed v_s without slipping occurring is given by

$$v_s = \sqrt{9{\cdot}81\, \mu r} \qquad \text{[eq. (8.4)]}$$

Putting $\mu = 0{\cdot}48$ then

$$v_s = \sqrt{(9{\cdot}81 \times 0{\cdot}48 r)}$$
$$= 2{\cdot}169\sqrt{r}$$

and the maximum speed v_t without overturning occurring is given by

$$v_t = \sqrt{\frac{9{\cdot}81\, ar}{h}} \qquad \text{[eq. (8.5)]}$$

Putting $a = 0{\cdot}75$ m and $h = 0{\cdot}7$ m then

$$v_t = \frac{9{\cdot}81 \times 0{\cdot}75r}{0{\cdot}7}$$

$$= 3{\cdot}297\sqrt{r}$$

Hence the car will slip before overturning.

For a bend of radius 20 m the maximum speed without slipping occurring is therefore

$$v_s = 2{\cdot}169\sqrt{20}$$

$$= 10{\cdot}245 \text{ m/s}$$

$$= 36{\cdot}88 \text{ km/h}$$

Example 8.4

A motor vehicle having a mass of 6000 kg starts the ascent of an incline, having a gradient of 1 in 40 and a length of 600 m, at a speed of 30 km/h. A constant tractive effort of 3·5 kN is maintained throughout the ascent and the road resistances are 0·3 N/kg.

After climbing the ascent the vehicle travels along a straight level road for a distance of 200 m prior to negotiating a bend of 50 m radius. The roadway is unbanked at the bend and the coefficient of friction between the tyres and the road is 0·45. Determine the maximum uniform tractive effort that can be applied over the 200 m straight if the vehicle is not to reach the bend at a speed at which skidding will occur.

Solution

A diagram is given in Fig. 8.8.

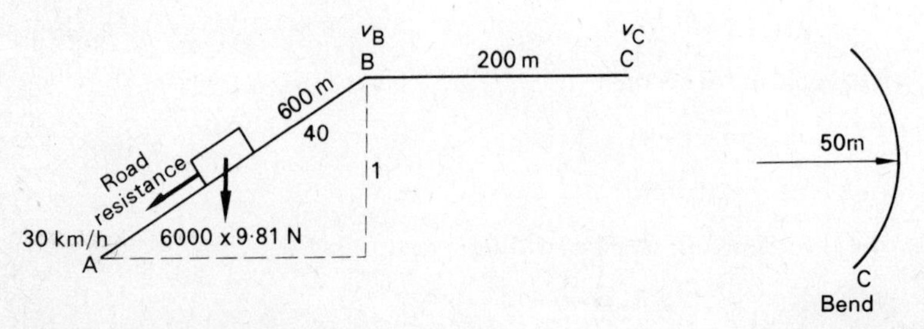

Fig. 8.8 Example 8.4

Consider the motion of the vehicle up the incline.

$$\text{Weight of vehicle} = 6000 \times 9{\cdot}81$$
$$= 58\,860 \text{ N}$$
$$\text{Component of this weight down incline} = 58\,860 \times \frac{1}{40}$$
$$= 1470 \text{ N}$$
$$\text{Road resistances to motion} = 0{\cdot}3 \times 6000$$
$$= 1800 \text{ N}$$
$$\therefore \quad \text{Total resistances to motion} = 1470 + 1800$$
$$= 3270 \text{ N}$$

Thus, the force F producing acceleration up the incline

$$= \text{tractive effort} - \text{resistance to motion}$$
$$= 3500 - 3270$$
$$= 230 \text{ N}$$

Applying $F = ma$ [eq. (8.1)]

$$230 = 6000a$$

where a = acceleration of vehicle up the incline.

Thus $a = 0{\cdot}0383 \text{ m/s}^2$

Applying $v_B^2 = v_A^2 + 2as$ [from eq. (7.6)]

to the motion up the incline gives

$$v_B^2 = \left(\frac{30 \times 1000}{3600}\right)^2 + 2 \times 0{\cdot}0383 \times 600$$
$$v_B = 10{\cdot}75 \text{ m/s}$$

This is the velocity at the top of the incline when the vehicle starts to travel along the straight 200-m stretch.

Now if the vehicle is to negotiate the bend at C without skidding then its maximum velocity at this position v_C is given by

$$v_C = \sqrt{9{\cdot}81\mu r} \qquad \text{[eq. (8.4)]}$$

where μ = coefficient of friction = 0·45

and r = radius of bend = 50 m

i.e. $v_C = \sqrt{(9{\cdot}81 \times 0{\cdot}45 \times 50)}$

$$= 14{\cdot}84 \text{ m/s}$$

The maximum acceleration a_h between B and C (200 m in length) can now be obtained from

$$v_C^2 = v_B^2 + 2a_h s$$

i.e., $$14{\cdot}84^2 = 10{\cdot}75^2 + 400a_h$$

$$a_h = 0{\cdot}261 \text{ m/s}^2$$

If the tractive effort on the horizontal is T_h, then

Force producing acceleration = $T_h - 1800$

Thus, applying $$F = ma$$

$$T_h - 1800 = 6000 \times 0{\cdot}261$$

$$T_h = 3366 \text{ N}$$

The maximum tractive effort on the horizontal if the vehicle is not to reach the bend at a speed which would cause skidding is 3366 N.

8.10 Vehicle on an inclined curved track

The case of a vehicle travelling around a curved track inclined at angle θ will now be considered.

A diagram is given in Fig. 8.9(*a*), N_1 and N_2, F_1 and F_2 again being the normal reactions and frictional forces at the wheels. The resultant reaction R of these four forces is still inclined at angle α to the normal to the track but is now inclined at $(\alpha + \theta)$ to the line of action of the weight of the vehicle, as shown in Fig. 8.9(*b*).

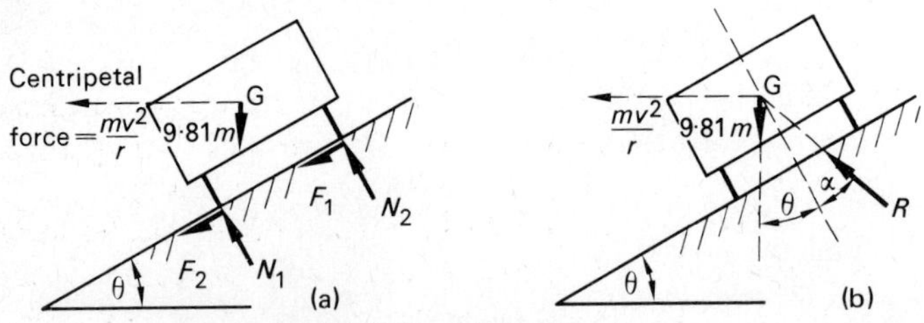

Fig. 8.9 Vehicle on an inclined curved track

Then, proceeding as for the horizontal track gives

$$R \sin(\alpha + \theta) = \frac{mv^2}{r}$$

and $$R \cos(\alpha + \theta) = 9{\cdot}81\, m$$

Thus, $$\tan(\alpha + \theta) = \frac{v^2}{9{\cdot}81\, r} \qquad (8.6)$$

This equation should be compared with eq. (8.3) obtained for a horizontal track.

Re-arranging eq. (8.6) and expanding $\tan(\alpha + \theta)$ gives

$$v^2 = 9{\cdot}81\, r \left(\frac{\tan\alpha + \tan\theta}{1 - \tan\alpha \tan\theta} \right) \tag{8.7}$$

In the case of a banked track there are two cases which need to be considered.

(*a*) When $\alpha = 0$ the reaction R will be normal to the track and hence there will be no road friction, i.e., $F_1 = F_2 = 0$ and $N_1 = N_2 = N$ as shown in Fig. 8.10.

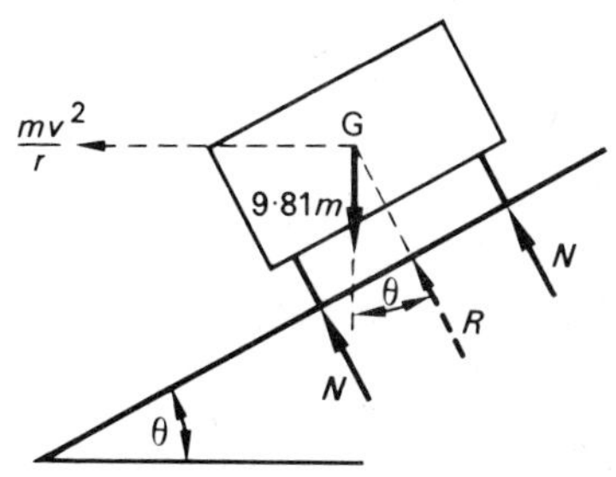

Fig. 8.10 Vehicle on an inclined curved track with no friction

Then from eq. (8.7)

$$v^2 = 9{\cdot}81\, r \tan\theta \qquad \text{since } \tan\alpha = 0$$

$$\tan\theta = \frac{v^2}{9{\cdot}81\, r}$$

The angle θ in this equation is therefore the angle of banking for which there is no tendency to skid, since there is no friction. Furthermore, as $N_1 = N_2$ the vehicle will not overturn.

It should also be noticed that for this condition the horizontal components of the normal reaction N supply the required centripetal force and that this situation can exist for one speed only, given by

$$v = \sqrt{(9{\cdot}81\, r \tan\theta)}$$

(*b*) When the vehicle is on the point of overturning there will be no reaction at the inside wheels and the resultant reaction R will therefore act entirely at the outside wheels, as shown in Fig. 8.11, such that

$$\tan\alpha = \frac{a}{h}$$

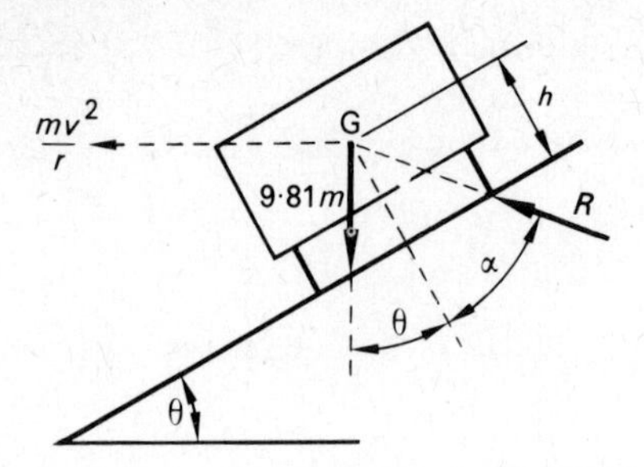

Fig. 8.11 Vehicle about to overturn on an inclined curved track

Substituting for tan α into eq. (8.7) gives

$$v^2 = 9{\cdot}81\, r\, \frac{(a/h) + \tan\theta}{1 - (a/h)\tan\theta}$$

This equation is only applicable providing that the friction is sufficient to allow the vehicle to rotate about its outer wheels:

i.e., $\tan\phi > \tan\alpha$

or $\mu > a/h$

If this is not the case and the friction is insufficient then skidding will occur before the overturning point is reached and so $\tan\phi < \tan\alpha$ or $\mu < a/h$.

The limiting condition for skidding is, therefore, when $\tan\alpha = \tan\phi = \mu$ and so the maximum speed for skidding is

$$v = \sqrt{\left\{9{\cdot}81r\left(\frac{\mu + \tan\theta}{1 - \mu\tan\theta}\right)\right\}} \tag{8.8}$$

Example 8.5

A car travels round a banked track having a mean radius of 40 m. If the angle of banking of the track is 30° and the coefficient of friction between the tyres and the ground is 0·6 determine the maximum speed at which the car can travel round the track without sideslip occurring.

Solution

When sideslip is about to occur then limiting friction conditions will exist at the wheels, i.e., $\tan\alpha = \tan\phi = \mu$.

Thus, applying eq. (8.8) gives

$$v^2 = 9{\cdot}81r\left(\frac{\mu + \tan\theta}{1 - \mu\tan\theta}\right)$$

$$= 9{\cdot}81 \times 40\left(\frac{0{\cdot}6 + \tan 30^\circ}{1 - 0{\cdot}6\tan 30^\circ}\right)$$

$$= 706{\cdot}9$$

$$\therefore \quad v = 26{\cdot}59 \text{ m/s}$$

$$= 93{\cdot}72 \text{ km/h}$$

The maximum possible speed for no sideslip is 93·72 km/h.

Example 8.6

A bicycle and rider have a total mass of 90 kg. Determine the angle which the machine and rider must make with the horizontal road when travelling round a curve of 12 m radius at 20 km/h without sideslip. What is the frictional force between the tyres and the road under these conditions?

Determine also the frictional force for the same speed and radius of curvature when the road is banked at 10° in favour of the cyclist.

Solution

When rounding the bend on a level road let the necessary angle between the cyclist and the road be θ, as shown in Fig. 8.12.

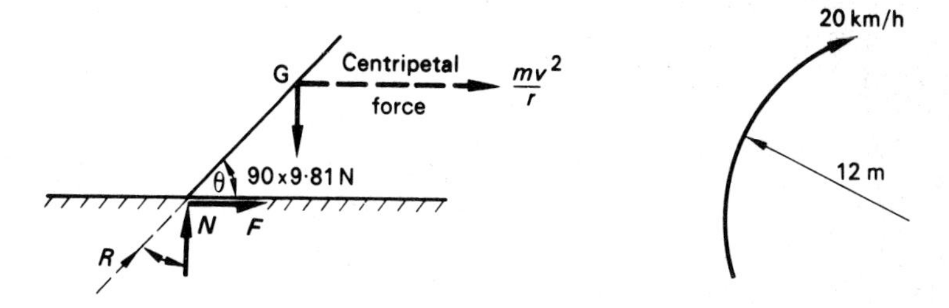

Fig. 8.12 Example 8.6

For vertical equilibrium the vertical reaction N must be equal to the weight of the rider and machine; i.e.,

$$N = 90 \times 9{\cdot}81$$

$$= 883 \text{ N}$$

The centripetal force necessary to maintain motion in a circular path can only be supplied by the frictional force F; i.e.,

$$F = \frac{mv^2}{r}$$

$$= \frac{90\ [\text{kg}]}{12\ [\text{m}]} \times \left(\frac{20}{3{\cdot}6}\right)^2\ [\text{m}^2/\text{s}^2]$$

$$= 231\ \text{N}$$

However, if the bicycle and rider are to be in rotational equilibrium the resultant reaction R must pass through the centre of mass G.

Thus $\tan\theta = \dfrac{N}{F}$

$$= \frac{883}{231} = 3{\cdot}8225$$

$$\theta = 75^\circ\ 20'$$

To negotiate a 12 m radius bend at 20 km/h, the cycle and rider should make an angle of 75° 20′ with the road and under these conditions the frictional force would be 231 N.

Consider now the motion on the banked road, as shown in Fig. 8.13.

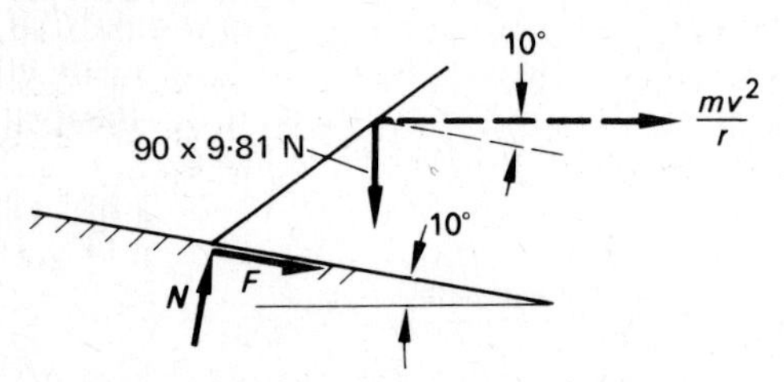

Fig. 8.13 Example 8.6

The horizontal centripetal force can be resolved into two components, parallel to and perpendicular to the road respectively.

Resolving parallel to the road:

$$F + 883 \sin 10^\circ = \frac{mv^2}{r} \cos 10^\circ$$

$$\therefore \quad F = 231 \cos 10^\circ - 883 \sin 10^\circ.$$

$$= 76{\cdot}2\ \text{N}$$

The frictional force at 20 km/h on a 10° banked track is 76·2 N.

8.11 The conical pendulum

A conical pendulum is a small mass m suspended on a light arm of length l which is made to revolve in a circular path of radius r with a constant angular velocity ω. This is illustrated in Fig. 8.14.

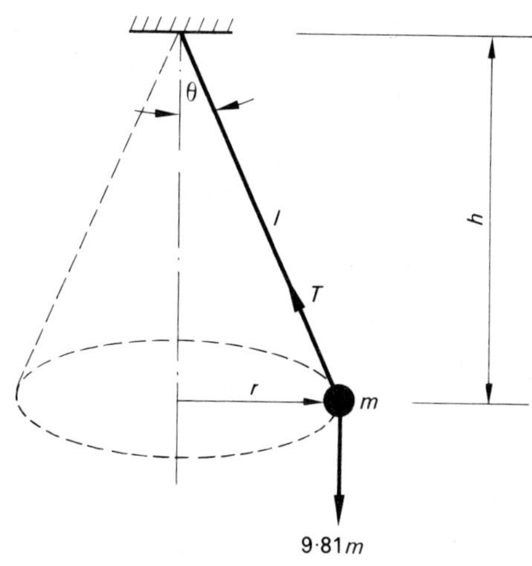

Fig. 8.14 The conical pendulum

For a certain speed of rotation the inclination θ of the arm to the vertical and the radius of rotation r will have definite values. If T is the tension in the arm then the centripetal force to maintain circular motion is provided by the horizontal component of T. Thus:

$$T \sin \theta = mr\omega^2$$

and for vertical equilibrium $$T \cos \theta = 9{\cdot}81\, m$$

Then $$\tan \theta = \frac{r\omega^2}{9{\cdot}81}$$

But from Fig. 8.14, $$\tan \theta = \frac{r}{h}$$

where h is the height of the pendulum from the point of suspension.

$\therefore$ $$\frac{r}{h} = \frac{r\omega^2}{9{\cdot}81}$$

$$h = \frac{9{\cdot}81}{\omega^2}$$

This indicates that the height of a conical pendulum from the point of suspension is independent of the length of the arm and depends only on the speed of rotation.

Also, from Fig. 8.14, $\sin\theta = \dfrac{r}{l}$

$$\therefore \quad T = \frac{mr\omega^2}{\sin\theta}$$

$$= ml\omega^2$$

Example 8.7

A simple governor has arms 0·25 m long which are pivoted on the axis of the governor spindle. Determine the speed of rotation at which the governor arms will be inclined at 25° to the axis of the spindle.

Solution

A diagram is given in Fig. 8.15.

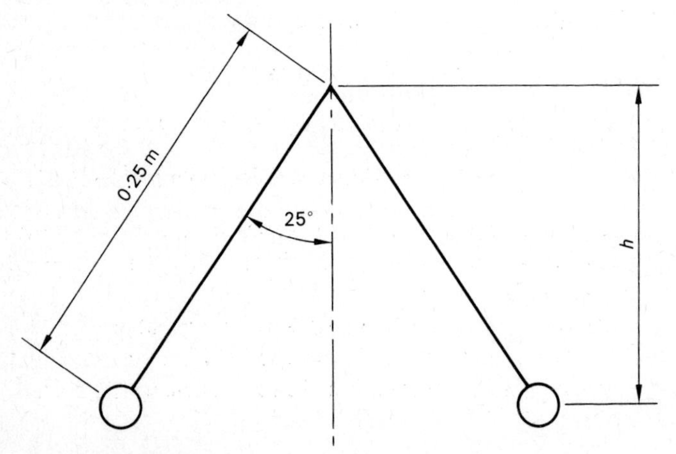

Fig. 8.15 Example 8.7

From this diagram:

$$\text{Cos } 25^\circ = \frac{h}{0{\cdot}25}$$

$$\therefore \quad h = 0{\cdot}2241 \text{ m}$$

Applying $h = \dfrac{9{\cdot}81}{\omega^2}$

gives $$\omega^2 = \frac{9{\cdot}81}{0{\cdot}2241} = 43{\cdot}57$$

$$\omega = 6{\cdot}616 \text{ rad/s}$$

$$= 63{\cdot}18 \text{ rev/min}$$

The governor arms will be inclined at 25° to the axis of the spindle when the speed of rotation is 63·18 rev/min.

8.12 Energy

The energy possessed by a body is a measure of the capacity of the body to do work or overcome resistance to motion. In mechanics we are concerned with two distinct forms of energy, namely potential energy and kinetic energy.

The potential energy of a body is a measure of the work required to raise it to a position above a given datum level, i.e.,

$$\text{Potential energy} = mgh \text{ joules} \tag{8.9}$$

where m = mass of body (kg)

g = gravitational acceleration (m/s^2)

h = height of body above datum level (m)

The kinetic energy of a body is due solely to the velocity of the body. Thus, a body may possess kinetic energy due to both translatory and rotational motion.

The *principle of conservation of energy* states that energy cannot be created or destroyed so that in the absence of frictional and other resistances the sum of the kinetic and potential energies remains constant.

8.13 Kinetic energy of translation

Suppose that a body of mass m that is initially at rest on a smooth level surface is acted upon by a force F such that it is displaced through a distance s as indicated in Fig. 8.16.

The acceleration a of the body is given by the relationship $F = ma$ and the velocity v after travelling a distance s is given by $v^2 = 2\,as$, since the initial velocity is zero.

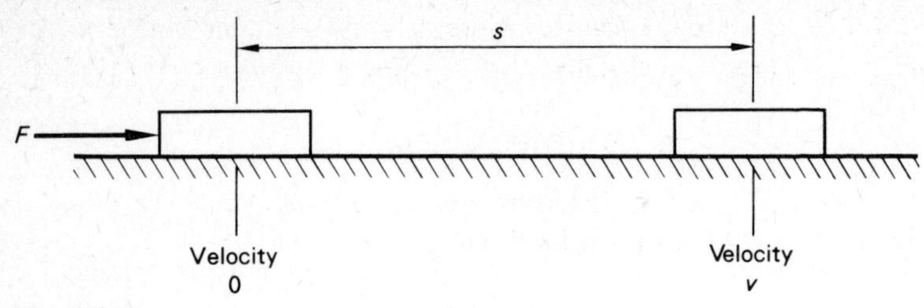

Fig. 8.16

Now, work done by force $F = Fs$

$$= ma \,.\, \frac{v^2}{2a}$$

$$= \tfrac{1}{2}mv^2$$

But the work done by the force is equal to the kinetic energy possessed by the body.

Thus, for a body of mass m [kg] moving with a velocity v [m/s]

$$\text{Linear kinetic energy} = \tfrac{1}{2}\, mv^2 \text{ joules} \tag{8.10}$$

8.14 Linear impulse

In section 8.4 it was shown that for a body having a constant mass m, Newton's second law could be written in the form

$$F = m\,\frac{\mathrm{d}v}{\mathrm{d}t}$$

where $\mathrm{d}v$ is the change in velocity occurring in a time $\mathrm{d}t$.

Re-arranging gives

$$F \,.\, \mathrm{d}t = m \,.\, \mathrm{d}v$$

If the velocity changes from v_1 to v_2 due to a constant force F acting for a time t, then by integration

$$F\int_0^t \mathrm{d}t = m\int_{v_1}^{v_2}\mathrm{d}v$$

$$Ft = m(v_2 - v_1) \tag{8.11}$$

If F is an impulsive force which acts for a short interval of time t then the product Ft is known as the linear impulse of force F. Also, $m(v_2 - v_1)$ is the change in linear momentum of mass m in the time t and so eq. (8.11) may be stated:

The linear impulse of a force which acts on a body for a given interval of time is equal to the change in linear momentum of the body during that time; i.e.,

Impulse = change in linear momentum

The units of impulse are therefore the same as those for momentum, namely kg m/s or newton seconds (N s).

8.15 Conservation of linear momentum

It has been shown that impulse is equal to the change of momentum and it follows that if there is no impulse then there cannot be a change in the linear momentum of a body. By coupling this statement with Newton's third law of motion (section 8.3) another important principle, known as the *principle of conservation of linear momentum*, is obtained. This states that:

The total momentum of a system, in a certain direction, remains constant unless an external force (and hence an impulse) is applied to the system in that direction.

This principle enables impact problems, where only internal forces are involved, to be readily solved.

Consider the firing of a gun which is free to move. The equal and opposite forces exerted by the gun on the projectile and the projectile on the gun will cause motion of each, but as no external forces are involved it follows that the change in momentum of the projectile due to firing must be equal and opposite to the change in momentum of the gun. The difference in the resulting velocities of the projectile and the gun is due entirely to their difference in mass.

8.16 Impact

A collision between bodies which occurs over a short interval of time is known as an impact.

Consider an impact between two bodies A and B as depicted in Fig. 8.17. On impact, the linear impulse of A on B is equal and opposite to the impulse of B on A (Newton's third law).

It therefore follows from the *principle of conservation of momentum*, that the total momentum of bodies A and B remains constant in the direction of the impulse; i.e.,

Momentum before impact = momentum after impact

$$m_A v_A + m_B v_B = m_A v_A' + m_B v_B'$$

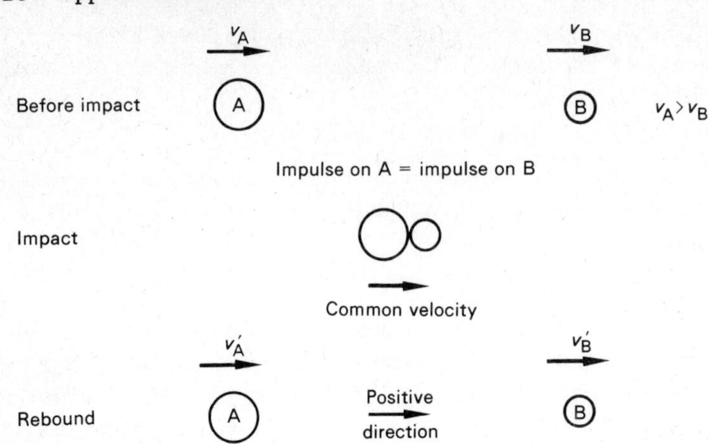

Fig. 8.17 Impulse

where v and v' represent the velocities before and after impact, respectively. When using this equation due regard must be given to the direction of motion.

During the short time of impact, the two bodies move together with a common velocity and then rebound with velocities v_A' and v_B' respectively.

It has been found by experiment that the ratio of the relative velocity of the two bodies after impact to their relative velocity before impact is constant, and depends upon the elastic properties of the two bodies.

This ratio is known as the coefficient of restitution *e*.

$$e = -\left(\frac{\text{relative velocity after impact}}{\text{relative velocity before impact}}\right) = -\left(\frac{v_A' - v_B'}{v_A - v_B}\right) \qquad (8.12)$$

The negative sign is introduced to take account of the fact that the relative velocities before and after impact are of opposite sense. The value of *e* is then positive.

The use of eq. (8.12) together with the principle of conservation of momentum enables the unknown velocities after impact to be determined.

In the case of an impact between two perfectly elastic bodies $e = 1$, whereas for completely inelastic bodies $e = 0$. For the latter case there would be no rebound and the bodies would move together after impact.

8.17 Loss of kinetic energy due to impact

Except in the case of impact between two perfectly elastic bodies ($e = 1$) there will always be some loss of energy due to impact. Thus, the total kinetic energy of the bodies after impact will always be less than the total kinetic energy of the bodies before impact and in the case of an inelastic impact ($e = 0$) the loss of kinetic energy will be a maximum.

Example 8.8

A small wooden block having a mass of 6 kg is moving along a smooth horizontal plane with a velocity of 15 m/s to the left when it is struck centrally by a bullet of mass 0·04 kg that is moving to the right with a velocity of 750 m/s. The bullet passes right through the block and emerges with a velocity of 250 m/s. Determine:

(*a*) the velocity of the block just after impact;
(*b*) the linear impulse on the block due to the impact;
(*c*) the loss of kinetic energy of the two bodies due to the impact.

Solution

The impact between the bullet and the block is depicted in Fig. 8.18.

(*a*) Let the velocity of the block after impact be v'_B in the positive direction, as shown. For a smooth horizontal plane there are no frictional forces acting.

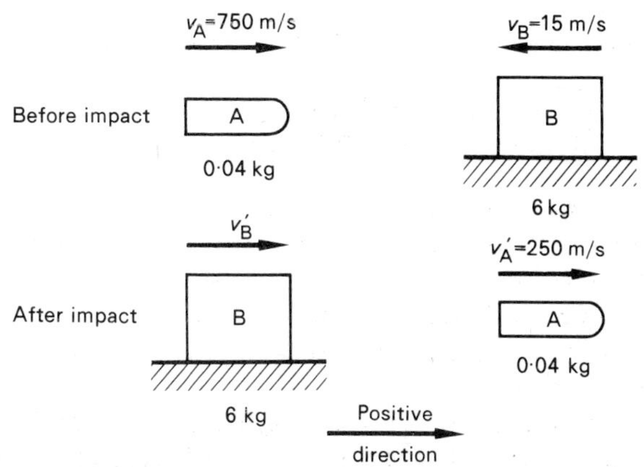

Fig. 8.18 Impact–Example 8.8

Then, by the *principle of conservation of linear momentum:*

Momentum before impact = momentum after impact

$$0{\cdot}04 \times 750 + 6(-15) = 0{\cdot}04 \times 250 + 6v'_B$$

$$v'_B = -11{\cdot}66 \text{ m/s}$$

The velocity of the block after impact is 11·66 m/s to the left.

(*b*) The linear impulse exerted on the block by the bullet is equal and opposite to that exerted on the bullet by the block. Thus, the change in momentum of either body may be used to determine the impact impulse.

Impulse on block = change in momentum of block

$$= m_B (v'_B - v_B)$$

$$= 6(-11{\cdot}66 - (-15))$$

$$= +20 \text{ kg m/s or } 20 \text{ N s}$$

Similarly, impulse on bullet $= m_A (v'_A - v_A)$

$$= 0{\cdot}04 (250 - 750)$$

$$= -20 \text{ N s}$$

The impulse on the block is 20 N s in the positive direction.

(*c*) Loss of kinetic energy = initial kinetic energy − final kinetic energy

$$= \tfrac{1}{2} m_A (v_A^2 - v_A'^2) + \tfrac{1}{2} m_B (v_B^2 - v_B'^2)$$

$$= \tfrac{1}{2} \,.\, 0{\cdot}04 (750^2 - 250^2) + \tfrac{1}{2} \,.\, 6 \,.\, [(-15)^2 - (-11{\cdot}66)^2]$$

$$= (10\,000 + 267) \text{ kg m}^2/\text{s}^2$$

$$= 10\,267 \text{ N m}$$

$$= 10\,267 \text{ J}$$

The loss of kinetic energy is 10 267 J.

8.18 Oblique impact

It was stated in section 8.15 that the total momentum of a system remains constant in a given direction unless an external impulse is applied in that direction.

Hence, for an oblique impact of smooth bodies, as illustrated in Fig. 8.19, the internal impulses are equal and opposite and the linear momentum along the line of centres – i.e., the **x** direction – remains

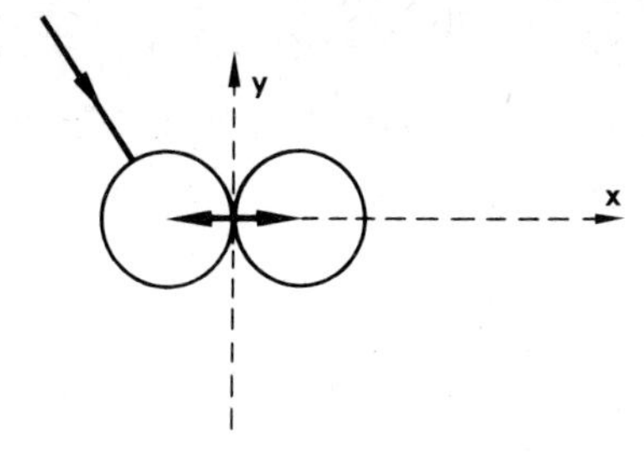

Fig. 8.19 Oblique impact

unchanged. Therefore, only component velocities in the **x** direction can be used in the momentum equation. This also applies to the equation giving e in terms of the relative velocities.

It should also be observed that there is no impact in the **y** direction and therefore the component velocities remain unchanged in this direction.

Example 8.9

A smooth ball A, having a mass of 1 kg, impacts with a similar ball B, having a mass of 2 kg, as shown in Fig. 8.20. The respective velocities and directions are as shown and the coefficient of restitution is 0·8. Determine the velocities of the balls after impact.
Neglect any rotational effects.

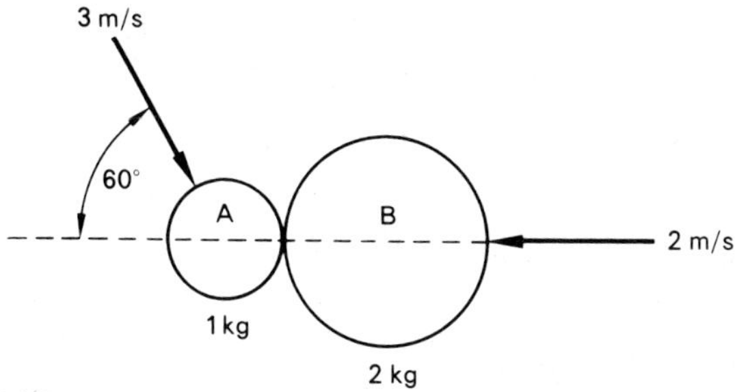

Fig. 8.20 Oblique impact–Example 8.9

Solution

The velocity of A can be resolved into

$$3 \cos 60^\circ = 1{\cdot}5 \text{ m/s along the line of centres}$$

and $3 \sin 60^\circ = 1{\cdot}5\sqrt{3}$ m/s perpendicular to the line of centres

Let v_A' and v_B' represent the component velocities, along the line of centres, after impact – considered positive.

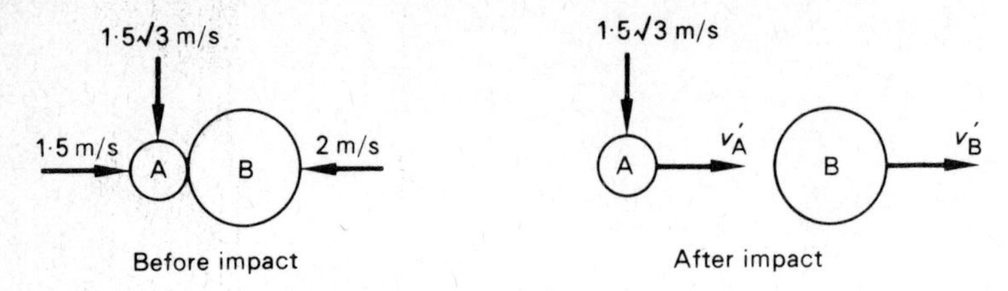

Internal impulses are equal and opposite along line of centres
∴ momentum unchanged in this direction

Fig. 8.21 Example 8.9

Applying the *principle of conservation of momentum* along line of centres gives:

$$1 \times 1{\cdot}5 + 2(-2) = 1 \times v_A' + 2 \times v_B'$$

$$v_A' + 2v_B' = -2{\cdot}5 \qquad (1)$$

Also,
$$e = -\left(\frac{\text{relative velocity after impact}}{\text{relative velocity before impact}}\right)$$

for the velocities along the line of centres.

Thus,
$$0{\cdot}8 = -\left(\frac{v_A' - v_B'}{1{\cdot}5 - (-2)}\right)$$

$$= -\left(\frac{v_A' - v_B'}{3{\cdot}5}\right)$$

and
$$v_A' - v_B' = -2{\cdot}8 \qquad (2)$$

Solving (1) and (2) gives:

$$v_A' = -2{\cdot}7 \text{ m/s} \qquad v_B' = 0{\cdot}1 \text{ m/s}$$

These are the component velocities, after impact, along the line of centres.

The $1{\cdot}5\sqrt{(3)}$ m/s component of the initial velocity of A is unchanged by the impact since there is no impact perpendicular to the line of centres.

Thus, referring to Fig. 8.22,

$$\text{Final velocity of A} = \sqrt{\{(2{\cdot}7)^2 + (1{\cdot}5\sqrt{3})^2\}}$$

$$= 3{\cdot}747 \text{ m/s}$$

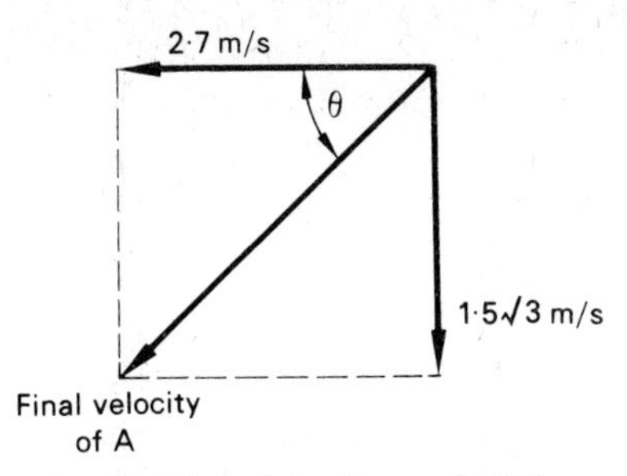

Fig. 8.22 Component velocities of A–Example 8.9

The direction of this velocity with the line of centres is given by

$$\tan\theta = \frac{1{\cdot}5\sqrt{3}}{2{\cdot}7} = 0{\cdot}9622$$

$$\theta = 43^\circ\ 54'$$

As B had no velocity component in the y direction before impact then it has no component in this direction after impact. Thus, the final velocity of B after impact is 0·1 m/s to the right, i.e., the direction of the velocity of B is reversed.

8.19 Impact of a fluid jet

The impact of a fluid jet upon a fixed vane is a good example of the application of the three laws of motion. If the jet is deflected from its line of motion then, from the first law, a force must be acting upon it. The second law tells us that the force which is acting must be proportional to the rate of change of momentum and act in the direction in which the change in momentum occurs. Finally, the third law states that the force exerted by the vane on the jet to change its momentum must be equal and opposite to the force exerted by the jet on the vane.

When dealing with problems on the impact of a fluid jet we are usually concerned with the mass flow rate of the jet. Thus, the second law of motion can be written in the form:

$$\begin{aligned}\text{Force} &= \text{rate of change of momentum}\\ &= \frac{\text{mass} \times \text{velocity change}}{\text{time taken}}\\ &= \text{mass flow rate} \times \text{change of velocity in the direction of the force}\end{aligned}$$

Example 8.10

A water jet having a diameter of 80 mm impinges perpendicularly upon a fixed plate with a velocity of 30 m/s. Calculate the force acting on the plate.

Solution

When the jet of water strikes the plate it will spray out radially, i.e., it is deflected through an angle of 90°. Thus, the jet will lose all its momentum at right angles to the plate, as illustrated in Fig. 8.23.

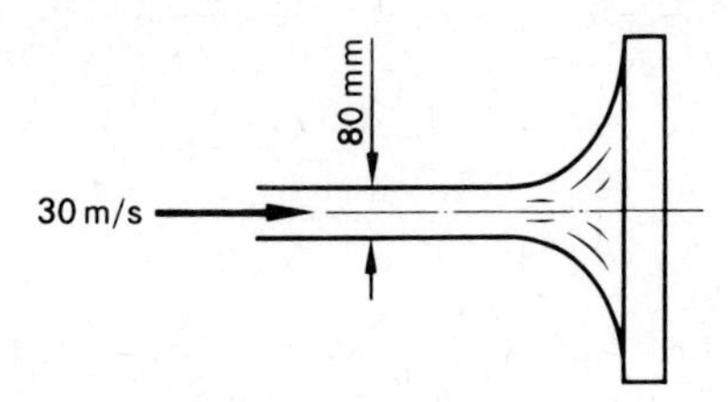

Fig. 8.23 Impact of a fluid jet–Example 8.10

$$\text{Area of jet} = \frac{\pi}{4} \times \left(\frac{80}{1000}\right)^2 = \frac{16\pi}{10^4}\ \text{m}^2$$

$$\text{Volume flow rate} = \text{area of jet} \times \text{velocity}$$

$$= \frac{16\pi}{10^4}\ [\text{m}^2] \times 30\ [\text{m/s}]$$

$$= 0{\cdot}048\pi\ \text{m}^3/\text{s}$$

$$\text{Mass flow rate} = \text{volume flow rate} \times \text{density}$$

$$= 0{\cdot}048\pi\ [\text{m}^3/\text{s}] \times 10^3\ [\text{kg/m}^3]$$

$$= 48\pi\ \text{kg/s}$$

Then, $$\text{Force on plate} = \text{mass flow rate} \times \text{change in velocity}$$

$$= 48\pi\ [\text{kg/s}] \times 30\ [\text{m/s}]$$

$$= 4520\ \text{N}$$

$$= 4{\cdot}52\ \text{kN}$$

The force on the plate is 4·52 kN.

Example 8.11

A jet of water moving at 20 m/s is deflected through an angle of 30° by a fixed curve blade. The mass flow rate of the jet, which enters the blade tangentially, is 5 kg/s. If friction between the water and the blade is negligible, determine the force of the water on the blade. What would the magnitude of this force be if the jet were deflected through an angle of 150°?

Solution

In the absence of frictional resistance the velocity of the water relative to the blade will remain unchanged. Therefore the exit velocity is 20 m/s, and this velocity can be resolved into components in the **x** and **y** directions (ref. Fig. 8.24).

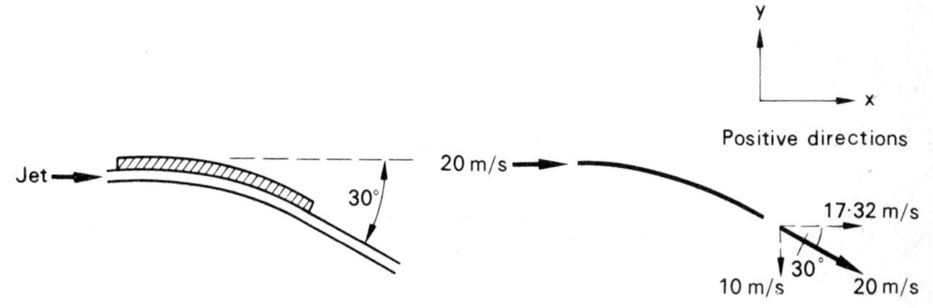

Fig. 8.24 Impact of a fluid jet on a curved blade–Example 8.11

The **x** component of the exit velocity, $v'_x = 20 \cos 30°$
$= 17{\cdot}32$ m/s

and the **y** component of the exit velocity, $v'_y = -20 \sin 30°$
$= -10$ m/s

The inlet velocity components are $v_x = 20$ m/s and $v_y = 0$

Then, **x** component of force on plate = mass flow rate x change in velocity in **x** direction

$$F_x = 5\ [\text{kg/s}] \times (20 - 17{\cdot}32)\ [\text{m/s}]$$
$$= 13{\cdot}40\ \text{N}$$

Similarly,
$$F_y = 5 \times (0 - (-10))$$
$$= 50\ \text{N}$$

The resultant force on the blade, $F = \sqrt{(F_x^2 + F_y^2)}$

$$= \sqrt{(13{\cdot}4^2 + 50^2)}$$

$$= 51{\cdot}76 \text{ N}$$

When the jet is deflected through 150° the components of the exit velocity are as shown in Fig. 8.25.

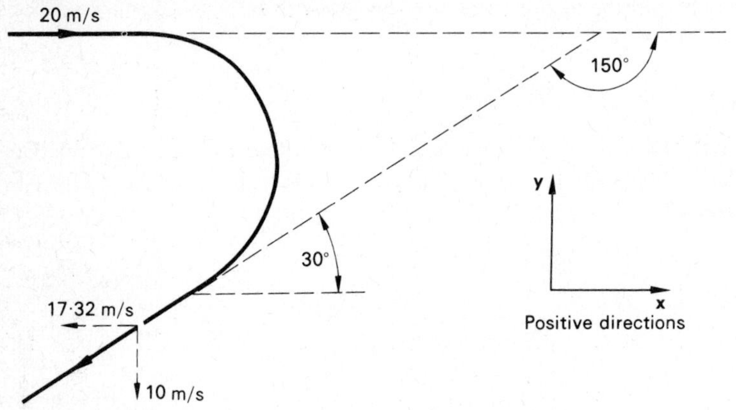

Fig. 8.25 Fluid jet deflected through 150°–Example 8.11

In this instance,

$$F_x = 5\,(20 - (-17{\cdot}32))$$

$$= 186{\cdot}6 \text{ N}$$

and $F_y = 50$ N as before

Resultant force on blade, $F = \sqrt{(186{\cdot}6^2 + 50^2)}$

$$= 193{\cdot}2 \text{ N}$$

The notation used in this example should be noted. Since (final velocity – initial velocity) would give the force acting on the jet, the reverse, as used, gives the force acting on the blade.

8.20 Motion under a varying force

(1) Force varying with time

For a body of constant mass Newton's second law can be written:

$$F = m \cdot \frac{dv}{dt} \qquad \text{(see section 8.4)}$$

In this instance F is a function of t and integrating with respect to time gives:

$$\int_0^t F\,.\,dt = \int_{v_1}^{v_2} m\,.\,dv$$

$$= m\,(v_2 - v_1)$$

or Area under force-time graph = change in linear momentum

(2) Force varying with displacement

F now becomes a function of s and Newton's second law can be written

$$F = m\,.\,\frac{dv}{dt}$$

$$= mv\,.\,\frac{dv}{ds} \qquad \text{(see section 7.3)}$$

Integrating with respect to s gives

$$\int_0^s F\,.\,ds = \int_{v_1}^{v_2} mv\,.\,dv$$

$$= \tfrac{1}{2}m(v_2^2 - v_1^2)$$

or Area under force-displacement graph = change in kinetic energy

Example 8.12

The cage of a lift has a mass of 400 kg and is pulled upwards by a wire rope. The cage starts from rest and the initial pull of the rope is 5924 N. The pull decreases uniformly with distance such that at a height of 30 m the pull is 2924 N. Determine:

(*a*) the velocity of the lift at this height;
(*b*) the maximum velocity of the lift and the height at which it occurs.

Solution

Weight of lift = 400 x 9·81

= 3924 N

(*a*) Thus the upward vertical accelerating force at the start of the motion

= 5924 – 3924

= 2000 N

and the upward vertical accelerating force at the end of the motion

$$= 2924 - 3924$$
$$= -1000\ \text{N}$$

This is shown in Fig. 8.26.

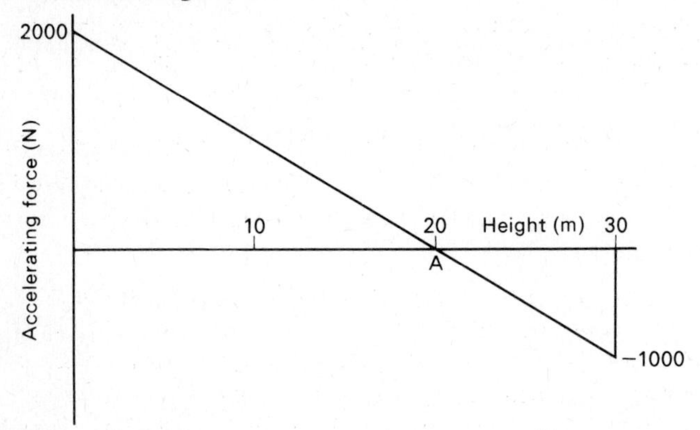

Fig. 8.26 Example 8.12

Applying:

Area under force-displacement graph = change in kinetic energy gives:

$$\left(\frac{2000 - 1000}{2}\right) \times 30 = \tfrac{1}{2} \,.\, 400\, v^2$$

$$\text{i.e. } v^2 = 75$$
$$v = 8{\cdot}66\ \text{m/s}$$

At a height of 30 m the lift will have a velocity of 8·66 m/s.

(*b*) The lift will only continue to accelerate upwards if there is an upward accelerating force. Thus as soon as point A is reached the lift will start to slow downward; the maximum velocity therefore occurs at A. By the ratios of two similar triangles, A is at a height of 20 m. Then applying

Area under force-displacement graph = change in kineuc energy for the motion to a height of 20 m gives:

$$\left(\frac{2000 + 0}{2}\right) \times 20 = \tfrac{1}{2} \,.\, 400\, v^2$$

$$v^2 = 100$$
$$v = 10\ \text{m/s}$$

The maximum velocity of the lift is 10 m/s and occurs when the lift is at a height of 20 m.

Problems

1. A locomotive can exert a maximum draw-bar pull of 150 kN. The locomotive has a mass of 80 Mg and is used to haul a train of mass 420 Mg, the rolling resistances to motion amounting to 0·05 N/kg for both the locomotive and the train. Determine:

(*a*) the maximum acceleration that can be achieved on an incline of 1 in 200;

(*b*) the greatest slope that could be climbed with an acceleration of 0·10 m/s^2;

(*c*) the gradient that can just be climbed at a constant speed.

2. (*a*) A hovercraft on the point of ascending a ramp has an initial speed up the ramp of 8 m/s and at this moment cuts its forward thrust to zero. The ramp has a length of 30 m followed by a horizontal surface. Calculate the angle of the ramp so that the craft will come to rest with its centre of mass exactly on the top of the slope. Calculate also the deceleration acting on the hovercraft.

(*b*) The same vehicle approaches the same ramp and starts to ascend it at 10 m/s again with no forward thrust during the time it is on the slope.
Calculate:

(*i*) its velocity at the top of the slope;

(*ii*) the horizontal distance in which the craft will come to rest if a constant reverse thrust of 5 kN is applied as soon as the top of the slope is reached.

The mass of the vehicle may be taken as 40 Mg (40×10^3 kg) and it may be assumed that there is no resistance due to wind or other air movement.

(U.E.I.)

3. A locomotive of mass 70 Mg is used to haul a train of mass 250 Mg. When negotiating an incline of 1 in 160 against a frictional resistance of 60 N/Mg the speed at a particular instant is 40 km/h but is increasing at the rate of 1 km/h in 6 s. Determine the draw-bar pull between the locomotive and the train and the power at which the locomotive is working.

4. The road surface of a humped back bridge has a radius of curvature of 9 m. Determine the maximum speed at which a car can cross the bridge without the wheels leaving the surface, if the centre of mass of the car is 1 m above the ground.

5. A car has a mass of 1·2 Mg and a track width of 1·6 m. Its centre of mass is centrally situated with respect to the wheels at a height of 700 mm above the ground. When traversing a bend of 80 m radius, calculate:
(*a*) the normal reaction at the outer wheels at a constant speed of 50 km/h;
(*b*) the maximum speed it can travel without either overturning or slipping laterally if the coefficient of friction between the wheels and the ground is 0·62. (N.C.T.E.C.)

6. A motor cyclist rounds a bend having a radius of 150 m at a speed of 60 km/h. Determine:
(*a*) the angle at which he must lean to negotiate the bend;
(*b*) the minimum coefficient of friction between the wheels and the ground to prevent sideslip.

7. A car having a 1·4 m track has a mass of 1000 kg and it may be assumed that the centre of mass is on the centre line of the vehicle, midway between the axles, at a height h above the ground. Determine the minimum coefficient of friction between the tyres and the road to allow the car to go round a level curved track of 55 m mean radius at 75 km/h without skidding. Calculate the maximum value of the height h of the centre of mass for there to be no overturning. (U.L.C.I.)

8. A bend is in the form of an arc of a circle of radius 80 m and is banked at an angle of Tan^{-1} 1/5. Determine the speed at which a vehicle can travel round this bend without there being any sidethrust on the tyres. Also find the speed at which skidding is about to occur if the coefficient of friction between the tyres and the ground is 0·6.

9. Derive an expression for the acceleration of a body rotating in a circular path at a constant speed.

A car travelling around a level bend of radius 20 m at a constant speed of 40 km/h is just on the point of slipping laterally. Working from first principles, calculate the coefficient of friction between the wheels and the ground.

If the bend had been banked at an angle of 10° to the horizontal so as to assist turning, calculate the speed at which there would be no tendency for the car to slip laterally. (N.C.T.E.C.)

10. Determine the greatest speed at which a car may traverse a banked road of 30 m radius without overturning, if its centre of mass is 1 m above the road surface and its track width is 1·5 m. The road is banked at 12° to the horizontal.

11. A conical pendulum has a bob of mass 2 kg attached to an inelastic cord of length 0·2 m. If the breaking load of the cord is 80 N determine the maximum speed of rotation. What would be the radius of rotation of the bob if the speed of rotation is half the maximum speed?

12. A conical pendulum has a bob of mass 2·5 kg suspended on a string 0·3 m long. If the pendulum rotates at 60 rev/min determine:

(*a*) the angle which the string makes with the vertical;
(*b*) the tension in the string.

13. Two wagons A and B are moving along the same straight track. A has a mass of 4 Mg and a velocity of 5 m/s while B has a mass of 6 Mg and a velocity of 1·5 m/s in the opposite direction to that of A. The two wagons collide and subsequently begin to move with a common velocity.

Determine the common velocity and the loss of kinetic energy.

How far will the wagons move together after the impact if their resistances to motion are 0·06 N/kg?

14. A truck having a mass of 10 Mg starts from rest and runs down an incline of 1 in 100 for 30 s, when it reaches a level track along which it runs for 15 s before colliding with a stationary truck of mass 16 Mg. The two trucks remain in contact after the impact and move with a common velocity.

Determine their velocity immediately after the impact and the distance that they move together before coming to rest.

Assume the track remains level and that the frictional resistance to the motion of the trucks is 0·05 N/kg throughout.

15. A railway truck of mass 9 Mg, travelling on a straight track at 50 km/h, overtakes and collides with another truck of mass 6 Mg travelling in the same direction at a speed of 10 km/h. Immediately after collision, they separate with a relative speed of 3 km/h. Calculate:

(*a*) the loss of kinetic energy caused by the collision;
(*b*) the distance between the two moving trucks 4 s after the collision if both trucks have the same uniform deceleration due to rolling resistance. (N.C.T.E.C.)

16. A pile-driver hammer of mass 1 Mg is used to drive a pile of mass 0·4 Mg into the ground. At a particular stage during the pile driving it is found that when the hammer is allowed to fall through a vertical height of 3 m the pile is driven 0·15 m into the ground.
Calculate:

(*a*) the common velocity of the hammer and pile immediately after impact, assuming no rebound occurs;
(*b*) the resisting force of the ground, assuming it to be constant.

17. A car having a mass of 1 Mg is travelling at 40 km/h when it runs head on into the side of a stationary van having a mass of 1·2 Mg. The two vehicles move together after impact with a common velocity and the van is moved sideways through a distance of 6 m. Determine:

(*a*) the common velocity immediately after impact;
(*b*) the loss of kinetic energy at impact.

18. Two smooth bodies A and B have masses of 2 kg and 3 kg respectively. Immediately prior to impact between them B is at rest and A has a velocity of 5 m/s. If the coefficient of restitution is 0·9 determine the magnitude and direction of the velocity after impact when

(*a*) the velocity of A is along the line of centres at impact;
(*b*) the velocity of A is at 45° to the line of centres at impact.

19. A jet of water having a diameter of 50 mm impinges perpendicularly on a flat plate. The force acting on the plate is 5·4 kN. Determine the velocity of the jet.

The density of water is 1 Mg/m^3.

20. A jet of liquid having a diameter of 40 mm, a relative density of 0·8 and a velocity of 20 m/s impinges tangentially on a curved vane which deflects the jet through an angle of 120°. Determine the magnitude of the resultant force acting on the vane.

The density of water is 1 Mg/m^3.

21. A pit cage, of mass 3 Mg, is drawn upwards by means of a vertical cable. The cage starts from rest with an initial tension in the cable of 40 kN. The tension decreases uniformly with the distance ascended until at a height of 20 m the tension in the cable is 24 kN. Calculate:

(*a*) the velocity of the cage at this height;
(*b*) the height at which the cage has its maximum velocity and hence the maximum kinetic energy of the cage. (N.C.T.E.C.)

22. A train of mass 200 Mg travelling on a track at 30 km/h is subjected to an accelerating force which varies uniformly from zero to 40 kN in a time of 2 min. Deduce an expression for the velocity of the train at a time *t* seconds after the acceleration starts. Hence calculate:

(*a*) the kinetic energy of the train after 2 min;
(*b*) the distance travelled by the train whilst accelerating.

(N.C.T.E.C.)

23. A car of mass 900 kg starts from rest on a straight level road and is accelerated to its maximum speed of 60 km/h in a time of 1 min. If the acceleration is one which decreases at a uniform rate determine the distance travelled from rest in 0·5 min and 1 min.

If the resistances to motion are 0·06 N/kg determine the power required after 0·25 min and 0·5 min from the start.

24. A train starts from rest with a constant acceleration of 0·06 m/s^2 which is maintained for 3 min. The power is then gradually reduced, thereby reducing the acceleration at a uniform rate until the train is running at full speed 5 min after starting. The brakes are then applied to produce a retardation which increases uniformly until the train is brought to rest in a further 2·5 min. Draw the acceleration-time graph and then calculate the maximum velocity of the train and the value of the retardation at the instant of stopping. Using the same time axis, sketch the velocity-time graph for the journey. (W.J.E.C.)

9

Kinetics — angular motion

9.1 Rotation of a body about a fixed axis

It has been seen previously (section 8.7) that when a body possesses translatory motion only, the resultant of the external forces which act upon it passes through the centre of mass of the body. The case when the resultant of the external forces does not pass through the centre of mass of the body will now be examined.

Consider the body shown in Fig. 9.1 which is subjected to a torque T and is rotating in a horizontal plane about a fixed axis through O, with an instantaneous angular velocity ω and angular acceleration α.

An elemental mass δm is subject to two acceleration components:

(1) a tangential component αr;

and (2) a centripetal component $\omega^2 r$ which passes through the centre of rotation O.

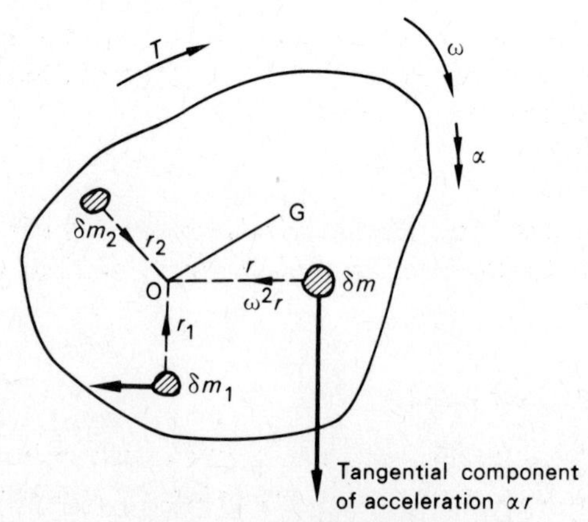

Fig. 9.1 Rotation of a body about a fixed axis

Thus, the force components acting on mass δm are $\delta m \,.\, \alpha r$ in the tangential direction and $\delta m \,.\, \omega^2 r$ in the radial direction.

Now, the centripetal component passes through O and therefore cannot provide a moment about this point.

Therefore, elemental turning moment about $O = \delta m \,.\, \alpha r \times r$

$$= \delta m \,.\, \alpha r^2$$

Summing for all such elemental masses gives:

$$\text{Total turning moment } T, \text{ about } O = \Sigma \delta m \,.\, \alpha r^2$$
$$= \alpha \Sigma\, \delta m \,.\, r^2 \text{ (since } \alpha \text{ is constant for all elemental masses)}$$

$\Sigma\, \delta m \,.\, r^2$ is the summation of all the elemental (mass x radius^2) terms about O and is known as the MOMENT OF INERTIA about O; i.e.,

$$I_O = \Sigma\, \delta m \,.\, r^2$$

Thus $$T = I_O \alpha \qquad (9.1)$$

If T has units of newtons x metres and I_O has units of kilogrammes x (metres)^2 then the units of α are rad/s^2.

9.2 Radius of gyration

The moment of inertia of a body about an axis is equal to the sum of all the elemental masses constituting a body multiplied by the square of their individual radii from the axis. Clearly, the sum of all the individual elemental masses must equal the total mass of the body, i.e., $m = \Sigma \delta m$.

If the total mass m is imagined to be concentrated at a distance k from the chosen axis such that $mk^2 = \Sigma \delta m \,.\, r^2$ then the moment of inertia can be expressed in the form

$$I_O = mk^2 \qquad (9.2)$$

The distance k is termed 'the radius of gyration' of the body about the chosen axis. Obviously, the axis about which k is taken must be clearly specified.

9.3 Centre of percussion

In many rotational problems it is frequently necessary to know the line of action of the resultant force F, shown in Fig. 9.2. This resultant force, which is equal to the mass of the body multiplied by the

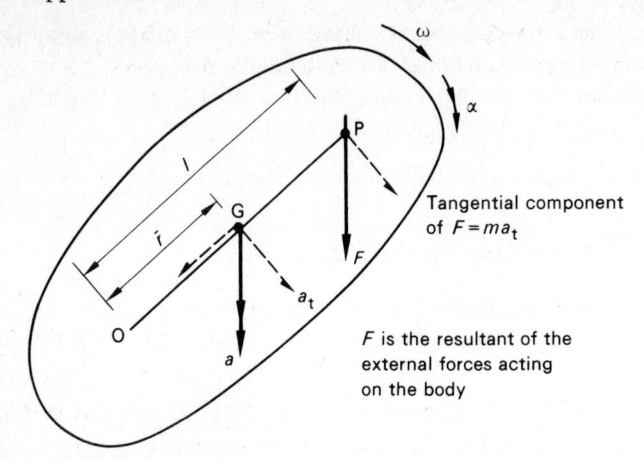

Fig. 9.2 Centre of percussion

acceleration of the centre of mass, will be assumed to pass through some point P on OG produced.

As in section 9.1, when dealing with a single elemental mass, only the tangential component of F contributes to the turning moment about O, since the centripetal component passes through O.

Thus, if a_t is the tangential component of the acceleration of the centre of mass G, then the tangential component of the resultant force acting through P is $F_t = ma_t$.

But $a_t = \alpha\bar{r}$ where $\bar{r}$ is the distance of the centre of mass G from the axis of rotation O.

$$\therefore \quad F_t = ma_t = m\alpha\bar{r}$$

The applied torque T is equal to the moment of F_t about O; i.e.,

$$T = F_t \times l$$

where l = moment arm of F_t about O.

But $$T = I_O\alpha$$

$$\therefore \quad m\alpha\bar{r} \times l = I_O\alpha$$

Putting $$I_O = mk_O^2$$

gives $$l = \frac{k_O^2}{\bar{r}}$$

which is the distance of P from the axis of rotation.

Now, by the parallel axis theorem (page 118) we know that if k_G is the radius of gyration about an axis through the centre of mass of the body, then:

$$I_O = I_G + m\bar{r}^2$$

$$k_O^2 = k_G^2 + \bar{r}^2$$

$$\therefore \quad \frac{k_O^2}{\bar{r}} = l = \frac{k_G^2}{\bar{r}} + \bar{r}$$

This shows that the resultant force F acts through a position P which is beyond G on the line OG produced and at a distance from G equal to $k_G^2/\bar{r}$.

Hence, point P is the position within the body where the moment of all the external forces is zero and is known as the *centre of percussion* (c.f. Centre of oscillation in section 10.6).

At this stage, however, we are generally concerned with bodies which are rotating about their centre of mass. For such cases O and G are coincident so that:

$$k_O^2 = k_G^2$$

since $\bar{r} = 0$

Then $T = I_G\alpha = mk_G^2\alpha$

It should now be appreciated why it is easier to spin a wheel about its central axis than about any other axis – the radius of gyration is least about the central axis and hence the required torque is a minimum.

Example 9.1

The total mass of a pair of uniform integral pulleys is 40 kg and the radius of gyration about the axis of rotation is 0·28 m. The effective diameters of the pulleys are 0·32 m and 1·10 m. Light ropes pass over the pulleys and masses of 15 kg and 25 kg are supported from the large and small pulleys respectively. If bearing friction is neglected and the system is released from rest, what will be movement of each mass during the first two seconds of motion?

Solution

The direction of motion of the system when released from rest can be obtained by considering the torques about the shaft axis when the system is at rest.

Referring to Fig. 9.3:

Clockwise torque due to 25-kg mass = 25 x 9·81 [N] x 0·16 [m]
= 39·24 N m

Anticlockwise torque due to 15-kg mass = 15 x 9·81 [N] x 0·55 [m]
= 80·93 N m

Thus, the pulley will rotate anticlockwise, the mass of 15 kg falling while the 25-kg mass rises.

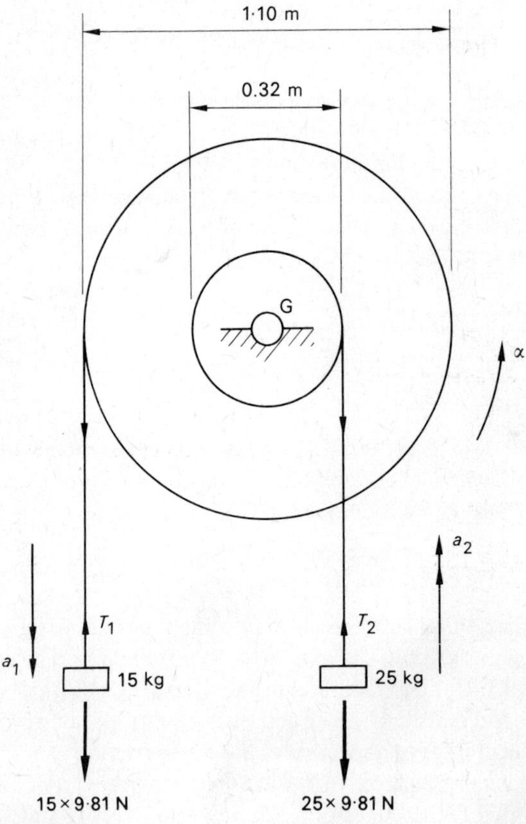

Fig. 9.3 Example 9.1

Note: When the system is in motion the resultant torque cannot be obtained from the above values. This is because when the system is at rest the tensions in the ropes are equal to the weight of the masses but this is not so when the system is in motion.

Let the tensions in the ropes be T_1 and T_2, the angular acceleration of the pulley system be α and the linear acceleration of the masses be a_1 and a_2 as shown in Fig. 9.3, the suffixes 1 and 2 referring to the 15-kg and 25-kg masses respectively.

Applying $T = I_G \alpha$

to the motion of the pulley system (since the axis of rotation is through the centre of mass of the pulleys) gives:

$$(T_1 \times 0{\cdot}55 - T_2 \times 0{\cdot}16)\ [\mathrm{N\,m}] = 40\ [\mathrm{kg}] \times 0{\cdot}28^2\ [\mathrm{m}^2] \times \alpha\ [\mathrm{rad/s^2}]$$

$$0{\cdot}55T_1 - 0{\cdot}16T_2 = 3{\cdot}136\,\alpha \qquad (1)$$

Applying $F = ma$

to the separate motion of the masses gives:

For 15-kg mass:

$$15 \times 9{\cdot}81 - T_1 = 15a_1 \qquad (2)$$

For 25-kg mass:

$$T_2 - 25 \times 9{\cdot}81 = 25a_2 \qquad (3)$$

A relationship can also be established between the linear accelerations of the masses a_1 and a_2, and the angular acceleration of the pulley block, α. Thus, applying eq. (7.12);

$$a_1 = 0{\cdot}55\,\alpha$$

and, $a_2 = 0{\cdot}16\,\alpha$

Then, $a_2 = \dfrac{0{\cdot}16}{0{\cdot}55}a_1 = 0{\cdot}2909\,a_1$

There are now sufficient equations to solve for a_1 and a_2.

From (2) $T_1 = -15a_1 + 147{\cdot}15$

From (3) $T_2 = 25a_2 + 245{\cdot}25 = 7{\cdot}2725a_1 + 245{\cdot}25$

Substituting for T_1, T_2, and α into eq. (1):

$$0{\cdot}55\,(-15a_1 + 147{\cdot}15) - 0{\cdot}16\,(7{\cdot}2725a_1 + 245{\cdot}25) = 3{\cdot}136 \times a_1/0{\cdot}55$$

Thus $a_1 = 2{\cdot}758\ \mathrm{m/s^2}$

and $a_2 = 0{\cdot}8023\ \mathrm{m/s^2}$

To determine the distance moved by the masses in 2 s from rest we can apply $s = v_1 t + \frac{1}{2}at^2$ to each with $v_1 = 0$:

For 15-kg mass: $s_1 = \frac{1}{2}a_1 t^2 = \frac{1}{2} \times 2{\cdot}758 \times 2^2 = 5{\cdot}516$ m downwards

For 25-kg mass: $s_2 = \frac{1}{2}a_2 t^2 = \frac{1}{2} \times 0{\cdot}8023 \times 2^2 = 1{\cdot}6046$ m upwards

9.4 Kinetic energy of rotation

Consider a body which is rotating about an axis through O with an angular velocity ω. For an elemental mass δm at a distance r from O the instantaneous linear velocity is $v = \omega r$, as shown in Fig. 9.4.

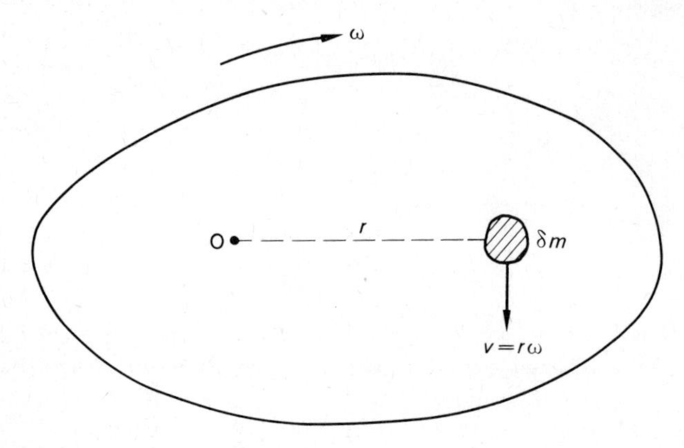

Fig. 9.4

Kinetic energy of mass $\delta m = \frac{1}{2} \cdot \delta m \cdot (\omega r)^2$.

The total kinetic energy of the body is given by the sum of all such terms, viz:

$$\text{Kinetic energy of body} = \Sigma \tfrac{1}{2} \cdot \delta m \cdot (\omega r)^2$$
$$= \tfrac{1}{2}\omega^2 \Sigma \delta m \cdot r^2$$

since ω is constant for all elements.

But from section 9.2, $\Sigma \delta m \cdot r^2 = I_O$, the moment of inertia about an axis through O.

$\therefore$ Kinetic energy of a rotating body $= \frac{1}{2} I_O \omega^2$

If I_O has units of [kg m^2] and ω has units of [rad/s] then the units of kinetic energy are [kg m^2/s^2] or joules (J), since 1 N = 1 kg m/s^2 (ref. section 8.4).

Example 9.2

A mass of 1 Mg is raised by means of a steel lifting cable which is coiled around a drum of effective diameter 2·5 m mounted on a horizontal shaft. The drum and shaft have a mass of 1·0 Mg and a radius of gyration of 1·10 m.

Determine the tension in the lifting cable and the torque required at the shaft to give the mass an acceleration of 0·8 m/s^2. What will be the kinetic energy possessed by the drum after 10 s?

Solution

Let the tension in the lifting cable be P, the shaft torque be T' and the angular acceleration of the drum be α, as shown in Fig. 9.5.

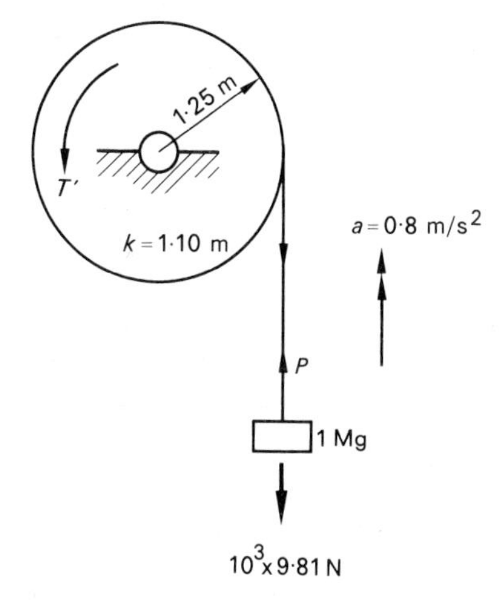

Fig. 9.5 Example 9.2

The tension in the lifting cable can be obtained by applying

$$F = ma$$

to the vertically upwards motion of the mass of 1 Mg.

Thus, $(P - 10^3 \times 9{\cdot}81) = 10^3 \times 0{\cdot}8$

$$P = 10\,610 \text{ N}$$

The tension in the cable is 10 610 N.

This cable tension results in a torque on the drum which opposes the applied torque T'. Thus, the torque T which produces the angular acceleration of the drum is

$$T = T' - 10\,610 \times 1{\cdot}25$$
$$= T' - 13\,262$$

Also $$I_O = mk_O^2$$
$$= 10^3\ [\text{kg}] \times 1{\cdot}10^2\ [\text{m}^2]$$
$$= 1210\ \text{kg m}^2$$

and $$\alpha = \frac{a}{r} = \frac{0{\cdot}8}{1{\cdot}25} = 0{\cdot}64\ \text{rad/s}^2$$

Applying $$T = I_O\alpha$$

gives $$T' - 13\,262 = 1210 \times 0{\cdot}64$$
$$T' = 14\,036\ \text{N m}$$

The torque that must be applied to the shaft is 14 036 N m.

Now the kinetic energy of a rotating body is given by $\frac{1}{2}I_O\omega^2$ and so we need to know the angular velocity of the drum after 10 s. As the drum starts from rest then

$$\omega = \alpha t$$
$$= 0{\cdot}64 \times 10 = 6{\cdot}4\ \text{rad/s}$$

$$\therefore\ \text{Kinetic energy after 10 s} = \tfrac{1}{2}\,.\,1210\ [\text{kg m}^2] \times 6{\cdot}4^2\ [(\text{rad/s})^2]$$
$$= 2478{\cdot}1\ \text{N m}$$
$$= 2478{\cdot}1\ \text{J}$$

The kinetic energy after 10 s is 2478·1 J.

Example 9.3

A flywheel has a mass of 800 kg and a radius of gyration of 0·6 m. Determine:

(*a*) the torque required to increase the speed of the flywheel from 100 rev/min to 300 rev/min in a time of 20 s, assuming a uniform acceleration;
(*b*) the average power absorbed by the flywheel in this time.

Solution

(*a*) To determine the required torque it is necessary to know the angular acceleration of the flywheel.

Thus, applying $\omega_2 = \omega_1 + \alpha t$

where, $\omega_2 = 300 \text{ rev/min} = \dfrac{300 \times 2\pi}{60} = 10\pi \text{ rad/s}$

$\omega_1 = 100 \text{ rev/min} = \dfrac{10\pi}{3} \text{ rad/s}$

and $t = 20 \text{ s}$

gives $$10\pi = \frac{10\pi}{3} + 20\alpha$$

$$\alpha = \frac{\pi}{3} \text{ rad/s}^2$$

Then
$$T = I_O \alpha$$
$$= 800 \text{ [kg]} \times 0{\cdot}6^2 \text{ [m}^2\text{]} \times \pi/3 \text{ [rad/s}^2\text{]}$$
$$= 301{\cdot}6 \text{ N m}$$

The torque required is 301·6 N m.

(*b*) The total angle turned through, θ, in 20 s is given by

$$\theta = \left(\frac{\omega_1 + \omega_2}{2}\right) t$$

But, the total work done by a torque = $T\theta$
Thus, as the power absorbed is equal to the rate of working we obtain:

$$\text{average power absorbed} = \frac{T\theta}{t} = \frac{T(\omega_1 + \omega_2)}{2}$$

$$= \text{torque} \times \text{mean angular velocity}$$

Also, $$T = I_O\alpha = I_O\left(\frac{\omega_2 - \omega_1}{t}\right)$$

Then, $$\text{average power absorbed} = I_O\left(\frac{\omega_2 - \omega_1}{t}\right)\left(\frac{\omega_1 + \omega_2}{2}\right)$$

$$= (\tfrac{1}{2} \,.\, I_O \,.\, \omega_2^2 - \tfrac{1}{2} \,.\, I_O \,.\, \omega_1^2)\,\frac{1}{t}$$

$$= \text{rate of change of kinetic energy}$$

Either of the above expressions may be used to determine the average power absorbed. However, they only apply to uniformly accelerated motion.

Using the first expression:

$$\text{Average power absorbed} = 301{\cdot}6 \text{ [N m]} \times \left(\frac{10\pi/3 + 10\pi}{2}\right) \text{ [rad/s]}$$

$$= 6307 \text{ W}$$

$$= 6{\cdot}307 \text{ kW}$$

The average power absorbed by the flywheel is 6·307 kW.

9.5 Kinetic energy of a body possessing translation and rotation

Now that the kinetic energy of a body possessing linear motion and one possessing rotational motion have been determined separately we can proceed to the case when a body possesses both translational and rotational motion.

Consider a body, such as a car wheel, that is rotating with a uniform angular velocity ω about its centre of mass in addition to possessing a linear velocity v. This is shown in Fig. 9.6.

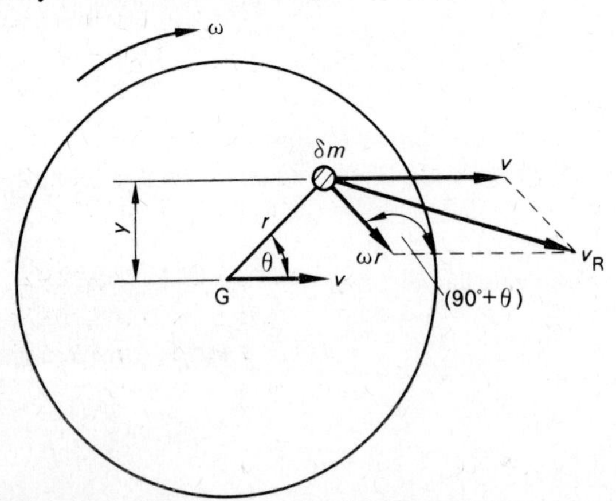

Fig. 9.6 Rotation of body possessing translation and rotation

An elemental mass δm at a distance r from the axis of rotation then possesses two velocity components. These are the linear velocity component v and a component of ωr at right angles to the radius r.

The resultant velocity is v_R, as shown, where

$$v_R^2 = v^2 + \omega^2 r^2 - 2v \cdot \omega r \cos(90 + \theta)$$
$$= v^2 + \omega^2 r^2 + 2v\omega r \sin\theta$$

The kinetic energy of the elemental mass $= \frac{1}{2} \cdot \delta m \cdot v_R^2$

Summing for the complete body gives:

$$\text{Total kinetic energy of body} = \Sigma \tfrac{1}{2} \cdot \delta m \cdot v_R^2$$
$$= \Sigma \tfrac{1}{2} \cdot \delta m (v^2 + \omega^2 r^2 + 2v\omega r \sin\theta)$$
$$= \Sigma \tfrac{1}{2} \cdot \delta m \cdot v^2 + \Sigma \tfrac{1}{2} \cdot \delta m \cdot r^2 \omega^2 + \Sigma \delta m \cdot v\omega r \sin\theta$$
$$= \frac{v^2}{2} \sum \delta m + \frac{\omega^2}{2} \sum \delta m \cdot r^2 + v\omega \sum \delta m \cdot y$$

since v and ω are constant and $r \sin\theta = y$

Now $m = \Sigma \delta m$

$I = \Sigma \delta m \cdot r^2$

and $\Sigma \delta m \cdot y = 0$ (being the first moment of mass about the centre of mass)

Thus, Total kinetic energy of body $= \frac{1}{2} m v^2 + \frac{1}{2} I \omega^2$

= kinetic energy of translation + kinetic energy of rotation

Example 9.4

Derive from first principles an expression for the moment of inertia of a solid circular bar about its longitudinal axis in terms of its mass and radius.

A solid cylinder having a mass of 2·5 kg and a diameter of 0·10 m rolls, without slipping, down a slope which is inclined at 10° to the horizontal. When the body is at a certain position on the incline its linear velocity is 1 m/s. How far will the cylinder have rolled down the plane when its velocity is 2 m/s?

Solution

A solid cylinder of radius R and length l is shown in Fig. 9.7.

Consider an elemental cylinder of the material having a radius r and thickness δr. If the density of the material is ρ then the mass of the elemental cylinder is $2\pi r \cdot \delta r \cdot l\rho$. As the mass of the element can be

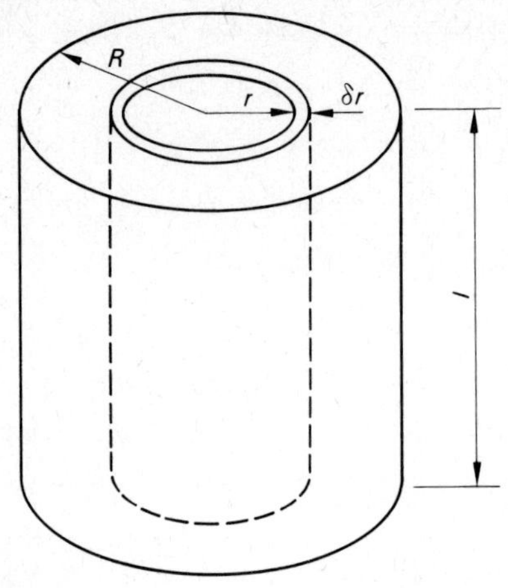

Fig. 9.7 Example 9.4

considered to be concentrated at a radius r from the longitudinal axis, we obtain:

Moment of inertia of elemental cylinder about longitudinal axis

$$= (2\pi\rho l r \,.\, \delta r) r^2$$

The total moment of inertia I of the solid will be the sum of all such elemental terms.

$$\therefore \quad I = \int_0^R 2\pi\rho l r^3 \,.\, \mathrm{d}r$$

$$= 2\pi\rho l \left[\frac{r^4}{4}\right]_0^R$$

$$= \pi\rho\frac{l}{2}R^4$$

$$= \pi\rho l R^2 \times \frac{R^2}{2}$$

$$= \frac{mR^2}{2}$$

where $m = \pi\rho l R^2$ = mass of solid bar.

A diagram for the cylinder rolling down the incline is given in Fig. 9.8. The body starts from position 1 with a linear velocity of 1 m/s and angular velocity ω_1, and after travelling a distance s it attains a linear velocity of 2 m/s and angular velocity ω_2, at position 2. In doing so the cylinder falls through a vertical height h and so loses

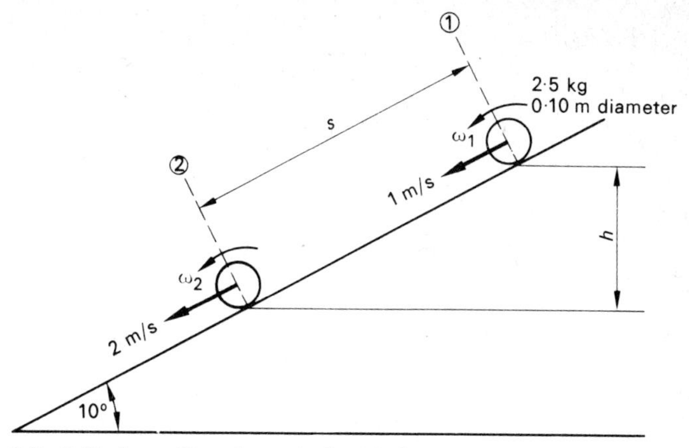

Fig. 9.8 Cylinder rolling down inclined plane

potential energy. As no slipping occurs this energy is completely converted into kinetic energy.

$$\text{Loss of potential energy} = 9{\cdot}81 \times 2{\cdot}5 \times h$$
$$= 24{\cdot}525 \times s \sin 10^\circ$$
$$= 4{\cdot}259\, s \text{ newton metres (joules)}$$

Gain in kinetic energy = gain in linear kinetic energy + gain in rotational kinetic energy

$$= \tfrac{1}{2}m(v_2^2 - v_1^2) + \tfrac{1}{2}I_G(\omega_2^2 - \omega_1^2)$$

But $I_G = \frac{1}{2}mr^2$ and $\omega = \dfrac{v}{r}$

$$\therefore \quad \text{Gain in kinetic energy} = \tfrac{1}{2}m(v_2^2 - v_1^2) + \tfrac{1}{2}(\tfrac{1}{2}mr^2)\left(\frac{v_2^2}{r^2} - \frac{v_1^2}{r^2}\right)$$
$$= \frac{3}{4} \cdot m(v_2^2 - v_1^2)$$
$$= \frac{3}{4} \cdot 2{\cdot}5\,(2^2 - 1^2)$$
$$= 5{\cdot}625 \text{ N m} = 5{\cdot}625J$$

Equating the energy expressions gives

$$s = \frac{5{\cdot}625}{4{\cdot}259} = 1{\cdot}344 \text{ m}$$

The cylinder will have rolled 1·344 m down the incline when its linear velocity is 2 m/s.

9.6 Angular impulse

The angular impulse of a force acting on a body about a certain axis is the moment of the linear impulse of the force about that axis.

If a constant force F acts at a distance d from a given axis for a period of time t, then since Ft is the linear impulse of the force,

$$\text{Angular impulse} = Ft \times d$$

But, Torque of force about axis, $T = Fd$

$\therefore$ $$\text{Angular impulse} = Tt$$

9.7 Angular momentum

The angular momentum of a body is the moment of its linear momentum.

Consider a body which is rotating about an axis through O with an angular velocity ω. The linear velocity v of an elemental mass δm at a distance r from O is $r\omega$, as shown in Fig. 9.9.

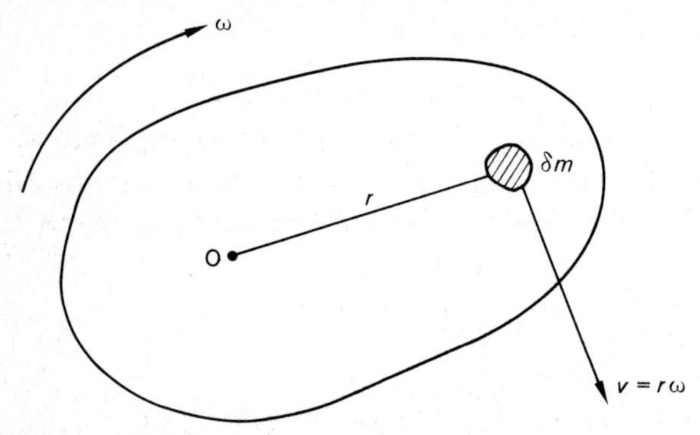

Fig. 9.9

$$\text{Linear momentum of elemental mass } \delta m = \delta m \, . \, v$$
$$= \delta m \, . \, r\omega$$

$$\text{Moment of linear momentum of } \delta m \text{ about O} = \delta m \, . \, r^2 \omega$$

Thus, for the complete body

$$\text{Moment of momentum about O} = \Sigma \delta m \, . \, r^2 \omega$$

i.e., Angular momentum about O $= \omega \Sigma \delta m \, . \, r^2$ (for constant ω)

$$= I\omega$$

where $I = \Sigma \delta m \, . \, r^2$ = moment of inertia

For a change in angular velocity from ω_1 to ω_2,

$$\text{Change in angular momentum} = I(\omega_2 - \omega_1)$$

The relationship between angular momentum and angular impulse is similar to that for the linear case; i.e.,

$$\text{Angular impulse} = \text{change in angular momentum}$$

$$Tt = I(\omega_2 - \omega_1)$$

The conservation of momentum can be stated for the angular case as follows:

If the resultant externally applied moment or torque on any system about a fixed axis is zero then the angular momentum of the system remains constant.

As in the linear case, there may be changes in the momentum of individual parts of the system but the total angular momentum remains constant.

Example 9.5

A and B are two separate clutch plates. Plate A has a mass of 40 kg, a radius of gyration of 0·16 m and rotates at 600 rev/min. Plate B has a mass of 50 kg, a radius of gyration of 0·12 m and is stationary. When the clutch plates are engaged slipping ceases after a time of 0·30 s. Determine:

(*a*) the common angular speed of the plates;
(*b*) the loss of kinetic energy;
(*c*) the angular acceleration of plate B;
(*d*) the torque transmitted by the clutch, assuming this to be constant during the 0·30-s period;
(*e*) the angle of slip between the two clutch faces.

Solution

(*a*) As there is no externally applied torque on the system the total angular momentum remains constant; i.e.,

$$\text{Angular momentum before engagement} = \text{angular momentum after engagement}$$

The angular momentum before engagement is due to plate A only, as plate B is stationary.

If the common angular speed of the plates after engagement is n rev/min, then

$$I_A\omega_A = (I_A + I_B)\omega$$

where ω is the common angular velocity.

$$\therefore \quad 40 \times 0{\cdot}16^2 \times \frac{600 \times 2\pi}{60} = (40 \times 0{\cdot}16^2 + 50 \times 0{\cdot}12^2) \times \frac{n \times 2\pi}{60}$$

$$n = \frac{40 \times 0{\cdot}16^2}{(40 \times 0{\cdot}16^2 + 50 \times 0{\cdot}12^2)} \times 600$$

$$= 352 \text{ rev/min}$$

The common angular speed of the plates is 352 rev/min.

(*b*) Loss of kinetic energy = kinetic energy before engagement − kinetic energy after engagement

$$= \tfrac{1}{2}I_A\omega_A^2 - \tfrac{1}{2}(I_A + I_B)\omega^2$$

$$= \tfrac{1}{2} \,.\, 40 \times 0{\cdot}16^2 \left(\frac{2\pi \,.\, 600}{60}\right)^2 - \tfrac{1}{2}(40 \times 0{\cdot}16^2 + 50 \times 0{\cdot}12^2)\left(\frac{2\pi \,.\, 352}{60}\right)^2$$

$$= (2021{\cdot}3 - 1186{\cdot}2) \text{ N m}$$

$$= 835{\cdot}1 \text{ J}$$

The loss of kinetic energy is 835·1 J.

(*c*) Plate B undergoes an acceleration from rest to 352 rev/min in a time of 0·30 s.

Applying $\omega_2 = \omega_1 + \alpha t$ to this motion

gives $$\frac{352 \times 2\pi}{60} = 0 + \alpha_B \times 0{\cdot}3$$

$$\alpha_B = 122{\cdot}9 \text{ rad/s}^2$$

The angular acceleration of plate B is 122·9 rad/s^2.

(*d*) The torque T transmitted by the clutch is given by

$$T = I_B\alpha_B$$

$$= 50 \times 0{\cdot}12^2 \times 122{\cdot}9$$

$$= 88{\cdot}5 \text{ N m}$$

As the impulse on the two plates is equal and opposite this torque could have been obtained from $T = -I_A\alpha_A$, where α_A is the angular acceleration of plate A.

Thus as plate A slows from 600 rev/min to 352 rev/min in 0·30 s we get:

$$\frac{352 \times 2\pi}{60} = \frac{600 \times 2\pi}{60} + 0{\cdot}3\alpha_A$$

$$\alpha_A = -86{\cdot}55 \text{ rad/s}$$

Then $$T = 40 \times 0{\cdot}16^2 \times 86{\cdot}55$$

$$= 88{\cdot}5 \text{ N m}$$

The torque transmitted by the clutch is 88·5 N m

(*e*) The loss of kinetic energy during the slip period will equal the work done by the torque during the same period;

i.e., $T\theta$ = loss in kinetic energy

where θ = angle of slip between plates (radians)

Then $88{\cdot}5\theta = 835{\cdot}1$

$$\theta = 9{\cdot}436 \text{ rad}$$

One face of the clutch moves through 9·436 rad relative to the other face.

This result could also be obtained by considering the motion of each plate during the slip period and determining the angle turned through, the slip angle then being the difference of the angles turned through by the individual plates. Solution by this method would be considerably longer.

Problems

1. Deduce from first principles the formula for the relationship between the torque and angular acceleration of a rotating drum.

A cage of mass 2 Mg is raised by means of a vertical rope which passes round a horizontal drum having an effective diameter of 1·5 m, a radius of gyration of 0·6 m and a mass of 4 Mg. Determine the torque that must be applied at the drum shaft to raise the cage with a uniform acceleration of 0·3 m/s^2.

2. Explain what is meant by the radius of gyration of a body.

A wheel and axle is mounted in horizontal frictionless bearings. The total mass is 90 kg and the radius of gyration about the axis of rotation is 250 mm. The diameter of the axle is 76 mm. A mass of 16 kg is attached to the hanging end of a light string wrapped round the axle and then released. Calculate the angular acceleration produced on the wheel.

When the wheel has turned through 8 rev from rest the string becomes detached from the axle and a retarding torque of 3 N m is then applied to the axle. Calculate the kinetic energy stored in the wheel and axle at this time and hence the subsequent number of revolutions made by the wheel in coming to rest. (N.C.T.E.C.)

3. In a flywheel experiment a mass of 4 kg is fastened to a cord that is attached to a peg on an axle. The axle has a diameter of 40 mm and the cord a diameter of 2 mm. The mass of 4 kg is released and falls through a distance of 2 m in a time of 6 s. The mass then becomes detached and the flywheel completes a further 100 revolutions before coming to rest. If the flywheel has a mass of 30 kg determine:

(*a*) the radius of gyration of the flywheel;
(*b*) the friction torque exerted by the bearings.

4. A hoist drum has a mass of 100 kg and a radius of gyration of 0·3 m. The drum is used to raise a lift of mass 1000 kg with an upward acceleration of 2 m/s^2. The drum diameter is 1 m. Determine:

(*a*) the torque required at the drum to accelerate the lift;
(*b*) the power required after accelerating for 3 s from rest.

(W.J.E.C.)

5. A mass of 6 kg is attached to a light cord that is wound round the axle of a flywheel. The flywheel, whose axle is 50 mm diameter and horizontal, has a mass of 25 kg and a radius of gyration of 150 mm. From tests it is known that a mass of 1·2 kg attached to the cord will just cause the flywheel to rotate uniformly.
Determine:

(*a*) the angular acceleration of the flywheel due to the 6 kg mass;
(*b*) the tension in the cord.

6. Explain the functions of a flywheel.

(*a*) A flywheel of mass 65 kg and having a radius of gyration of 110 mm runs at a uniform speed of 10 rev/s. Calculate its store of kinetic energy.

(*b*) If the machine to which this flywheel is attached absorbs 280 J of this energy, calculate the final flywheel speed.

(*c*) Find the mean angular deceleration of the flywheel and the number of revolutions it makes if the speed change in (*b*) takes place in 5 s.

(U.L.C.I.)

7. Explain what is meant by moment of inertia and show that the kinetic energy of a body rotating about a fixed axis is given by $\frac{1}{2}I\omega^2$ where I is the moment of inertia and ω the angular velocity.

A rope hanging over a pulley of radius 0·5 m has attached to its two ends masses of 9 kg and 10 kg respectively. The moment of inertia about its axis of rotation is 16 kg m^2. Determine the velocity of the masses after they have moved a distance of 2 m from rest. Assume that there is no slip between the rope and the pulley and that the pulley runs on frictionless bearings.

8. A lift cage and its contents have a total mass of 1400 kg and are raised by a steel cable which is wound round a drum of 1 m diameter. The drum has a mass of 200 kg and a radius of gyration of 0·25 m. The drum drive is from an electric motor which produces a maximum torque of 8 kNm at the drum. If the resisting torque amounts to 5% of the applied torque determine the maximum acceleration of the cage and the tension in the lifting cable.

9. A flywheel has a moment of inertia of 8 kg m^2 about its axis of rotation. Two small dense masses, each being one-quarter of the mass of the flywheel, are attached symmetrically to the flywheel at a radius of 200 mm and increase the moment of inertia to 9·2 kg m^2. Find the radius of gyration of the flywheel alone.

A tangential force of 18 N is applied to the loaded flywheel at a radius of 400 mm. Neglecting friction, calculate:

(*a*) the angular acceleration produced;

(*b*) the angular distance moved by the flywheel from rest in 2 min.

Hence, or otherwise, calculate the kinetic energy stored in the loaded flywheel at this time. (N.C.T.E.C.)

10. A truck is carried by four wheels each of which may be regarded as a solid disc 0·8 m diameter and 40 kg mass. Apart from its wheels the mass of the truck is 200 kg. Calculate the total kinetic energy of the truck and wheels when travelling at 30 m/s. What braking force applied to the coupling of the truck will reduce the speed to 20 m/s in 40 s? (E.M.E.U.)

11. Two separate clutch plates, A and B, are mounted on a common axis of rotation. Plate A has a moment of inertia about the axis of 0·24 kg m^2. Plate B has a moment of inertia about the axis of 0·09 kg m^2. With plate A rotating at a speed of 10 rev/s and plate B rotating at 1·2 rev/s in the opposite sense, the plates are engaged and slipping ceases after 0·4 s. Calculate:

(*a*) the common angular speed of the plates when slipping has ceased;
(*b*) the loss of kinetic energy of plate A due to engagement;
(*c*) the average angular deceleration of plate A during slip and hence the mean torque transmitted during slip.

(N.C.T.E.C.)

12. The flywheel of a power press has a mass of 200 kg and a radius of gyration of 0·24 m. The flywheel is rotating at 600 rev/min when the ram is operated. The press is used to blank a material having a thickness of 10 mm and the average ram force during this operation is 250 kN. Calculate the reduction in the speed of the flywheel during the blanking operation.

13. A truck of mass 7 Mg is drawn up an incline of 1 in 20 by means of a light cable. The cable, which is parallel to the incline, passes around a winding drum at the top of the incline at a radius of 1·3 m. The drum has a mass of 5 Mg and a radius of gyration of 1·1 m about its axis of rotation. The rolling resistance to the truck is 620 N and the frictional torque on the drum is 190 N m.

Calculate the driving torque required on the drum to give uniform acceleration to the truck so as to reach a speed of 48·3 km/h from rest while travelling a distance of 50 m up the incline. (N.C.T.E.C.)

14. The flywheel of an engine has a mass of 4 Mg and a radius of gyration of 0·7 m. After the fuel supply is cut off when the engine is running at 80 rev/min on no load the engine comes to rest in 3 min 20 s. Determine the power required to run the engine on no load at 80 rev/min and the kinetic energy of the flywheel at this speed.

15. The two halves of a discharged friction clutch are mounted on their independent co-axial shafts, neither being driven from an external source. One element A and its shaft is rotating freely at 420 rev/min while the other element B is rotating at 300 rev/min in the opposite direction. The clutch is now engaged. Determine the common speed after engagement if A has a mass of 15 kg and a radius of gyration of 150 mm, and B has a mass of 20 kg and a radius of gyration of 120 mm. Determine also the loss of kinetic energy due to the clutch engagement.

16. A punching machine is used to punch 40 mm diameter holes in a steel plate of 12 mm thickness. The work done by the machine is 6 MJ per square metre of sheared area. The machine flywheel has a mass of 3600 kg and a radius of gyration of 0·6 m.

(*a*) If the normal running speed of the machine on no load is 100 rev/min determine the loss of speed due to punching one hole assuming that the energy for punching the hole is supplied entirely by the flywheel.

(*b*) If the flywheel is supported in two 150 mm diameter bearings calculate the number of revolutions that the flywheel would make in coming to rest from 100 rev/min with the power off if the coefficient of friction at the bearings is 0·12.

17. A truck has a total mass of 4000 kg including its four wheels each of which has a mass of 250 kg, a radius of gyration of 200 mm, and is 900 mm diameter. Initially the truck is moving on a level track with a speed of 48 km/h and after travelling against a constant tractive effort for a distance of 240 m the speed becomes 24 km/h. At this point the truck begins to ascend a gradient of 1 in 200 (measured as a sine). Calculate the initial kinetic energy of the truck including the wheels, the tractive resistance, and the distance the truck travels up the incline before coming to rest. (E.M.E.U.)

10

Simple harmonic motion

10.1 Periodic motion

Motion which repeats itself after a time interval is said to be *periodic.* A simple example of this kind of motion can be obtained by holding a scale rule over the edge of a table and tapping its overhanging end. If the disturbing force is not re-applied then the end of the rule will vibrate, or oscillate, freely about a mean position. This type of repeated to and fro motion occurs in elastic structures whenever they are disturbed from their equilibrium positions, and if the resulting vibration is within the audible range a musical note is produced. It is for this reason that the motion is described as harmonic.

With such a periodic motion the vibrations do not continue indefinitely, but are eventually 'damped' out by the internal resistance of the body and the external resistance of the air. In later studies the effect of this damping, or friction, must be taken into account but at this stage it will be assumed that the damping effects are negligible and we shall restrict our analysis to those common types of free vibration which approximate to simple harmonic motion.

10.2 Simple harmonic motion

Simple harmonic motion is defined as the motion of a body whose acceleration is always directed *towards* a fixed point in its path and is proportional to its displacement from that point. The general terms employed in the analysis of simple harmonic motion may be explained by reference to Fig. 10.1 in which an imaginary point P moves round a circle of radius r with a constant angular velocity ω. (The circle shown in Fig. 10.1 is sometimes called the *auxiliary circle.*)

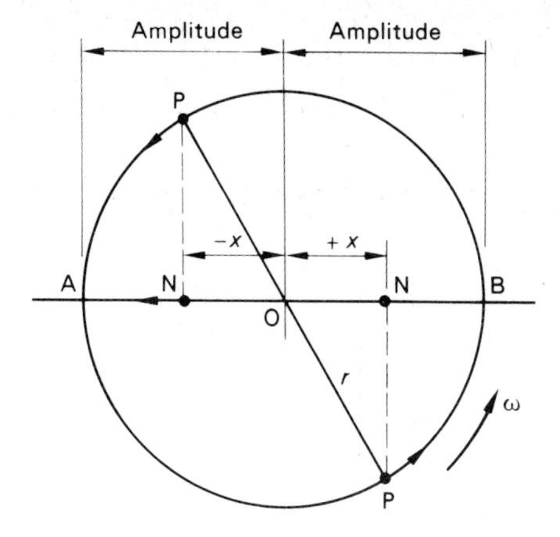

Fig. 10.1

The oscillation of the point N – the projection of P on the diameter AB – will be shown to occur with simple harmonic motion across the diameter AB.

It will be appreciated that whilst P makes one complete revolution the point N will make one complete oscillation across the diameter AB, i.e., from A to B and back again to A. Hence, the time for one revolution of P is equal to the time for one complete oscillation or period of N and this is called the periodic time T.

Now, the distance moved by P in one revolution = 2π radians

But as $\omega = \dfrac{2\pi}{T}$

then $T = \dfrac{2\pi}{\omega}$ seconds

The frequency f, or number of oscillations per second, is then given by

$$f = \frac{1}{T} = \frac{\omega}{2\pi} \text{ oscillations or cycles per second}$$

The unit of frequency is the hertz (Hz) which is one cycle/second (c/s).

The amplitude of the oscillation of N is the radius of the circle and is simply the maximum displacement of N from the centre of oscillation O.

From Chapter 7 we know that, throughout its circular motion, P will be subjected to a constant centripetal acceleration of $\omega^2 r$, directed towards O. Thus, at the instant when N is at a distance $+x$ from O, as shown in Fig. 10.2, the horizontal component of the acceleration of P will be $\omega^2 r \cos\theta$ in a direction towards O.

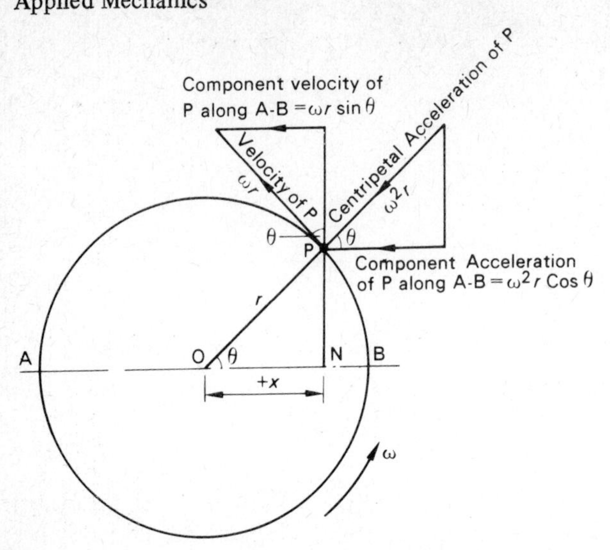

Fig. 10.2

Since N has the same horizontal motion as P but no vertical motion this horizontal component is the actual acceleration of N.

Therefore, the acceleration of N directed towards O is $-\omega^2 r \cos\theta$ (the negative sign indicating that the acceleration is in the opposite direction to that of increasing x); i.e.,

$$\text{Acceleration of N} = -\omega^2 r \cos\theta$$

But from Fig. 10.2, $r \cos\theta = x$

$$\therefore \quad \text{Acceleration of N} = -\omega^2 x \qquad (10.1)$$

Equation (10.1) indicates that the two necessary conditions for S.H.M. have been established, viz. the acceleration of N is directed towards the fixed point O and is proportional to the x displacement from O. That these conditions are satisfied for every position of P on the circumference is evident from the fact that the acceleration of P is always radially inwards.

Equation (10.1) may therefore be written:

$$\text{Acceleration of N} = -\text{ constant} \times x \text{ displacement}$$

It is important to note that in practice we are concerned with the motion of N – representing the vibrating body – and not with the imaginary point P. We have merely used the idea of an imaginary radius vector OP rotating at constant angular speed ω in order to establish the motion of N.

As the acceleration of N continually varies [ref. eq. (10.1)] then the velocity of N will also vary continually.

From Fig. 10.2 the velocity of P perpendicular to the radius vector OP is ωr. The horizontal component of this velocity, $\omega r \sin\theta$, is therefore the actual velocity of N at the instant shown.

From the right-angled triangle OPN,

$$\sin\theta = \frac{\sqrt{r^2 - x^2}}{r}$$

Therefore, for any displacement x from the centre of oscillation O,

$$\begin{aligned}\text{Velocity of N} &= -\omega r \sin\theta \\ &= -\omega\sqrt{(r^2 - x^2)} \qquad (10.2)\end{aligned}$$

(the negative sign indicating that the velocity is towards O).

Equations (10.1) and (10.2) give the acceleration and velocity of N which moves about O with simple harmonic motion.

$$\begin{aligned}\text{Acceleration of N} &= -\omega^2 x \\ &= -\omega^2 r\cos\theta \text{ (i.e., directed towards O)} \qquad (10.3)\end{aligned}$$

$$\text{Velocity of N} = -\omega r \sin\theta \quad \text{(i.e., directed towards O)} \qquad (10.4)$$

As both ω and r are constant, these equations show that the variation of acceleration with respect to time can be represented by a cosine curve and the variation of velocity by a sine curve. Fig. 10.2 and eq. (10.4) clearly indicate that point N passes through O, with maximum velocity and zero acceleration ($\theta = 90^\circ$). It then approaches A with reducing velocity and increasing acceleration until at A it has zero velocity and a maximum acceleration, again directed towards the centre of oscillation.

Thus, at the extremities of the vibration, A and B, the velocity of N is zero and its acceleration has a maximum value.

Most free (undamped) vibrations found in engineering have the characteristics just described so that the first step in any problem is to investigate the relationship which exists between acceleration and displacement. If it can be shown that:

$$\text{Acceleration} = -\text{ constant} \times \text{displacement}$$

then we can infer that simple harmonic motion conditions exist. This approach is illustrated in the examples which follow.

10.3 Mass on a vertical spring

Figure 10.3 illustrates a mass m suspended by a spring from a fixed rigid support.

It is assumed that the spring is perfectly elastic and obeys Hooke's law and also that the mass of the spring is very small compared with that of the suspended mass.

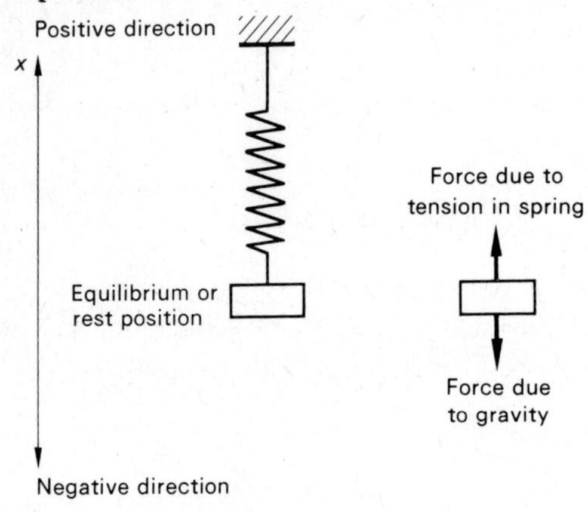

Fig. 10.3

Let the stiffness of the spring be k, i.e., k is the force required to produce a unit extension of the spring.

The static equilibrium position is where the gravitational force on the mass is balanced by the force in the spring. This position is indicated in Fig. 10.4(a).

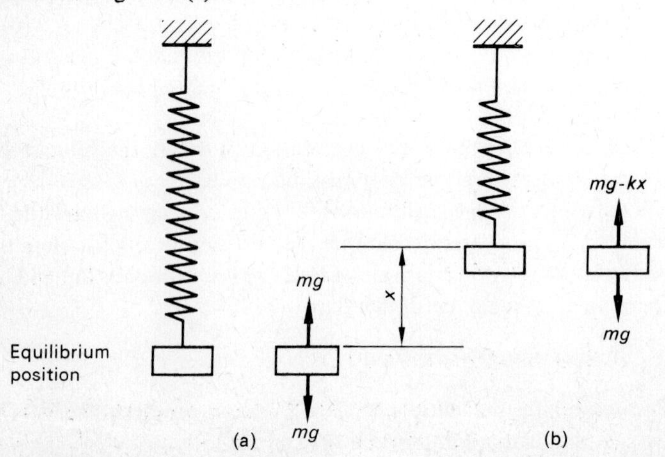

Fig. 10.4

Now, if the mass is displaced upwards and released then at some instant it will be in the position shown in Fig. 10.4(*b*). (It is assumed that the mass is constrained such that only motion in the vertical plane is possible.)

The forces acting on the mass will then be:

(*i*) the downward gravitational force mg

(*ii*) the upward force of the spring, $mg - kx$

$$\therefore \quad \text{Net upward force on mass} = (mg - kx) - mg$$
$$= -kx$$

The net force is therefore $-kx$, the negative sign indicating that it acts in a downward direction tending to restore the mass to its static equilibrium position.

Note that this restoring force is independent of the gravitational force (mg) acting on the mass.

Applying Newtons' second law [eq. (8.1)] gives:

$$-kx = ma$$

$$a = -\frac{k}{m}x \qquad (10.5)$$

Now the ratio k/m is clearly a constant quantity and therefore the acceleration of the vibrating mass is a negative constant times the displacement from its equilibrium position.

It is thus shown that the vibrating mass moves with simple harmonic motion.

Comparing eq. (10.5) and eq. (10.1) gives

$$\omega^2 = \frac{k}{m} \quad \text{or} \quad \omega = \sqrt{\frac{k}{m}}$$

Hence, periodic time, $T = \dfrac{2\pi}{\omega} = 2\pi\sqrt{\dfrac{m}{k}}$ (10.6)

and frequency, $f = \dfrac{1}{T} = \dfrac{1}{2\pi}\sqrt{\dfrac{k}{m}}$ (10.7)

10.4 Static deflection of a spring

When a mass m is attached to an unloaded spring of stiffness k the spring deflects to its equilibrium position by an amount mg/k. This is called the static deflection of the spring δ_s (Fig. 10.5).

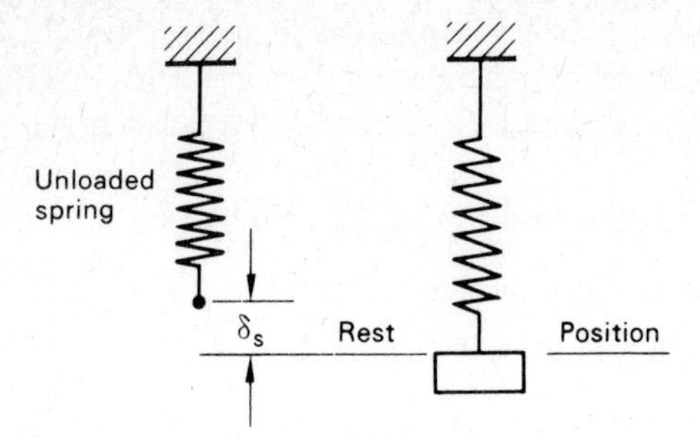

Fig. 10.5

The relationships (10.6) and (10.7) for T and f can now be expressed in terms of this static deflection δ_s.

Since $\delta_s = \dfrac{mg}{k}$ then $\dfrac{m}{k} = \dfrac{\delta_s}{g}$

Hence
$$T = 2\pi\sqrt{\frac{\delta_s}{g}} \tag{10.8}$$

and
$$f = \frac{1}{2\pi}\sqrt{\frac{g}{\delta_s}} \tag{10.9}$$

It is important to note that the periodic time, and hence the frequency, depend only on the magnitude of the static deflection which is itself only dependent on the spring stiffness and the magnitude of the suspended mass.

Therefore, since the periodic time is independent of the amplitude of the motion, the actual initial displacement of the spring is of no consequence – providing that Hooke's law is still satisfied. In practice, friction or damping will be present and therefore the amplitude of oscillation will diminish with time but the periodic time will remain unaltered. This explains why simple harmonic motion is also termed isochronous, i.e., repeated motion in the same time interval.

Example 10.1

A helical spring of negligible mass is required to support a mass of 70 kg. If the spring requires a force of 600 N in order to give an extension of 10 mm what will be the frequency of natural vibration when the mass is displaced 20 mm and then released? Also determine the velocity of the mass when it is 10 mm below the rest position.

Solution

$$\text{Stiffness of spring, } k = \frac{600\ [\text{N}]}{0{\cdot}01\ [\text{m}]} = 60 \times 10^3\ \text{N/m}$$

$$\omega = \sqrt{\frac{k}{m}} = \sqrt{\frac{60 \times 10^3}{70}}$$

$$= 29{\cdot}3\ \text{rad/s}$$

$$\text{Frequency } f = \frac{\omega}{2\pi} = \frac{29{\cdot}3}{2\pi} = 4{\cdot}66\ \text{c/s}$$

$$= 4{\cdot}66\ \text{Hz}$$

The velocity of the mass when it is 10 mm below its rest position, shown diagrammatically in Fig. 10.6, can be obtained by applying eq. (10.2).

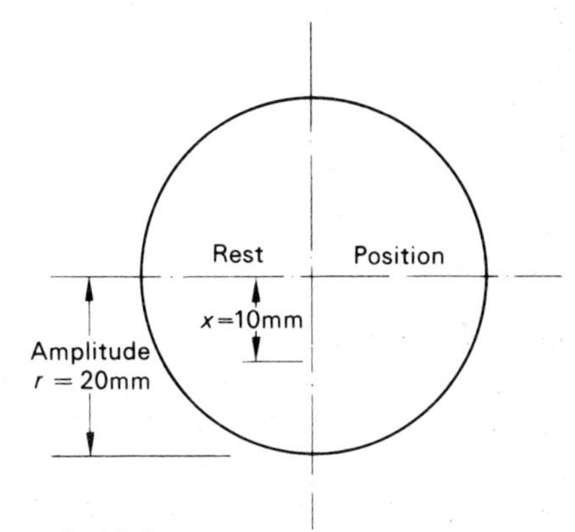

Fig. 10.6 Example 10.1

Thus, $v = \omega\ \sqrt{(r^2 - x^2)}$

$$= 29{\cdot}3\sqrt{\left\{\left(\frac{20}{1000}\right)^2 - \left(\frac{10}{1000}\right)^2\right\}}$$

$$= 0{\cdot}51\ \text{m/s}$$

Note: When the mass on any spring is caused to oscillate it is evident that the spring must also oscillate with it. The effect of the spring on the frequency of oscillation need not be considered when its mass is negligible, as in this problem. However, when this is not the case it can be shown that the effect of the spring can be allowed for by adding one third of its mass to the attached mass and then using the relationships for a light spring as given above.

Example 10.2

A mass is supported by two springs and is constrained to oscillate freely in a vertical direction. When the springs are connected in parallel the frequency of oscillation is found to be 2·05 Hz and when they are connected in series the frequency is 1 Hz. If the stiffness of one of the springs is 20 kN/m what is the stiffness of the other?

Solution

Let the spring stiffnesses be k_1 and k_2.

PARALLEL CASE

When the mass is displaced, clearly each spring must be deflected by the same amount, as shown in Fig. 10.7.

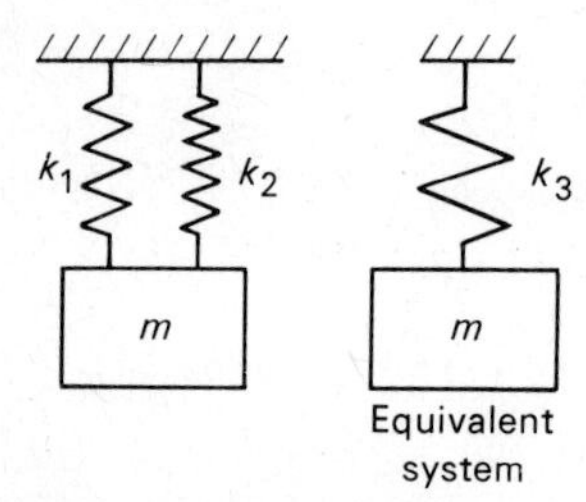

Fig. 10.7 Springs in parallel – Example 10.2

Hence for a displacement of mass m of x:

$$\text{Restoring force on mass} = -(k_1 + k_2)x$$

Then, from Newton's second law:

$$-(k_1 + k_2)x = ma$$

and,

$$a = -\frac{(k_1 + k_2)}{m}x \quad *$$

Hence, by comparison with eqs. (10.5) and (10.7)

$$f_p = \frac{1}{2\pi}\sqrt{\frac{(k_1 + k_2)}{m}} \tag{1}$$

* *Note:* The stiffness k_3 of a single spring that would replace the two springs in parallel and produce the same effect is

$$k_3 = k_1 + k_2$$

SERIES CASE

In this case each spring supports the mass m and the total deflection will therefore equal the sum of the separate static deflections of each spring, as shown in Fig. 10.8. Thus,

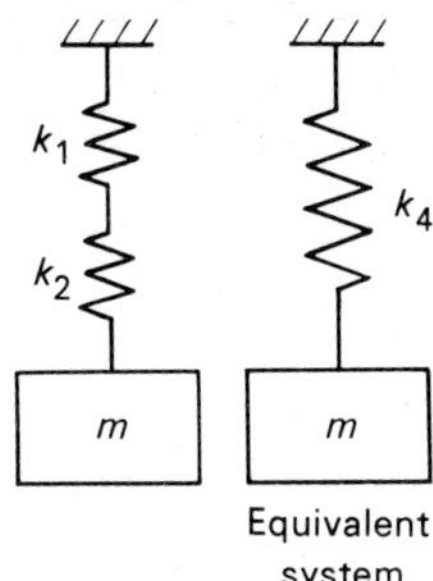

Fig. 10.8 Springs in series – Example 10.2

$$\delta_s = \frac{mg}{k_1} + \frac{mg}{k_2} = mg\left(\frac{1}{k_1} + \frac{1}{k_2}\right)$$

$$= mg\left(\frac{k_1 + k_2}{k_1 k_2}\right)^*$$

Hence, from eq. (10.9)

$$\text{Frequency}, f_s = \frac{1}{2\pi}\sqrt{\left\{\frac{1}{m}\left(\frac{k_1 k_2}{k_1 + k_2}\right)\right\}} \quad (2)$$

From eqs. (1) and (2) the ratio of the frequencies is given by

$$\frac{f_p}{f_s} = \frac{\frac{1}{2\pi}\sqrt{\left\{\frac{1}{m}(k_1 + k_2)\right\}}}{\frac{1}{2\pi}\sqrt{\left\{\frac{1}{m}\left(\frac{k_1 k_2}{k_1 + k_2}\right)\right\}}}$$

$$\left(\frac{f_p}{f_s}\right)^2 = \frac{(k_1 + k_2)^2}{k_1 k_2}$$

Hence $$\left(\frac{2{\cdot}05}{1}\right)^2 = \frac{k_1^2 + 2k_1 k_2 + k_2^2}{k_1 k_2}$$

$$\therefore \quad k_1^2 - 2{\cdot}2\, k_1 k_2 + k_2^2 = 0$$

* *Note:* The stiffness k_4 of a single spring that would replace the two springs in series and produce the same effect is given by

$$\frac{1}{k_4} = \frac{1}{k_1} + \frac{1}{k_2} \quad \text{or} \quad k_4 = \frac{k_1 k_2}{(k_1 + k_2)}$$

Putting $k_1 = 20$ kN/m

gives $400 - 44\,k_2 + k_2^2 = 0$

Solving the quadratic:

$$k_2 = 31{\cdot}15 \text{ or } 12{\cdot}83 \text{ kN/m}$$

Thus, in combination with a spring of stiffness 20 kN/m the stiffness of the other spring must be either 31·15 kN/m or 12·83 kN/m.

Example 10.3

The static deflection at the end of a light uniform cantilever when supporting a mass at its free end is given by $Wl^3/3EI$ where W is the weight of the mass, l the length of the cantilever and EI is the flexural rigidity of the cantilever. If the mass is displaced from its equilibrium position and then released show that it will undergo simple harmonic motion.

A cantilever of length 0·50 m has a static deflection of 30 mm when carrying a given end-mass. What length of cantilever of the same flexural rigidity will be necessary to give a periodic time of 1 s when carrying the same end mass?

Solution

Ler k be the beam stiffness, i.e., the end-load necessary to produce a unit deflection at the end of the cantilever.

If the end-load is displaced by a distance x from its·equilibrium position δ_s, then

Restoring force acting on mass = $-kx$

But from Newton's second law [eq. (8.1)]

$$-kx = ma$$

$$\therefore \quad a = -\frac{k}{m}x$$

Hence, the cantilever will act in the same manner as a spring if the end-load is given an initial displacement and then released and allowed to oscillate.

Then $\omega = \sqrt{\dfrac{k}{m}}$ or $\omega = \sqrt{\dfrac{kg}{W}}$

But $k = \dfrac{W}{\delta_s}$ by definition.

Therefore $\omega = \sqrt{\dfrac{g}{\delta_s}}$

Now, $\delta_s = \dfrac{Wl^3}{3EI}$

$= Cl^3$

where $C = \dfrac{W}{3EI}$ = a constant

$\therefore \quad \omega = \sqrt{\dfrac{g}{Cl^3}}$

and $T = \dfrac{2\pi}{\omega} = 2\pi\sqrt{\dfrac{Cl^3}{g}}$

Now for the cantilever of length 0·50 m:

$$\delta_s = Cl^3$$

$$\frac{30}{10^3} = C\,(0{\cdot}50)^3$$

$$\therefore \quad C = 0{\cdot}24$$

The length of cantilever to give a periodic time of 1 s is obtained from eq. (1); i.e.,

$$1 = 2\pi\sqrt{\frac{0{\cdot}24 \times l^3}{9{\cdot}81}}$$

$$l^3 = \frac{1}{4\pi^2} \times \frac{9{\cdot}81}{0{\cdot}24}$$

$$l = 1{\cdot}011 \text{ m}$$

The required length of cantilever for a periodic time of 1 s is 1·011 m.

Example 10.4

A glass U-tube is mounted with its limbs vertical and contains a liquid. A small pressure is applied to the liquid surface in one of the limbs and then released causing the column of liquid in the tube to oscillate. Show that the liquid moves with simple harmonic motion and that the periodic time is independent of the density of the liquid and of the bore of the tube, but is a function of the overall length of the liquid column.

Hence, calculate the period of oscillation of a column of mercury of 1 m overall length.

Solution

Let l = total length of liquid column;

d = diameter of tube;

ρ = density of liquid.

Figure 10.9(*a*) shows the liquid column at rest – the common liquid surface being represented by the line XX. If the liquid in the right-hand limb is now displaced a distance h below XX, then, since the tube is uniform, the liquid in the left-hand limb will rise a distance h above XX, as shown in Fig. 10.9(*b*).

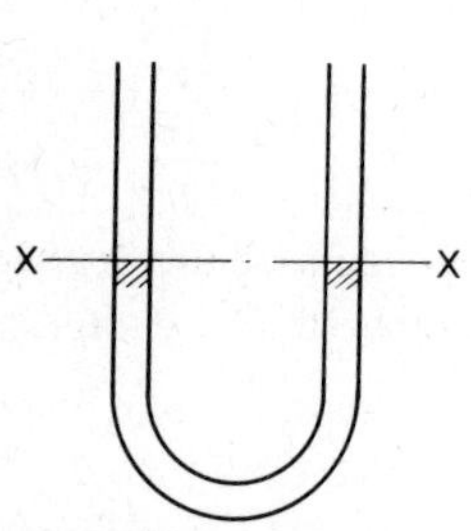

XX- Position of equilibrium before pressure applied

(a)

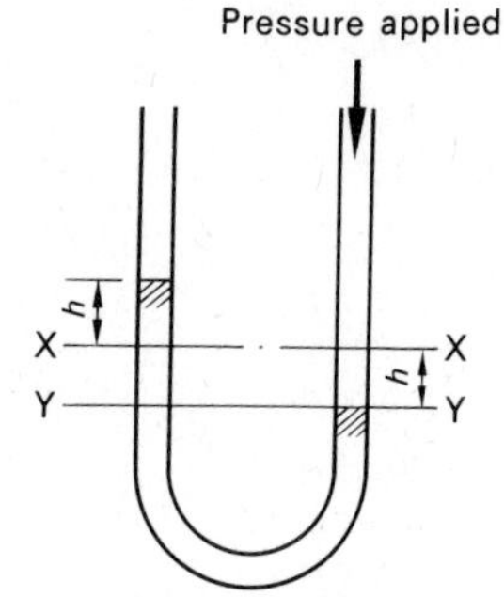

YY- Position of equilibrium with pressure applied

(b)

Fig. 10.9 Example 10.4

Consider now the line YY through the surface of the liquid in the right-hand limb. Whilst the small pressure is acting, equilibrium will exist across YY – the applied force just balancing the force due to the mass of liquid of length $2h$ above YY in the left-hand limb.

When the pressure is removed the force due to the now unbalanced mass of liquid in the left limb will cause the whole liquid column of length l to move.

The restoring force causing motion will be the gravitational force on the mass of liquid column of length $2h$.

$$\text{Mass of liquid column of length } 2h = \text{volume x density}$$

$$= \frac{\pi d^2}{4} \cdot 2h\rho$$

$$\therefore \quad \text{Restoring force} = -\left(\frac{\pi d^2}{4} \cdot 2h\rho\right) g$$

$$\text{Mass of liquid column which is moved by this force} = \frac{\pi d^2}{4} \cdot l\rho$$

Applying Newton's second law:

$$\frac{-\pi d^2}{4} \cdot 2h\rho g = \frac{\pi d^2}{4} \cdot l\rho \times \text{acceleration}$$

$$\therefore \quad \text{Acceleration} = -\left(\frac{2g}{l}\right)h$$

$$= -\text{ constant} \times \text{displacement}$$

This relationship between acceleration and displacement indicates that the oscillating liquid column moves with simple harmonic motion.

Hence, $\omega = \sqrt{\frac{2g}{l}}$

and $T = \frac{2\pi}{\omega} = 2\pi\sqrt{\frac{l}{2g}}$

This proves that the periodic time is independent of the density ρ and the bore d but is a function of the overall length of the liquid column l.

For a column of 1 m length:

$$\text{Periodic time, } T = 2\pi\sqrt{\frac{1}{2 \times 9{\cdot}81}}$$

$$= 1{\cdot}42 \text{ s}$$

10.5 The simple pendulum

A simple pendulum consists of a concentrated mass at one end of a light cord, the other end of which is fixed. If the mass is given a lateral displacement and released it will oscillate in a vertical plane under the action of gravity.

A simple pendulum having a length of cord l is shown in Fig. 10.10 and at a certain instant let the cord be inclined at angle θ to the vertical.

As the mass m moves in a circular path of radius l, the only forces acting on it are the gravitational force mg and the tension in the cord T. For the position shown the gravitational force produces a torque about O tending to restore the mass to its equilibrium position. Thus

$$\text{Restoring torque} = -mgl \sin\theta$$

$$= -mgl\,\theta \quad \text{(for small values of } \theta)$$

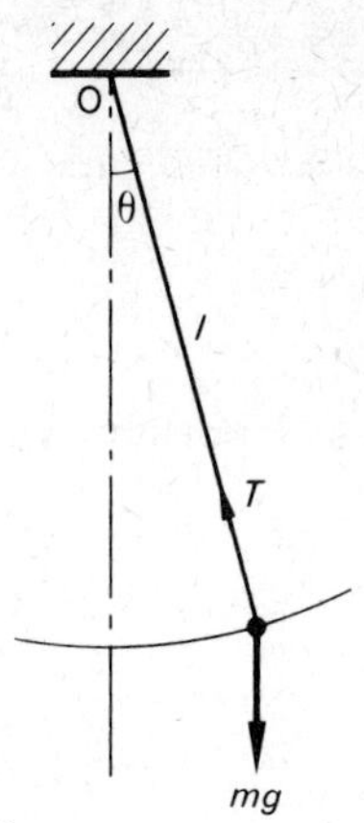

Fig. 10.10 A simple pendulum

For mass m, the moment of inertia about O is

$$I_O = ml^2$$

Applying Newton's second law to the rotation about O:

$$T = I_O\alpha \qquad \text{[eq. (9.1)]}$$

gives

$$-mgl\theta = ml^2\alpha$$

$$\alpha = -\frac{g}{l}\theta \qquad (10.10)$$

This relationship is clearly of the form

Acceleration = – constant x displacement

provided that the angle of swing is small. Therefore the motion of the pendulum is simple harmonic.

Comparing eq. (10.10) with eq. (10.1) gives $\omega = \sqrt{g/l}$.
Then:

$$\text{Periodic time, } T = \frac{2\pi}{\omega} = 2\pi\sqrt{\frac{l}{g}}$$

and

$$\text{frequency, } f = \frac{1}{T} = \frac{1}{2\pi}\sqrt{\frac{g}{l}}$$

These relationships indicate that the periodic time, and hence the frequency, of a simple pendulum is independent of the mass of the suspended body and depends only upon the length of the pendulum and the value of the local gravitational acceleration.

10.6 The compound pendulum

A body which is suspended vertically such that it is capable of free rotation under the action of gravity is termed a compound pendulum.

In Fig. 10.11, O is the 'centre of suspension' and k_O the radius of gyration of the body about an axis through O perpendicular to the plane of motion.

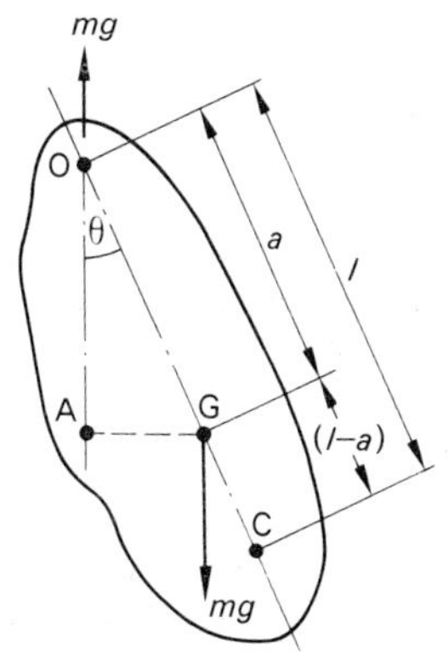

Fig. 10.11 A compound pendulum

G is the centre of mass which is at a distance a from O. Then, as for the simple pendulum:

$$\text{Restoring torque, } T = -mga \sin\theta$$
$$= -mga\,\theta \quad \text{(for small values of } \theta\text{)}$$

Applying $$T = I_O \alpha$$

gives $$-mga\theta = mk_O^2 \alpha$$

$$\alpha = -\frac{ga}{k_O^2}\theta$$

This is again simple harmonic motion, for small values of θ, and comparison with eq. (10.1) gives

$$\omega = \sqrt{\frac{ga}{k_O^2}}.$$

Hence $$T = 2\pi\sqrt{\frac{k_O^2}{ga}}$$

and $$f = \frac{1}{2\pi}\sqrt{\frac{ga}{k_O^2}}$$

If k_G is the radius of gyration of the body about an axis through its centre of mass, perpendicular to the plane of motion then from the parallel axis theorem (section 4.5):

$$I_O = I_G + ma^2$$

or, $mk_C^2 = mk_G^2 + ma^2$

and $k_O^2 = k_G^2 + a^2$

Substituting for k_O into the above equations for T and f gives:

$$T = 2\pi\sqrt{\frac{k_G^2 + a^2}{ga}} \quad \text{and} \quad f = \frac{1}{2\pi}\sqrt{\frac{ga}{k_G^2 + a^2}} \tag{10.11}$$

Comparison with the expression for a simple pendulum gives the length l of a simple pendulum which would have the same periodic time and frequency as the compound pendulum, i.e.,

$$l = \frac{k_G^2 + a^2}{a}$$

$$= \frac{k_G^2}{a} + a$$

Hence, if point C is taken on Fig. 10.11 such that the length OC is equal to that of a simple pendulum which has the same periodic time, then

$$OC = l = \frac{k_G^2}{a} + a \tag{10.12}$$

The point C is known as the 'centre of oscillation' or 'centre of percussion' (see section 9.3).

If the centres of oscillation and suspension are now interchanged and the pendulum is allowed to oscillate about an axis through C parallel to the original axis through O then the length CG now corresponds to the length a in the original equations (see Fig. 10.11).

Thus $(l - a)$ must now be substituted for a.

Then $\quad T = 2\pi\sqrt{\dfrac{k_G^2 + (l - a)^2}{g(l - a)}}$

But from eq. (10.12)

$$l - a = \frac{k_G^2}{a}$$

$$\therefore \quad T = 2\pi\sqrt{\frac{k_G^2 + (k_G^2/a)^2}{g(k_G^2/a)}}$$

$$= 2\pi\sqrt{\frac{a^2 + k_G^2}{ag}}$$

which is identical with the periodic time for oscillation about the axis through O. This shows that the centres of oscillation and suspension are interchangeable.

Example 10.5

Show that the periodic time for a compound pendulum is a minimum when the centre of suspension is at a distance equal to k_G from the centre of mass.

A connecting rod AB oscillates at 1 Hz when suspended from a horizontal knife-edge at A (see Fig. 10.12). The centre of mass of the

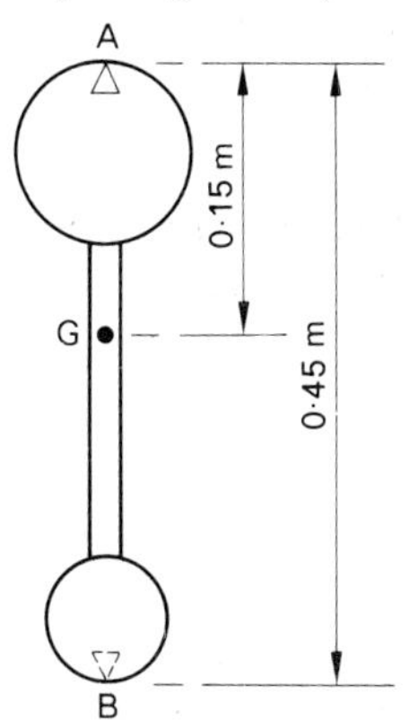

Fig. 10.12 Example 10.5

rod is 0·15 m from the knife-edge at A. Determine the radius of gyration of the rod about an axis through the centre of mass parallel to the knife-edge. Also calculate the frequency of small angular oscillations when the rod is suspended at B. The length of the connecting rod AB is 0·45 m.

Solution

A diagram is given in Fig. 10.12
From eq. (10.11)

$$T = 2\pi\sqrt{\frac{k_G^2 + a^2}{ga}}$$

where a represents the distance of the centre of mass from the centre of suspension.

If T is to be a minimum then

$$\frac{k_G^2 + a^2}{a}$$

must be a minimum (since g is a constant).

Then by differentiation of

$$\frac{k_G^2 + a^2}{a} \text{ with respect to } a$$

$$\frac{dT}{da} = \frac{-k_G^2}{a^2} + 1$$

$$= \text{zero for a minimum}$$

i.e., $k_G = a$

Hence the periodic time will be a minimum when the centre of suspension is at a distance equal to k_G from the centre of mass.

From eq. (10.11)

$$\text{frequency}, f = \frac{1}{2\pi}\sqrt{\frac{ga}{k_G^2 + a^2}}$$

Now $a = 0{\cdot}15$ m when suspended from A and the frequency is 1 Hz.

$$\therefore \qquad 1 = \frac{1}{2\pi}\sqrt{\frac{9{\cdot}81 \times 0{\cdot}15}{k_G^2 + 0{\cdot}15^2}}$$

Hence $k_G^2 = 0{\cdot}01477$

and $k_G = 0{\cdot}121$ m

When suspended from B, $a = 0{\cdot}30$ m (Fig. 10.12)

$$\therefore \qquad f = \frac{1}{2\pi}\sqrt{\frac{9{\cdot}81 \times 0{\cdot}3}{0{\cdot}01477 + 0{\cdot}3^2}}$$

$$= 0{\cdot}844 \text{ Hz}$$

Problems

1. The ram of a pump may be assumed to move with simple harmonic motion. Find the amplitude or half travel of the ram if, for velocities of 2·5 m/s and 2 m/s, the displacements from the mid-position are 50 mm and 60 mm respectively.

2. A body of mass 50 kg moves along a straight line with simple harmonic motion. At distances of 0·3 m and 0·6 m from the mid-point of oscillation the velocities of the body are 0·9 m/s and 0·6 m/s respectively. Determine:

(*a*) the amplitude of the motion;
(*b*) the time for one complete oscillation of the body;
(*c*) the force acting on the body when it is 0·6 m from the extremity of the oscillation.

3. Part of a machine has a reciprocating motion which is simple harmonic making 200 complete oscillations in one minute. The mass of the part is 5 kg and it has a stroke of 220 mm. Find:

(*a*) the accelerating force acting upon it and its velocity when it is 60 mm from mid-stroke;
(*b*) the maximum accelerating force;
(*c*) the maximum velocity.

4. A piston moving with simple harmonic motion passes through two points A and B, 400 mm apart with the same velocity having taken 2 seconds in passing from A to B. Three seconds later it returns to point B. Determine:

(*a*) the period and amplitude of the oscillation;
(*b*) the maximum acceleration of the piston.

5. A body of mass 25 kg moves with simple harmonic motion of frequency 2 Hz. The maximum velocity attained by the body is 4·5 m/s. Determine:

(*a*) the amplitude of the motion;
(*b*) the displacement of the body from the mid-position when the force acting on the body is 750 N;
(*c*) the time taken by the body to move from the extremity of the motion to a point mid-way between the extreme and the mid-position of the motion.

6. An engine valve is operated by a cam which imparts simple harmonic motion to the valve. The mass of the moving parts is 0·6 kg and the cam speed is 300 rev/min. The valve opens and closes completely in 60° of cam rotation and the valve lift is 12 mm. Determine:

(*a*) the maximum velocity of the valve;
(*b*) the velocity when the valve has lifted 3 mm;
(*c*) the maximum acceleration of the valve;
(*d*) the acceleration when the valve has lifted 3 mm;
(*e*) the maximum accelerating force acting on the valve.

7. A vertical helical spring having a stiffness of 800 N/m is clamped at its upper end and carries a mass of 3 kg attached to the lower end. The mass is displaced vertically a distance of 30 mm from the equilibrium position and released. Determine the period of the resultant vibration and the maximum velocity attained by the mass.

8. A helical spring is suspended from a rigid support and a mass of 0·9 kg is attached to its lower end causing the spring to extend 100 mm. This mass is then removed and replaced by a body of unknown mass. When this body is allowed to oscillate vertically it is found to complete 72 vibrations in one minute. Calculate the value of the unknown mass and determine its maximum acceleration given that its maximum displacement from the equilibrium position is 150 mm.

9. A mass of 15 kg is supported by two close coiled helical springs connected in series. The stiffness of one spring is 1800 N/m. When the mass is displaced 0·05 m from its equilibrium position and then released 30 oscillations are made in 28 s. Determine the stiffness of the other spring and the maximum velocity of the oscillating mass.

10. A cantilever spring of negligible weight is 0·35 m long and carries a mass 2·5 kg at the free end. When an additional mass of 10 kg is applied to the free end the cantilever is deflected downwards 24 mm from the equilibrium position. If this additional mass is suddenly removed determine the period of vibration of the 2·5 kg mass and also its maximum velocity.

11. Show that the period of vibration T, of a simple pendulum of length l, is given by $T = 2\pi\sqrt{l/g}$. Hence or otherwise deduce the length of a simple pendulum having a period of 2 s.

12. A mass is suspended from the lower end of a cord of length 1 m which is attached to a rigid support at its upper end. If the mass is displaced through an angle of 5° from its vertical position and then released determine:

(*a*) the periodic time of oscillation;
(*b*) the acceleration when its velocity is zero;
(*c*) the acceleration when its displacement is zero;
(*d*) the length of the pendulum necessary to increase the periodic time by 15%.

13. A small flywheel of mass 90 kg was suspended in a vertical plane as a compound pendulum. When the distance to the centre of mass from the knife edge was 250 mm the flywheel made 100 oscillations in 127 s. Determine the value of the moment of inertia of the flywheel about an axis through the centre of mass.

14. A connecting rod of mass 40 kg and 0·8 m between centres is suspended vertically and allowed to oscillate. The time for 60 oscillations is found to be 92 s when the axis of oscillation coincides with the small end centre and 88 s when it coincides with the big end centre. Find:

(*a*) the distance of the centre of mass from the small end centre;
(*b*) the moment of inertia of the rod about an axis through the centre of mass.

15. Show that the radius of gyration of a uniform rod of length l about an axis through the centre of mass and perpendicular to its length is $l/2\sqrt{3}$. A uniform rod is 2 m long and is oscillated as a compound pendulum about axes perpendicular to its length. Calculate:

(*a*) the periodic time of swing when the axis is 0·7 m from one end;
(*b*) the position of the axis for the periodic time to be a minimum;
(*c*) the minimum periodic time.

11

Fluids at rest

11.1 Fluids

Fluids are divided into liquids and gases. Liquids offer great resistance to change of volume, their shape being that of the containing vessel. If the volume of liquid is less than that of the vessel there will be a horizontal free surface. Gases, on the other hand, can be compressed relatively easily and will always completely fill the vessels which contain them. If free to expand, gases will expand indefinitely.

11.2 Liquids and solids

When a solid is subjected to a shear force, as in the case of a loaded beam, a definite distortion occurs and the internal resistance offered by the beam material balances the external shear force. Unless overstrained, the beam will return to its original shape when the external force is removed. It is well known that this is not the case for a liquid. The application of a shear force, however small, will cause the liquid to flow and it does not regain its original shape when the shear force is removed. Hence the distinction between solid and liquid – a solid can support shear forces whereas a liquid is incapable of withstanding tangential forces. It is for this latter reason that the force exerted on any surface in a liquid at rest is always perpendicular to that surface.

11.3 Density and relative density

The density ρ (rho) of a liquid is its mass per unit volume. The basic unit of density is therefore the kilogramme per cubic metre (kg/m^3). Other forms frequently used are kg/litre and $tonne/m^3$, where
1 litre = 10^{-3} m^3 and 1 $tonne/m^3$ = 1 Mg/m^3

The density of water is 1000 kg/m^3 or 1 Mg/m^3.

The relative density of a liquid is the ratio of the density of that liquid to the density of water. Thus relative density has no units.

11.4 Intensity of pressure and thrust

If a force or thrust F is applied uniformly over an area A to a liquid in a container the intensity of pressure p is defined as

$$p = \frac{F}{A}$$

(compare with the definition of stress given in chapter 3). When F is not applied uniformly across the area this relationship gives an average value.

In engineering, the term intensity of pressure is often abbreviated to pressure and the term pressure is used when total force is clearly intended. To avoid any misunderstanding in this text the term **pressure** will be used in place of intensity of pressure to mean force per unit area, and the term **thrust** will denote force only.

11.5 Pressure at a point in a liquid

It has already been mentioned that the force exerted on any surface in a liquid at rest can only act normal to that surface. This fact can be used to show that the pressure at a point in a liquid is the same in all directions.

Consider a small right-angled triangular prism of liquid within a body of liquid at rest as shown in Fig. 11.1.

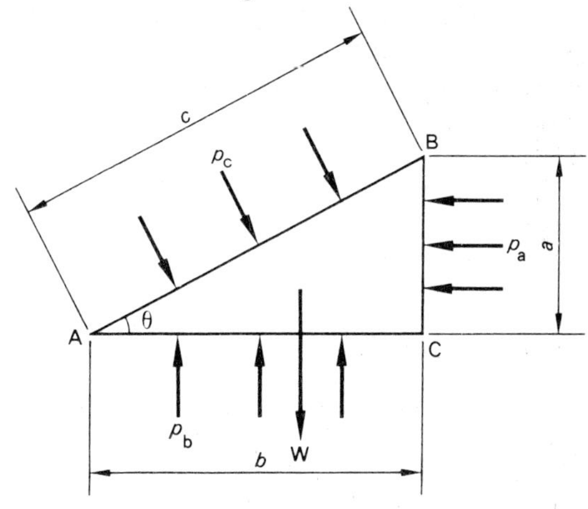

Fig. 11.1

Let the lengths of the faces of the triangle be a, b, and c and the average pressures on these faces be p_a, p_b, and p_c respectively as shown.

The forces on the faces will then be $p_a a$, $p_b b$ and $p_c c$ per unit length of the prism.

As the liquid is at rest the element is in equilibrium under its gravitational force and the forces acting normally to the three faces.
The mass of the prism is $\frac{1}{2}\rho ab$ per unit length.
The gravitational force is therefore $\frac{1}{2}\rho gab$.

Resolving forces horizontally:

$$p_c c \sin\theta - p_a a = 0$$

But $$a = c \sin\theta$$

$\therefore$ $$p_c = p_a$$

Resolving forces vertically:

$$p_b b - p_c c \cos\theta - \tfrac{1}{2}\rho gab = 0$$

But $$b = c \cos\theta$$

$\therefore$ $$p_b = p_c + \tfrac{1}{2}\rho ga$$

Now if the prism is made smaller until in the limit, $a \to 0$, the term $\frac{1}{2}\rho ga$ tends to zero. Thus in the limit

$$p_a = p_b = p_c$$

Thus the pressure at a point in a liquid is the same in all directions.

11.6 Variation of pressure with depth

To determine the pressure at a depth h below the free surface of a liquid at rest, consider an imaginary cylinder of liquid as shown in Fig. 11.2(*a*).

For vertical equilibrium the weight of the liquid column must be balanced by the pressure difference across the ends of the cylinder.

Weight of column of liquid, $W = \rho g Ah$

Thus for vertical equilibrium:

$$pA = \rho g Ah + p_a A$$

where p_a is the atmospheric pressure at the surface.

$\therefore$ $$p = p_a + \rho gh \qquad (11.1)$$

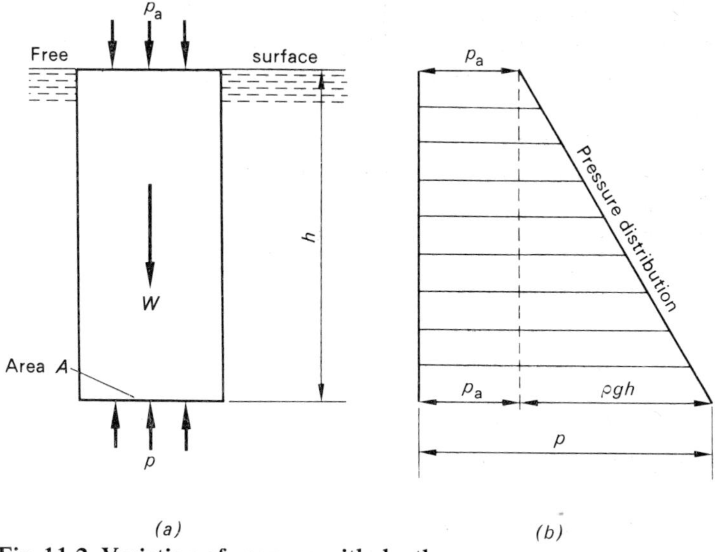

Fig. 11.2 Variation of pressure with depth

As liquids are generally regarded as being incompressible, their density is assumed to remain constant and therefore the pressure p within the liquid is shown to increase linearly with depth h.

Equation (11.1) indicates that the atmospheric pressure p_a is transmitted uniformly throughout the depth h as shown in Fig. 11.2(*b*).

11.7 Measurement of pressure

Engineers are usually interested in pressures relative to the atmospheric pressure and instruments of the Bourdon gauge type* record the pressure above atmospheric, i.e., the gauge pressure.

Thus, the ρgh term in eq. (11.1) represents the gauge pressure. The absolute pressure is the sum of the atmospheric pressure and the gauge pressure; i.e.,

$$\text{Absolute pressure} = \text{atmospheric pressure} + \text{gauge pressure}$$

$$p = p_a + p_g$$

where $$p_g = \text{gauge pressure} = \rho gh$$

* See Jackson, A., *Mechanical Engineering Science 01* (Chapter 10), Longman, Harlow, 1971.

Not all pressure measurements are made with Bourdon gauges, and extensive use is made of manometers or U-tube type instruments. These employ the principle of balancing a static liquid column against the pressure to be measured. When pressure is measured in this way it is usually expressed in terms of the height or head h of the fluid which produces the pressure.

Hence from $\quad p = \rho g h$

$$h = \frac{p}{\rho g}$$

h being referred to as the pressure head.

The piezometer is the simplest type of manometer and is illustrated in Fig. 11.3. It consists of an open-ended transparent tube inserted into the apparatus at the point where the fluid pressure is to be determined.

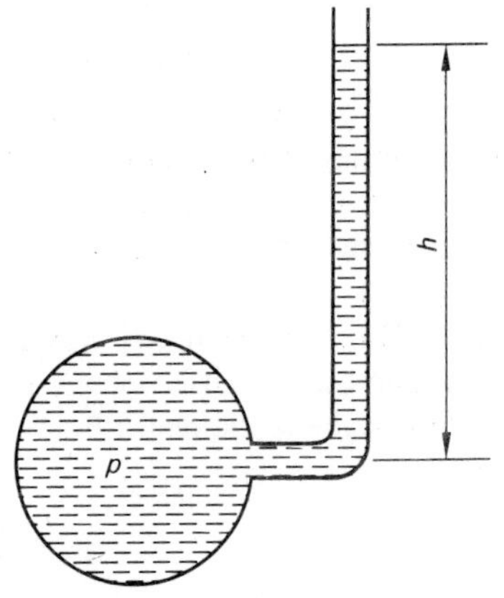

Fig. 11.3 The Piezometer

The fluid in the tube rises to a height h and thus the pressure p is equivalent to the head h of the fluid.

This simple instrument is not suitable for gases or for high pressures which would clearly require a long length of tube.

An alternative arrangement for the measurement of higher pressure is shown in Fig. 11.4.

The manometer contains a different fluid from that whose pressure is required. Usually water is used if air pressure is to be measured or mercury for the measurement of water pressure.

Before the manometer is connected to the tapping point on the apparatus the fluid in both legs of the manometer will be at the same level XX.

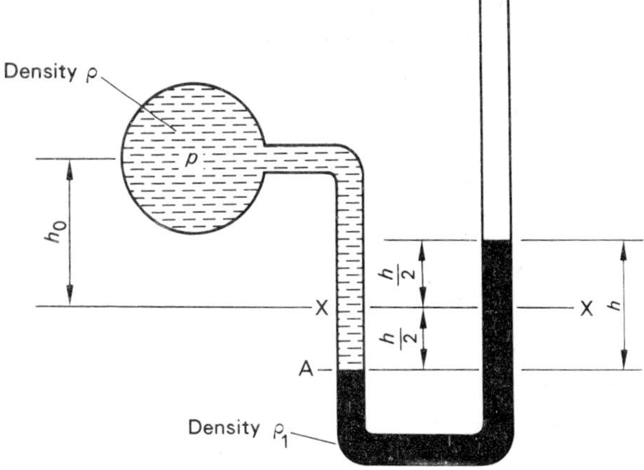

Fig. 11.4 A manometer

The application of the pressure p will depress the fluid in the left leg of the manometer by $h/2$ and produce a corresponding rise in the right-hand leg. The difference in level h of the two surfaces in the manometer then gives the gauge pressure at surface A; i.e.,

$$p_A = \rho_1 gh$$

This pressure is greater than the pressure p in the pipe since the point A is below the centre of the pipe.

Thus $\quad p = \rho_1 gh - \rho g(h/2 + h_0)$

Then $\quad p = gh\left(\rho_1 - \frac{\rho}{2}\right) - \rho g h_0$

or $\quad \frac{p}{\rho g} = h\left(\frac{\rho_1}{\rho} - \frac{1}{2}\right) - h_0$

The pressure p in the pipe is therefore given in terms of the difference in level h – the other terms in the equation being constants.

This manometer arrangement can also be used to determine the difference in pressures between two points in a piece of apparatus as illustrated in Fig. 11.5.

For this arrangement

$$p_1 - p_2 = gh(\rho_1 - \rho) + \rho g h_0$$

or differential pressure head is

$$\frac{p_1 - p_2}{\rho g} = h\left(\frac{\rho_1}{\rho} - 1\right) + h_0 \tag{11.2}$$

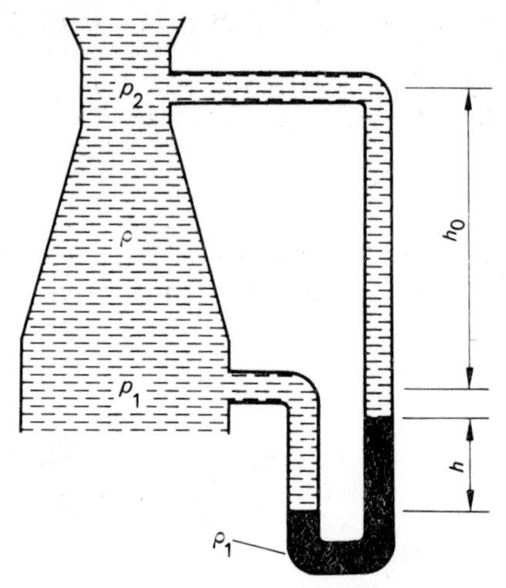

Fig. 11.5 Difference in pressures measured by a manometer

In this case h_0 is again a constant and is the difference in height between the tapping points.

Note: If a mercury filled manometer is used to measure the pressure of water then the ratio

$$\frac{\rho_1}{\rho} = 13{\cdot}6$$

Equation (11.2) then becomes

$$\frac{p_1 - p_2}{\rho g} = 12{\cdot}6\, h + h_0$$

11.8 Atmospheric pressure and equivalent head

Standard atmospheric pressure is 101·3 kN/m^2. This may be expressed as an equivalent head of water by applying the equation $p = \rho g h$ where ρ, the density of water, is 1000 kg/m^3.

Then $101{\cdot}3 \times 10^3\ [\mathrm{N/m^2}] = 10^3\ [\mathrm{kg/m^3}] \times 9{\cdot}81\ [\mathrm{m/s^2}] \times h$

$\therefore \quad h = 10{\cdot}33$ m of water

Atmospheric pressure can also be expressed as an equivalent head of mercury. Since mercury has an approximate relative density of 13·6, then 10·33 m of water is equivalent to

$$\frac{10{\cdot}33}{13{\cdot}6} = 0{\cdot}76 \text{ m} = 760 \text{ mm of mercury}$$

Therefore we have the following alternative methods of expressing atmospheric pressure:

$101{\cdot}3\ \text{kN/m}^2$, or $10{\cdot}33$ m of water or 760 mm of mercury.

Example 11.1

A uniform upright tube, 10 mm diameter, contains a length of 180 mm of water above a 50 mm length of mercury. What is the pressure at the base of the tube? The relative density of mercury is 13·6.

Solution

The pressure at the common surface OO is due to the 180 mm column of water and this pressure is transmitted through the mercury as shown by the pressure distribution in Fig. 11.6.

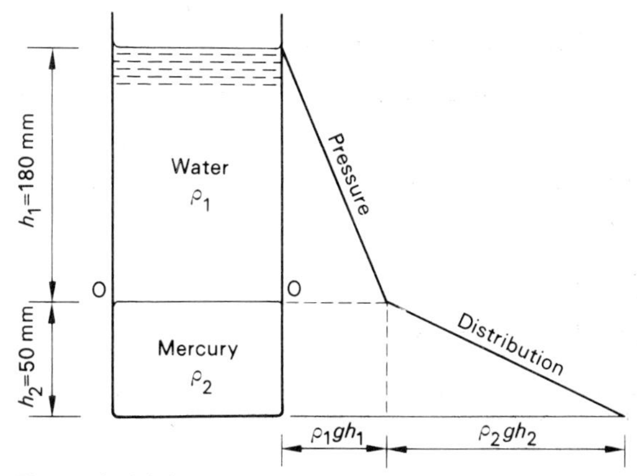

Fig. 11.6 Example 11.1

$$\text{Total pressure at the base} = \rho_1gh_1 + \rho_2gh_2$$

$$= 10^3 \times 9{\cdot}81 \times \frac{180}{10^3}$$

$$+ 13{\cdot}6 \times 10^3 \times 9{\cdot}81 \times \frac{50}{10^3}$$

$$= 8437\ \text{N/m}^2$$

It should be noted that the diameter of the tube does not appear in the calculation. This is because pressure is concerned with height only and hence the diameter is irrevelant to this question.

Example 11.2

When one leg of an open mercury U-tube is connected to the centre of a pipe containing water under pressure, the mercury column is deflected 200 mm. If the centre of the pipe is 350 mm below the level of the lower mercury meniscus, what is the pressure head in the pipe? If the lower mercury meniscus is now made to coincide with the centre of the pipe, how is the manometer reading affected?
Assume the atmospheric pressure to be 760 mm of mercury.

Solution

Let the pressure head in the pipe be h [m] of water.

Referring to Fig. 11.7(a) the equilibrium condition (since the mercury column is continuous) is

pressure at A = pressure at B

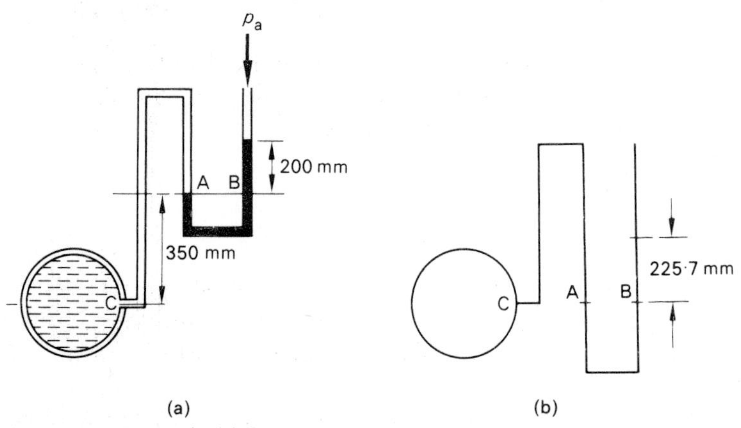

Fig. 11.7 Example 11.2

But the pressure at B is that due to a column of mercury (Hg) 200 mm long plus atmospheric pressure.

$$\therefore \quad \text{pressure head at A} = 200 + 760 = 960 \text{ mm Hg}$$

$$= \frac{960}{1000} \times 13{\cdot}6 = 13{\cdot}06 \text{ m of water}$$

Now the pressure at the centre of the pipe supports the pressure at A together with a column of water, CA of height 350 mm.

$$\text{Hence,} \quad \text{pressure head in pipe} = 13{\cdot}06 + \frac{350}{1000}$$

$$= 13{\cdot}41 \text{ m of water}$$

Note: 13·41 m of water is equivalent to a pressure of

$$\frac{13{\cdot}41}{10{\cdot}33} \times 101{\cdot}3 \times 10^3 = 131{\cdot}5 \times 10^3 \text{ N/m}^2$$

Since atmospheric pressure has been included in the calculation this is the absolute pressure in the pipe. If the pressure has been measured by a Bourdon gauge placed central with the pipe the gauge reading would be

$$131{\cdot}5 \times 10^3 - 101{\cdot}3 \times 10^3 = 30{\cdot}2 \times 10^3 \text{ N/m}^2$$

If point A is now made to coincide with the centre line of the pipe, as in Fig. 11.7(*b*), the pressure in the pipe would still remain the same. However, the mercury column would now have a total deflection of 200 mm plus the deflection due to a 350 mm length of water column.

Hence, $\quad$ manometer reading $= 200 + \dfrac{350}{13{\cdot}6} = 225{\cdot}7$ mm Hg

This example illustrates that the position of the fluid level in a manometer must be considered in relation to the position of the tapping point in the container if a true pressure measurement is required.

11.9 Thrust on an immersed surface

Having established the variation of pressure with depth and also that the thrust exerted on any surface in a liquid is always perpendicular to that surface we can now determine the total thrust exerted on an immersed surface.

Consider a uniform plate immersed in a liquid as shown in Fig. 11.8. OO is the line of intersection of the plane of the plate continued with the free surface, the two planes being inclined at angle θ.

For an elemental strip of area δA distance y from OO,

Normal pressure p on elemental area $= \rho g h$

$$\text{Thrust on element} = \rho g h\,.\,\delta A \text{ (pressure} \times \text{area)}$$

But

$$h = y \sin \theta$$

$\therefore$

$$\text{Thrust on elemental area} = \rho g y \sin \theta\,.\,\delta A \qquad (11.3)$$

The total thrust exerted on the whole plate will be the sum of all such elemental thrusts.

Hence, $\quad$ total thrust on plate $F = \Sigma \rho g y \sin \theta\,.\,\delta A$

$$= \rho g \sin \theta\, \Sigma y\,.\,\delta A$$

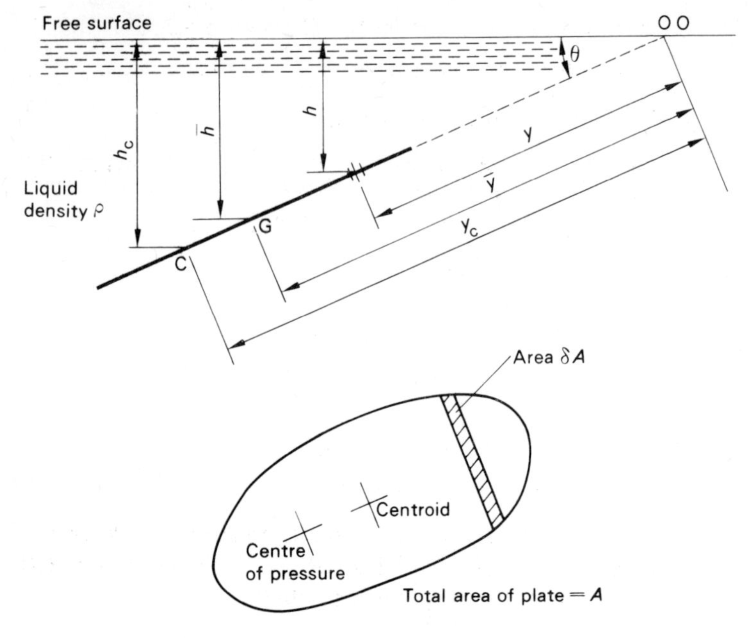

Fig. 11.8 Thrust on an immersed surface

But $\Sigma y \cdot \delta A$ = first moment of area of the whole plate about OO.

$$= A\bar{y}$$

where $\bar{y}$ is the distance of the centroid from OO.

$$\therefore \qquad F = \rho g \sin\theta \cdot A\bar{y} \qquad (11.4)$$

Replacing $\bar{y} \sin\theta$ by $\bar{h}$, the depth of the centroid below the free surface, gives:

Total thrust $F = \rho g A\bar{h}$

It is thus shown that the total thrust on an immersed surface is given by the product of the pressure at the centroid and the total area of surface. The thrust is therefore independent of the inclination of the immersed surface provided that its centroid remains at the same depth.

11.10 Centre of pressure

The position where the total thrust on an immersed surface acts will now be determined.

From eq. (11.3) the thrust on the elemental area $\delta A = \rho g y \sin\theta \cdot \delta A$. The moment of this thrust about OO $= \rho g y \sin\theta \cdot \delta A \times y$

Then the total moment of thrust for the whole plate $= \rho g \sin\theta \Sigma y^2 . \delta A$

But $\Sigma y^2 . \delta A$ is the second moment of area I_O of the plate about OO.

$\therefore$ Total moment of thrust $= \rho g \sin\theta . I_O$

Now for equilibrium about OO this moment must be equal to the moment of F about OO.

Let F act at the centre of pressure distance y_c from OO.

Then $$F . y_c = \rho g \sin\theta I_O$$

From eq. (11.4) $$F = \rho g \sin\theta . A\bar{y}$$

$\therefore$ $$y_c = \frac{I_O}{A\bar{y}}$$

$$= \frac{\text{second moment of area about OO}}{\text{first moment of area about OO}}$$

It is important to remember that in this expression both the first and second moments of area are calculated about the axis through OO.

Alternatively, by using the parallel axis theorem:

$$I_O = I_G + A\bar{y}^2$$

where I_G is the second moment of area about the centroid,

$$y_c = \frac{I_G + A\bar{y}^2}{A\bar{y}} = \frac{I_G}{A\bar{y}} + \bar{y}$$

This clearly shows that the centre of pressure is below the centroid of the section by an amount $I_G/A\bar{y}$.

A further simplification, using $I_G = Ak_G^2$, where k_G is the radius of gyration, gives:

$$y_c = \frac{k_G^2}{\bar{y}} + \bar{y}$$

But $$\bar{h} = \bar{y} \sin\theta$$

$\therefore$ $$y_c = \frac{k_G^2}{\bar{h}} \sin\theta + \frac{\bar{h}}{\sin\theta}$$

If the plate is vertical then $\sin\theta = 1$ and $y_c = h_c$.

Then $$h_c = \frac{k_G^2}{\bar{h}} + \bar{h}$$

where $\bar{h}$ is the vertical depth of the centroid below the free surface.

Two important cases will now be considered.

1. A vertical rectangular plate with one edge at the free surface.
For a rectangular vertical surface with its top edge at the water surface $\theta = 90°$ and $y_c = h_c$ (see Fig. 11.8).

Then from the equation $y_c = I_O/A\bar{y}$, I_O corresponds to I about the base of a rectangle.

Hence, for the plate shown in Fig. 11.9(*a*).

$$I_O = \frac{bd^3}{3}, \quad A = bd, \quad \bar{y} = \bar{h} = \frac{d}{2}$$

Then $$y_c = h_c = \frac{bd^3}{3} \Big/ bd \cdot \frac{d}{2} = \frac{2d}{3}$$

The depth of the centre of pressure is $\frac{2}{3}$ of the length of the vertical below the free surface.

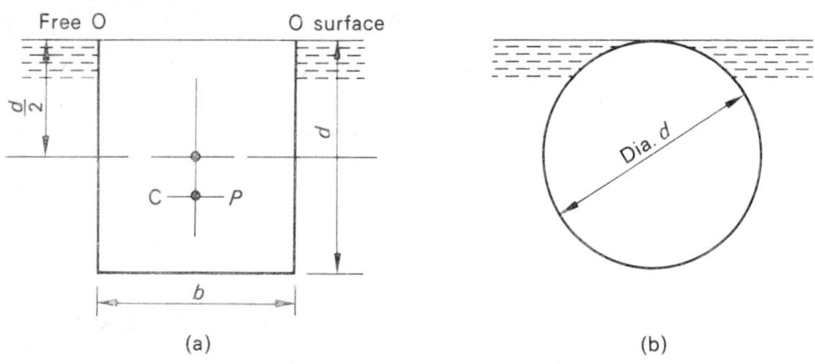

Fig. 11.9 Centre of pressure

2. A vertical circular plate with edge in free surface, Fig. 11.9(*b*).
Using the parallel axis equation

$$y_c = \frac{I_G}{A\bar{y}} + \bar{y}$$

$$I_G = \frac{\pi d^4}{64}, \quad A = \frac{\pi d^2}{4}, \quad \bar{y} = \frac{d}{2}$$

Then $$y_c = h_c = \left[\frac{\pi d^4}{64} \Big/ \left(\frac{\pi d^2}{4} \cdot \frac{d}{2}\right)\right] + \frac{d}{2} = \frac{5}{8} d$$

The depth of the centre of pressure is $\frac{5}{8}d$ below the free surface.

Example 11.3

A sluice gate is 6 m wide and has water to a depth of 8 m on one side and 3 m on the other. Find the resultant horizontal thrust on the gate and the position of its line of action from the bottom.
The density of water is 1000 kg/m^3.

Solution

Referring to Fig. 11.10:

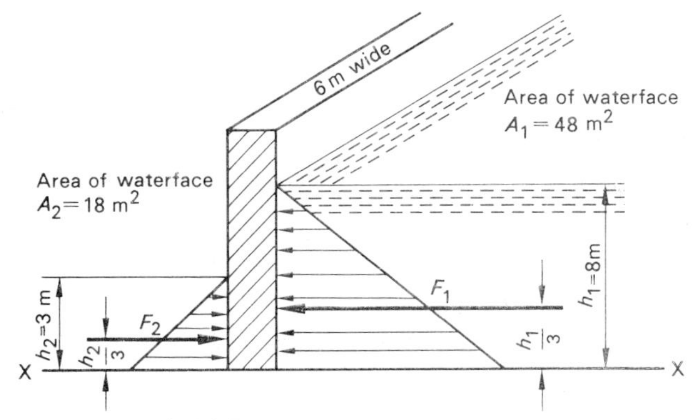

Fig. 11.10 Example 11.3

Horizontal thrust on deep side, $F_1 = \rho g A_1 \bar{h}_1$

$$= 10^3 \times 9{\cdot}81 \times (6 \times 8) \times 4$$

$$= 1\,883\,520 \text{ N}$$

$$= 1883{\cdot}52 \text{ kN}$$

Horizontal thrust on shallow side, $F_2 = \rho g A_2 \bar{h}_2$

$$= 10^3 \times 9{\cdot}81 \times (6 \times 3) \times 1{\cdot}5$$

$$= 264\,870 \text{ N}$$

$$= 264{\cdot}87 \text{ kN}$$

Therefore resultant horizontal thrust on the gate

$$= 1883{\cdot}52 - 264{\cdot}87$$

$$= 1618{\cdot}65 \text{ kN}$$

Thrust F_1 acts at $\frac{h_1}{3} = \frac{8}{3}$ m from XX

$$\text{Thrust } F_2 \text{ acts at } \frac{h_2}{3} = 1 \text{ m from XX}$$

$$\text{Moment of } F_1 \text{ about XX} = 1883{\cdot}52 \times \frac{8}{3} \text{ kN m}$$

$$\text{Moment of } F_2 \text{ about XX} = 264{\cdot}87 \times 1 \text{ kN m}$$

Let the resultant thrust act at distance y from XX.

Then, $$1618{\cdot}65 \times y = 1883{\cdot}52 \times \frac{8}{3} - 264{\cdot}87$$

$$y = \frac{4757{\cdot}85}{1618{\cdot}65} = 2{\cdot}94 \text{ m}$$

The resultant thrust is 1618·65 kN and acts at a distance of 2·94 m from the bottom of the gate.

Example 11.4

A flap valve (Fig. 11.11) 0·6 m diameter, closes the end of a horizontal pipe against internal water pressure, the head of water over the centre line of the pipe being 2 m. The valve is held by a hinge, whose pin is

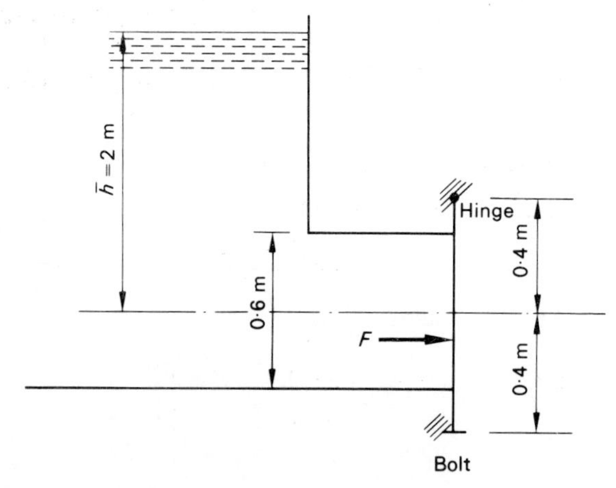

Fig. 11.11 Example 11.4

0·4 m above the centre of the flap, and by a bolt which is 0·4 m below the centre of the flap. Determine the forces acting on the hinge and the bolt.
The density of water is 1000 kg/m^3.

Solution

The flap valve will be in equilibrium under the forces at the hinge and bolt and the total thrust F acting at the centre of pressure.

$$\text{Total thrust } F = \rho g A\bar{h}$$
$$= 10^3 \times 9{\cdot}81 \times \frac{\pi}{4}(0{\cdot}6)^2 \times 2$$
$$= 5547 \text{ N}$$

This thrust acts at the centre of pressure which is at a distance below the free surface given by

$$\bar{h} + \frac{I_G}{A\bar{h}},$$

or alternatively, $I_G/A\bar{h}$ below the centre line of the pipe.
For a circular disc $I_G = \pi d^4/64$.

Hence the depth of the centre of pressure below the centre line of the pipe $= \dfrac{I_G}{A\bar{h}} = \dfrac{\dfrac{\pi d^4}{64}}{\dfrac{\pi d^4}{4} \times 2}$

$$= \frac{d^2}{32} = \frac{0{\cdot}36}{32} = 0{\cdot}0112 \text{ m}$$

The force on the bolt, F_B, can be found by taking moments about the hinge; i.e.,

$$F_B \times 0{\cdot}8 = F \times (0{\cdot}4 + 0{\cdot}0112)$$

But the thrust on the valve, $F = 5547$ N

$$\therefore \qquad F_B = \frac{5547 \times 0{\cdot}4112}{0{\cdot}8}$$
$$= 2851 \text{ N}$$

and $\qquad$ Force on hinge, $F_H = 5547 - 2851$
$$= 2696 \text{ N}$$

Example 11.5

A storage tank is to have sides sloping at 30° to the horizontal and retain water to a depth of 1·8 m. The sides of the tank are to be supported at 3-m intervals by timber struts normal to them. These struts are to meet the tank at a point 2 m from the ground along the sloping side. Determine the thrust carried by each strut.

Solution

A diagram is given in Fig. 11.12.

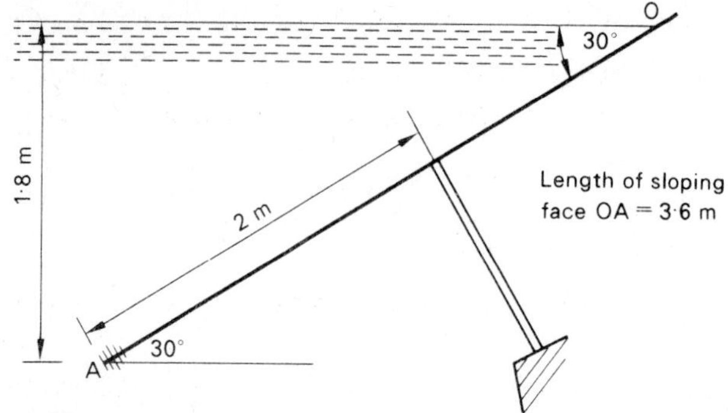

Fig. 11.12 Example 11.5

Area of sloping face of tank supported by each strut = 3·6 x 3

= 10·8 m²

Thrust on side of tank, $F = \rho g A \bar{h}$

where $\bar{h}$ = 0·9 m below free surface

$\therefore$ $F = 10^3 \times 9{\cdot}81 \times 10{\cdot}8 \times 0{\cdot}9$

= 95 353 N

This thrust will act at the centre of pressure distance y_c from O given by,

$$y_c = \frac{I_G}{A\bar{y}} + \bar{y}$$

where $\bar{y}$ is the distance to the centroid from O measured along the slope.

$$\therefore \quad y_c = \left[\frac{3 \times 3{\cdot}6^3}{12} \Big/ 10{\cdot}8 \times 1{\cdot}8 \right] + 1{\cdot}8$$

$= 0{\cdot}6 + 1{\cdot}8 = 2{\cdot}4$ m from O

The thrust taken by each strut may be found by taking moments about A.

Thrust on strut x 2 = $F \times (3{\cdot}6 - 2{\cdot}4)$

$$\text{Thrust on strut} = \frac{95\,353 \times 1{\cdot}2}{2}$$

= 57 212 N

Problems

The following values should be used:

Density of water 1000 kg/m^3. Relative density of mercury 13·6.

1. A tube is filled with water to a depth of 600 mm and then 450 mm of oil of relative density 0·72 is poured in and allowed to come to rest. Determine the pressure:

(*a*) at the common liquid surface;
(*b*) at the base of the tube.

2. One leg of a vertical U-tube containing mercury is connected to the bottom of an open vertical tank, the other leg being open to the atmosphere. When the tank is empty the level of the mercury in the U-tube is 1 m below a fixed mark on the tank. If water is now poured into the tank up to the level of the fixed mark determine the resulting difference in the mercury levels in the U-tube.

3. A mercury U-tube manometer is used to measure the gauge pressure of water in a pipe. The water from the pipe is in contact with the mercury in the left-hand limb of the manometer and the mercury surface in this limb is 300 mm below the centre-line of the pipe. The mercury surface in the right-hand limb is 200 mm above the centre-line of the pipe.

Determine the gauge pressure in the pipe.

4. The difference in height between two points A and B in a water main is 0·5 m, point A being above point B. It is known that the pressure at A is greater than that at B and a mercury manometer is used to determine the pressure difference.

The left-hand limb of the manometer is connected to point A and the right-hand limb to point B. The manometer reading is 200 mm, water being above the mercury surface in each limb. Sketch the arrangement of the manometer and determine the pressure difference between A and B.

5. A rectangular tank, 2 m long, 1·5 m high and 1 m wide is filled with water to a height of 1·3 m.
Determine the force:

(*a*) on the base of the tank;
(*b*) on one side;
(*c*) on one end of the tank.

6. A lock gate 6 m wide has water on both sides, the depth of water on one side being 4·5 m and on the other 3 m. Determine the magnitude of the resultant thrust on the gate and the position at which it acts.

7. A hollow right circular cylinder is submerged in water so that its axis is horizontal and 3 m beneath the surface. The ends of the cylinder are closed. To what pressure must the air inside the cylinder be raised in order that there shall be no resultant thrust on the circular ends of the cylinder?

If the diameter of the cylinder is 1 m calculate the resultant couple on one end. The thickness of the cylinder material is negligible.

8. An open tank with vertical sides contains water. In one of the sides a hole 1 m diameter, with its centre 1·5 m below the water surface, is covered by a plate. Determine the resultant force on the plate and the point at which it may be assumed to act.

9. A cylindrical tank 1 m in diameter has its axis horizontal. At the middle of the tank, on top, is a pipe 50 mm in diameter which extends vertically upwards. The tank and pipe are filled with oil of relative density 0·97 with the free surface in the 50 mm pipe at a level of 4 m above the tank top. Determine the total thrust on one end of the tank.

10. A tank is divided into two compartments by a vertical partition plate. One compartment contains water to a depth of 2·1 m the other holds oil of relative density 0·94 to a depth of 1·2 m. Determine, per metre width of plate, the magnitude and position of the resultant force on the partition.

11. A vertical lock gate has a rectangular aperture in it 0·9 m high and 1·2 m wide closed by a door which overlaps the aperture by 150 mm all round. The door is hinged at its top edge. Calculate the force on the hinge and on the fastening at the bottom edge of the door when the water stands 1·8 m above the top edge of the aperture, the door and the water being on opposite sides of the lock gate.

12. A square trap door of side 1 m in the vertical side of a water tank has its lower edge horizontal and hinged. It is kept closed by a normal force of 2·7 kN applied to the upper edge. Determine the greatest depth of water above the lower edge which the tank can contain.

13. One side of a water tank is inclined at 60° to the horizontal. Incorporated in this side is a plate glass inspection window 1 m wide and 2 m deep (measured down the side). The centroid of the window is

2 m vertically below the water surface. Determine the thrust on the window and the vertical distance below the water surface at which it may be assumed to act.

14. A circular plate 1·5 m in diameter is immersed in water such that the distance of its perimeter measured vertically below the water surface varies between 0·5 m and 1·5 m. Determine the thrust of the water acting on one side of the plate and the vertical distance of the centre of pressure below the water surface.

12

Liquids in motion

12.1 Introduction

When a solid body is in motion the individual particles which make up the body all undergo similar movements. This is so because the positions of all the particles are fixed relative to each other. In the case of a liquid, the individual particles are not fixed relative to each other and when the liquid is in motion the particles can move in a variety of patterns. The manner in which the particles of liquid move will depend not only on the magnitude of the pressure forces causing motion, and whether the liquid is being accelerated or decelerated, but also on the characteristic properties of the liquid itself. For example, it is well known that a liquid such as water will flow more easily than a liquid such as a heavy oil and this suggests that although liquids offer very little resistance to shear, each liquid has a characteristic property which determines its ability to flow. This particular property of a liquid is termed **viscosity** and its effect is to introduce frictional forces which oppose motion.

It is clearly not possible at this stage to take into account all the complex factors which affect the type and rate of flow of a liquid. However a useful introduction to the study of hydrodynamics can be obtained by considering the conditions which appertain to the steady flow of a frictionless incompressible liquid.

Steady flow occurs when the velocity and pressure of the liquid at a particular section in a pipe or channel do not vary with time. Two types of steady flow arise – uniform and non-uniform.

(i) STEADY UNIFORM FLOW

This is illustrated in Fig. 12.1(*a*). The liquid particles are assumed to travel in parallel paths and the pressure and velocity conditions at

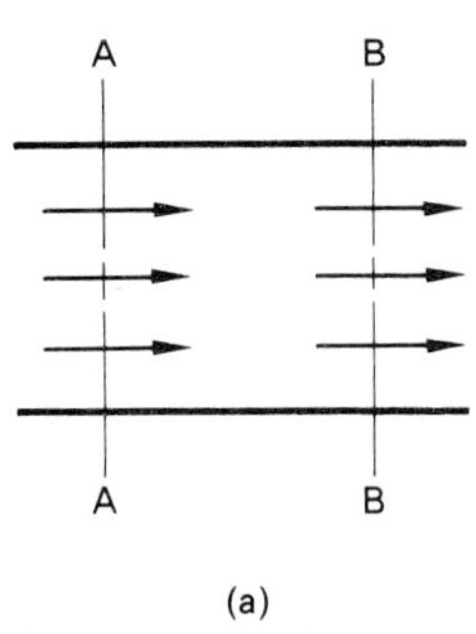

Fig. 12.1 Steady flow

section AA are identical with those at section BB and do not vary with time. Such conditions are usually found in parallel pipes at low rates of flow.

(ii) STEADY NON-UNIFORM FLOW

The flow of a liquid in a converging pipe produces this type of flow, as illustrated in Fig. 12.1(*b*). The velocities of the liquid particles will increase as they pass along the pipe and therefore the conditions at the sections AA and BB are no longer the same. However, as the velocity and pressure are constant at a particular section the flow, while non-uniform along the pipe, is still classified as steady flow.

12.2 Rate of flow

When a liquid flows in a pipe the volume of liquid passing any normal cross-section of pipe in unit time is referred to as the volumetric rate of flow or discharge Q.

If the cross-sectional area of the pipe is A and the uniform velocity of flow is v then

$$Q = Av$$

$$\text{The mass flow rate} = \rho Av = \rho Q$$

where ρ is the density of the liquid.

Q is measured in m^3/s and mass flow rate in kg/s.

12.3 Equation of continuity

For steady flow in a converging pipe, as shown in Fig. 12.2, the mass flow rate passing any cross-section is constant. Thus for sections (1)

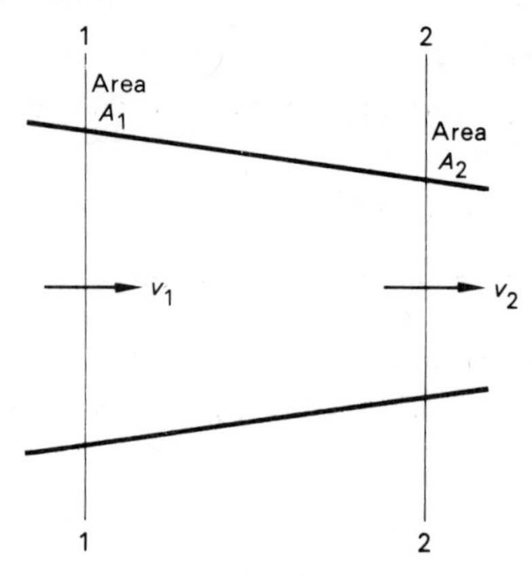

Fig. 12.2 Steady flow in a converging pipe

and (2) where the cross-sectional areas are A_1 and A_2 and the velocities are v_1 and v_2,

$$\text{Mass flow rate} = \rho_1 A_1 v_1 = \rho_2 A_2 v_2$$

This is known as the equation of continuity and applies for all fluids, i.e. liquids and gases.

For liquids, which are incompressible, the density ρ is constant so that

$$A_1 v_1 = A_2 v_2 = Q$$

It is important to note that in practice v_1 and v_2 represent the mean value of all the individual particle velocities at the given sections. Since it is clearly impractical to determine the value of the mean velocity by direct measurement it is necessary to measure the discharge and cross-sectional area and calculate the mean velocity from the relationship

$$Q = Av$$

12.4 Energy of a liquid in motion

An incompressible liquid can possess energy in three forms, of which two, namely potential energy and kinetic energy, are determined in the same way as for a solid body. The third, pressure energy, which has no counterpart in solid body dynamics, must be possessed by a liquid if flow is to occur against opposing hydrostatic pressure.

(i) POTENTIAL ENERGY

This is the energy possessed by a mass of liquid m by virtue of its height Z above a given datum.

$$\text{Potential energy} = mgZ$$

and potential energy per unit mass = gZ (J/kg)

(ii) KINETIC ENERGY

This is energy due to motion and for a mass m moving with uniform velocity v,

$$\text{Kinetic energy} = \tfrac{1}{2}mv^2$$

and kinetic energy per unit mass = $\tfrac{1}{2}v^2$ (J/kg)

(iii) PRESSURE ENERGY

When a liquid flows in a pipe under pressure work is done in moving the liquid through each section of pipe against the hydrostatic pressure existing at each section.

If at a given section in a pipe the cross-sectional area is A and the uniform pressure p, then the force on the liquid is pA in the direction of flow.

If the volumetric flow through the section is $Q = Av$, the force pA will move the liquid a distance $v = Q/A$ in unit time.

Then, work done on the liquid per unit time = force x distance moved

$$= pAv$$

$$= pA \cdot \frac{Q}{A} = pQ$$

Now in unit time the mass flowing is ρQ.

$$\therefore \quad \text{Work done per unit mass} = \frac{pQ}{\rho Q} = \frac{p}{\rho} \text{ (J/kg)}$$

p/ρ is called the pressure energy of unit mass of the liquid.

In general the energy possessed by unit mass of liquid will be the sum of its pressure, kinetic and potential energies.

$$\text{Therefore total energy per unit mass} = \frac{p}{\rho} + \frac{v^2}{2} + gZ \qquad (12.1)$$

Each term has units of J/kg.

12.5 Bernoulli's equation

In the case of the steady flow of an incompressible liquid the division of energy between the three forms pressure, kinetic and potential may vary at different sections of the flow but according to the principle of the conservation of energy the total energy will remain constant.

Then for frictionless flow in the pipe shown in Fig. 12.3 we have:

Total energy at section (1) = total energy at section (2) = a constant.

Considering unit mass of liquid this gives, from eq. (12.1):

$$\frac{p_1}{\rho} + \frac{v_1^2}{2} + gZ_1 = \frac{p_2}{\rho} + \frac{v_2^2}{2} + gZ_2 \qquad (12.2)$$

The energy equation expressed in this form is known as Bernoulli's equation.

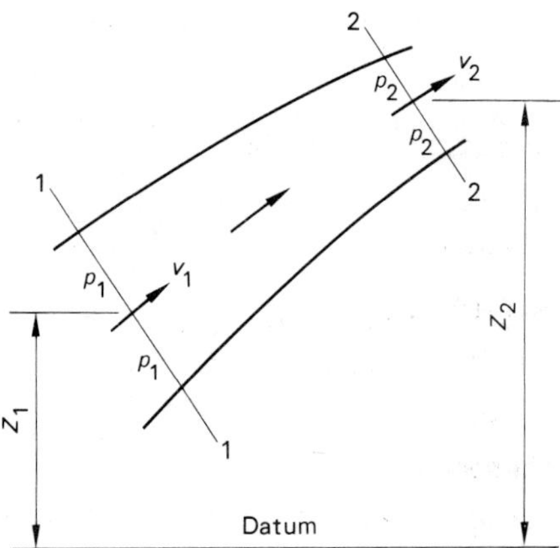

Fig. 12.3

It must be appreciated that the conditions expressed by Bernoulli's equation represent a perfect interchange of energy as no account is taken of the loss of energy which must occur in the direction of motion whenever a liquid flows between two points in a system.

To allow for this loss of energy Bernoulli's equation must be modified and expressed in the form:

Total energy at (1) = total energy at (2)
+ loss of energy occurring between (1) and (2)

Thus for actual flow conditions we have:

$$\frac{p_1}{\rho} + \frac{v_1^2}{2} + gZ_1 = \frac{p_2}{\rho} + \frac{v_2^2}{2} + gZ_2 + \text{losses (between (1) and (2))} \qquad (12.3)$$

Even allowing for losses, eq. (12.3) is still an over-simplification of the conditions found in practice. From the introduction to this chapter sufficient has been said to indicate that the main assumptions of eq. (12.3), namely the existence of uniform velocity and pressure, can hardly be expected in actual liquid motion.

Under practical conditions, we must therefore be content with calculations for kinetic energy based on the average flow velocity and the knowledge that even if the pressure is not uniform, the sum of the pressure and potential energy terms is likely to be constant at any section of a straight pipe. In a curved pipe we should expect centripetal forces to be developed (Section 8.8) and although eq. (12.2) does not allow for flow conditions in a curved path this is a factor which must not be overlooked when considering the limitations of Bernoulli's equation.

For many problems it is often more convenient to express the energy terms of Bernoulli's equation as equivalent heads (in metres) of the flowing liquid.

Thus from
$$\frac{p_1}{\rho} + \frac{v_1^2}{2} + gZ_1 = \frac{p_2}{\rho} + \frac{v_2^2}{2} + gZ_2$$

we obtain
$$\underset{\text{Pressure head}}{\frac{p_1}{\rho g}} + \underset{\text{Kinetic head}}{\frac{v_1^2}{2g}} + \underset{\text{Potential head}}{Z_1} = \frac{p_2}{\rho g} + \frac{v_2^2}{2g} + Z_2$$

Example 12.1

Water is discharged from a constant head tank at a point 20 m below the tank level. If the diameter of the pipe at outlet is 50 mm determine the discharge. Find also the pressure and velocity at a point 7 m below the tank level at which the pipe diameter is 100 mm. Neglect all losses. Density of water 1000 kg/m^3.

Solution

A diagram is given in Fig. 12.4

The position of the datum is quite arbitrary and is therefore chosen to coincide with the outlet point C. Hence the potential head term Z_C is zero.

Applying Bernoulli's equation between A and C:

$$\frac{p_A}{\rho g} + \frac{v_A^2}{2g} + Z_A = \frac{p_C}{\rho g} + \frac{v_C^2}{2g} + Z_C$$

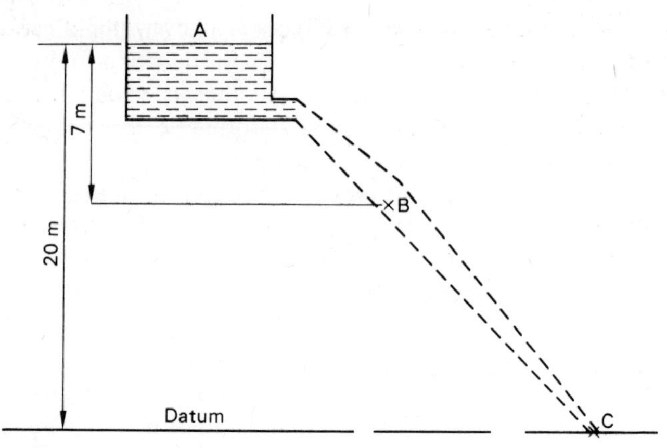

Fig. 12.4 Example 12.1

The pressure at the free surface of the tank and at the outlet is atmospheric – i.e. equivalent to 10·33 m of water.

The water at A may be regarded as at rest, i.e. $v_A = 0$

Hence $\quad 10{\cdot}33 + 0 + 20 = 10{\cdot}33 + \dfrac{v_C^2}{2g} + 0$

$$\therefore \qquad \frac{v_C^2}{2g} = 20$$

$$v_C = 19{\cdot}8 \text{ m/s}$$

Discharge $\quad Q = A_C v_C = 19{\cdot}8 \times \dfrac{\pi}{4}(0{\cdot}05)^2$

$$= 0{\cdot}039 \text{ m}^3/\text{s}$$

$$= 39 \text{ litres/s}$$

The velocity at B, 7 m below the free surface can be obtained from the continuity equation

$$A_C v_C = A_B v_B$$

or

$$v_B = v_C \left(\frac{50}{100}\right)^2$$

$$= 4{\cdot}95 \text{ m/s}$$

To find the pressure at B the datum may be taken either through B or C.

Applying Bernoulli's equation between A and B with the datum through C gives:

$$\frac{p_A}{\rho g}+\frac{v_A^2}{2g}+Z_A=\frac{p_B}{\rho g}+\frac{v_B^2}{2g}+Z_B$$

$$10{\cdot}33+0+20=\frac{p_B}{\rho g}+\frac{(4{\cdot}95)^2}{2\times 9{\cdot}81}+13$$

$$\frac{p_B}{\rho g}=16{\cdot}08 \text{ m of water}$$

Alternatively $p_B = 16{\cdot}08 \times 1000 \times 9{\cdot}81$

$= 157{\cdot}7 \text{ kN/m}^2$ (abs.)

The discharge is $0{\cdot}039 \text{ m}^3/\text{s}$ and the velocity and pressure at a point 7 m below the tank level are 4·95 m/s and $157{\cdot}7 \text{ kN/m}^2$ respectively.

Example 12.2

Oil of relative density 0·85 flows along a diverging pipe AB at the rate of $7 \text{ m}^3/\text{s}$ without loss of energy. The areas of cross-section at A and B are $0{\cdot}5 \text{ m}^2$ and $1{\cdot}0 \text{ m}^2$ respectively and the centroid of the section at A is 8 m above the centroid of the section at B. If the pressure at A is 140 kN/m^2, determine the energy content of the oil per unit mass at A taking the centroid of B as datum. Hence determine the pressure at B.

Solution

A diagram is given in Fig. 12.5.

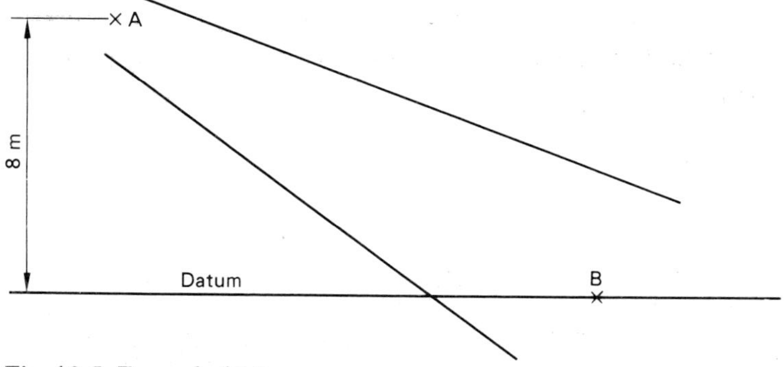

Fig. 12.5 Example 12.2

The density of the oil = 0·85 x 1000 = 850 kg/m^3

From $\quad Q = A_A v_A = A_B v_B$

$$v_A = \frac{7}{0\cdot5} = 14 \text{ m/s}, \quad v_B = \frac{7}{1\cdot0} = 7 \text{ m/s}$$

Total energy per unit mass at A with reference to datum through B

$$= \frac{p_A}{\rho} + \frac{v_A^2}{2} + Z_A$$

$$= \frac{140 \times 10^3}{850} + \frac{14^2}{2} + 8 = 270\cdot7 \text{ J/kg}$$

The pressure at B is obtained by equating the energy at A to that at B, since there are no losses.

Thus $\quad 270\cdot7 = \frac{p_B}{\rho} + \frac{v_B^2}{2} + Z_B$

or $\quad 270\cdot7 = \frac{p_B}{850} + \frac{7^2}{2} + 0$

Hence $\quad p_B = 209\cdot3 \times 10^3 \text{ N/m}^2 = 209\cdot3 \text{ kN/m}^2$

12.6 Flow measurement

The two types of flow measuring instruments in common use are the orifice and the Venturi meter.

(i) THE ORIFICE

This is simply a sharp edged hole in a plate or the side of a tank. Its use as a flow measuring device can be demonstrated by considering the large tank in Fig. 12.6 from which liquid is discharged to the atmosphere under a constant head h above the orifice centre line.

In order to calculate the quantity flowing through the orifice it is necessary to determine the velocity of flow.

The connection between velocity and head can be obtained from the energy equation by considering the points (1) and (2) at the liquid surface in the tank and near the orifice outlet.

Since the liquid approaches and passes through the orifice in a curved path Bernoulli's equation cannot be applied when section (2) is taken in the actual plane of the orifice (see page 305). However the jet is found to converge to a minimum area at a distance roughly half the orifice diameter away from the face of the orifice. This minimum

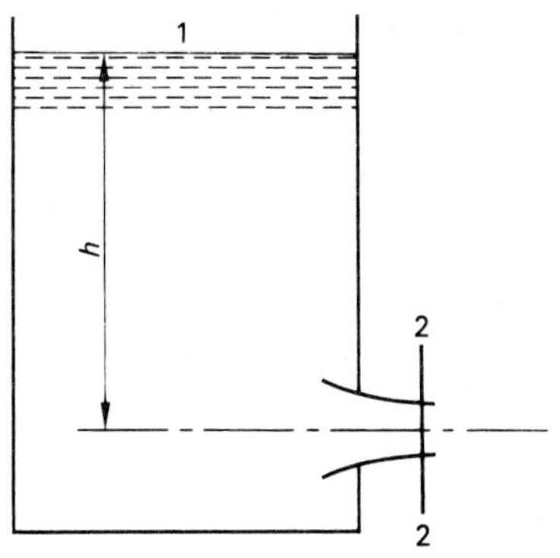

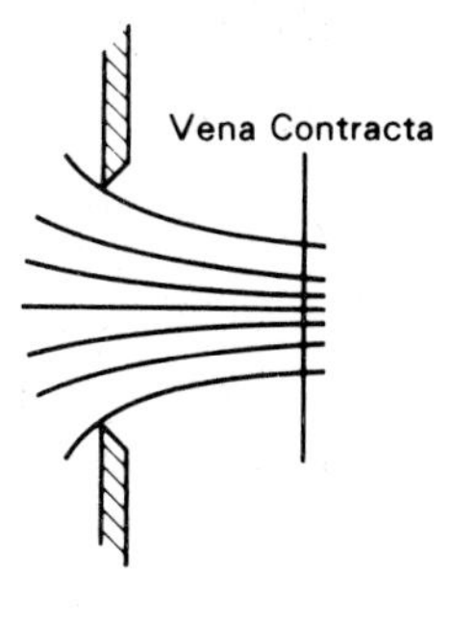

Fig. 12.6 Orifice in the side of a tank

or contracted area is called the **vena contracta** and is the first section over which the velocity is almost wholly axial. Hence the reason for choosing section (2) at the position shown.

If the tank is large, the velocity at the surface of the liquid is negligible and the pressure at (1) and (2) is atmospheric.

Applying the equivalent head equation to sections (1) and (2)

$$\frac{p_1}{\rho g} + \frac{v_1^2}{2g} + Z_1 = \frac{p_2}{\rho g} + \frac{v_2^2}{2g} + Z_2 + \text{Losses}\,(1-2)$$

Taking the datum level through the centre-line of the orifice ($Z_2 = 0$) and noting that $p_1 = p_2, v_1 = 0$ and $Z_1 = h$ gives

$$h = \frac{v_2^2}{2g} + \text{losses} \qquad (12.4)$$

If there were no losses, then clearly $h = \dfrac{v_2^2}{2g}$

and $v_2 = \sqrt{2gh}$ which may be recognised as the velocity attained by a solid body falling freely from a height h.

Thus for no energy loss, v_2 can be considered as the ideal jet velocity. However, as indicated by equation 12.4 the actual value of v_2 will be less than $\sqrt{2gh}$ on account of losses, which in this case are mainly due to friction. While this friction is minimised by making the orifice sharp-edged on the upstream side as shown in Fig. 12.6 it cannot be eliminated entirely.

The ratio of the actual velocity to ideal jet velocity at the vena contracta is called the coefficient of velocity and is given the symbol C_v.

i.e. $$C_v = \frac{\text{Actual velocity } v_2}{\sqrt{2gh}}$$

$$\therefore \quad v_2 = C_v\sqrt{2gh}$$

The quantity of liquid flowing is then obtained from

$$Q = Av \qquad \text{(ref. Section 12.2)}$$

Hence if A_2 is the area of flow at the vena contracta, then

$$Q = A_2 v_2 = A_2 C_v \sqrt{2gh} \qquad (12.5)$$

Now the ratio of the area of the jet A_2 at the vena contracta to the actual area of the orifice A is called the coefficient of contraction C_c.

$$\therefore \quad C_c = \frac{A_2}{A}$$

and $\quad A_2 = A \,.\, C_c$

Substituting this value in eq. (12.5) gives

$$Q = C_c C_v \times A\sqrt{2gh}$$

Because of the difficulty in determining the coefficients C_v and C_c it is more convenient to replace the product $C_c C_v$ by a single coefficient C_d called the coefficient of discharge.
This gives:

$$Q = C_d \,.\, A \,.\, \sqrt{2gh} \qquad (12.6)$$

The discharge coefficient C_d is the ratio of the actual discharge to the ideal one based on no contraction or loss of energy.

i.e. $\quad C_d = \dfrac{Q}{A\sqrt{2gh}}$

The value of C_d can be determined for a given orifice and head by measuring the discharge over a suitable period. The value so obtained can then be used to determine discharges under other values of head.*

Figure 12.7 shows an alternative arrangement of a sharp-edged orifice as used for flow measurement in a pipe.

In this case the discharge is given by

$$Q = \frac{C_d A \sqrt{2gH}}{\sqrt{1 - \left(\dfrac{C_c A}{A_1}\right)^2}}$$

* Actually C_d is not constant, particularly at low heads. It varies with the ratio of head to orifice diameter. This effect is considered further in later studies but a commonly accepted value is in the region of 0·62.

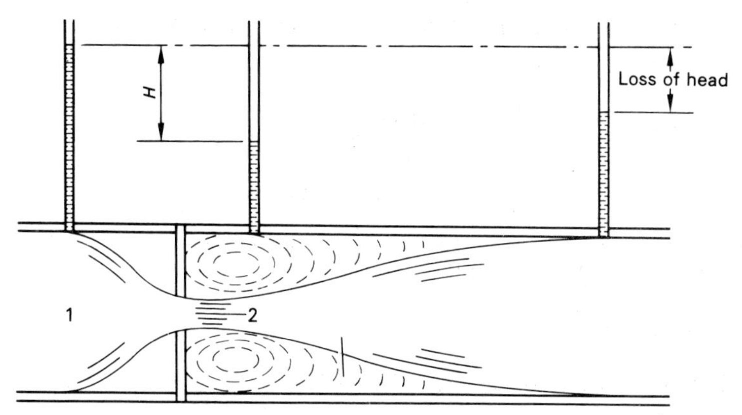

Fig. 12.7 Orifice in a pipe

where A is again the cross-sectional area of the orifice as in eq. (12.6),
A_1 is the cross-sectional area of the pipe,
and C_d and C_c are the orifice coefficients.

h in eq. (12.6) is replaced by H which is the difference in pressure head measured between a suitably placed upstream tapping and a downstream tapping made to coincide approximately with the vena-contracta as in Fig. 12.7.

Example 12.3

When water is discharged through a 50 mm diameter sharp-edged orifice under a head of 5 m the measured rate of flow is found to be 2100 litres in 3 minutes. Determine the coefficient of discharge and the probable discharge when the head over the orifice is 7 m.

Solution

Applying $Q = C_d A\sqrt{2gh}$ [eq. (12.6)]

where discharge $Q = \dfrac{2100}{3} = 700$ litres/min $= \dfrac{700}{60 \times 10^3}$ m^3/s

$$= 11{\cdot}7 \times 10^{-3} \text{ m}^3/\text{s}$$

and $A = \dfrac{\pi}{4}(0{\cdot}05)^2 = 1{\cdot}97 \times 10^{-3}$ m^2

gives $$C_d = \frac{Q}{A\sqrt{2gh}} = \frac{11{\cdot}7 \times 10^{-3}}{1{\cdot}97 \times 10^{-3}\sqrt{2 \times 9{\cdot}81 \times 5}}$$

$$= 0{\cdot}6$$

Assuming that C_d remains constant over the given head range then Q is proportional to $\sqrt{h}$.

$$\therefore \quad \text{Discharge at 7 m head} = 700\sqrt{\frac{7}{5}}$$

$$= 828 \text{ litres/min}$$

The coefficient of discharge is 0·6 and the probable discharge under a 7 m head is 828 litres/min.

Example 12.4

In an experiment to determine the coefficients of contraction, velocity and discharge for a circular orifice of 9 mm diameter, water was discharged through the orifice located in the vertical side of a tank. The constant head of water above the centre-line of the orifice was 1·2 m.

The jet was discharged into a collecting tank in which the water surface was 155 mm below the centre-line of the orifice and the jet was found to strike the water surface at a horizontal distance of 850 mm from the vena-contracta.

If the measured discharge from the orifice was 90·7 kg in 470 seconds, determine the values of the coefficients for the orifice.

Density of water = 1000 kg/m^3.

Solution

A diagram is given in Fig. 12.8

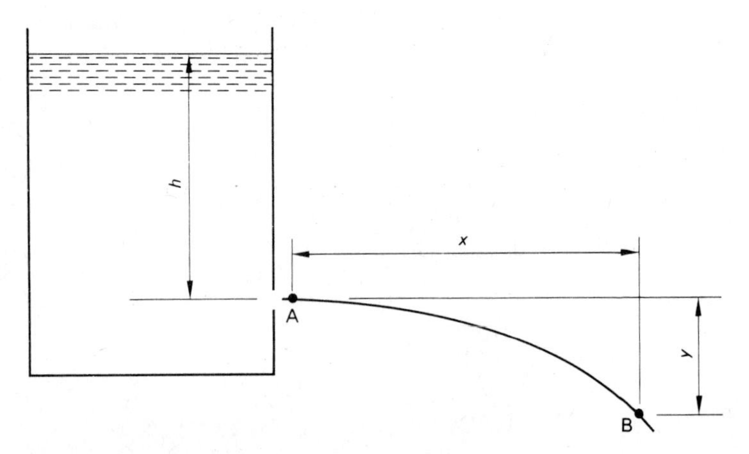

Fig. 12.8 Example 12.4

Determination of C_v.

The jet issues from the orifice as illustrated in Fig. 12.8.

If the actual velocity of the jet is v and t is the time for a particle of water to travel from the vena-contracta at A to the point B, then

$$x = vt \text{ and } y = \tfrac{1}{2}gt^2$$

Rearranging $\quad v = \dfrac{x}{t}$ and $t = \sqrt{\dfrac{2y}{g}}$

Substituting for t gives

$$v = \frac{x\sqrt{g}}{\sqrt{2y}} = \sqrt{\frac{gx^2}{2y}}$$

Now C_v is the ratio of the actual velocity v to the theoretical velocity $\sqrt{2gh}$.

$$\therefore \quad C_v = \frac{v}{\sqrt{2gh}}$$

Substituting for v gives

$$C_v = \sqrt{\frac{x^2}{4yh}}$$

For the given problem:

$$x = 0{\cdot}85 \text{ m}, \quad y = 0{\cdot}155 \text{ m and } h = 1{\cdot}2 \text{ m}$$

$$\therefore \quad C_v = \sqrt{\frac{(0{\cdot}85)^2}{4 \times 0{\cdot}155 \times 1{\cdot}2}} = 0{\cdot}985$$

Determination of C_d.

$$C_d = \frac{\text{actual discharge}}{\text{theoretical discharge}}$$

$$\text{Mass flow rate} = \frac{90{\cdot}7}{470} = 0{\cdot}193 \text{ kg/s}$$

$$\therefore \quad \text{Actual discharge} = \frac{0{\cdot}193}{1000} = 0{\cdot}193 \times 10^{-3} \text{ m}^3\text{/s}$$

$$\text{Theoretical discharge} = A\sqrt{2gh}$$

where A = area of orifice $= \dfrac{\pi}{4}(0{\cdot}009)^2 = 0{\cdot}0635 \times 10^{-3} \text{ m}^2$

$$\therefore \quad \text{Theoretical discharge} = 0{\cdot}0635 \times 10^{-3}\sqrt{2 \times 9{\cdot}81 \times 1{\cdot}2}$$

$$= 0{\cdot}308 \times 10^{-3} \text{ m}^3\text{/s}$$

Then

$$C_d = \frac{0{\cdot}193 \times 10^{-3}}{0{\cdot}308 \times 10^{-3}} = 0{\cdot}627$$

Determination of C_c.

Since by definition $C_d = C_c \times C_v$

Then
$$C_c = \frac{C_d}{C_v}$$
$$= \frac{0{\cdot}627}{0{\cdot}985} = 0{\cdot}637$$

(ii) THE VENTURI METER

Whichever instrument is used for flow measurement, some loss of energy will occur and in this respect the pipe orifice shown in Fig. 12.7 is not entirely satisfactory. It certainly has the advantages of being cheap and relatively easy to install, but in cases where such factors are unimportant the use of a Venturi meter is preferable.

A Venturi meter, shown in Fig. 12.9, is usually made a permanent installation in the pipe-work and consists simply of a tube which converges to a minimum area called the throat and then diverges

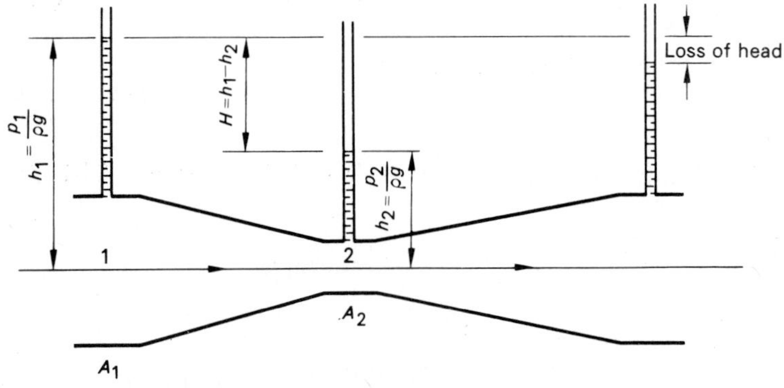

Fig. 12.9 The Venturi meter

gradually back to the normal pipe diameter. Its operation, therefore, is like that of the orifice in Fig. 12.7 and is based on the conversion of pressure energy in the pipe to kinetic energy in the throat. However, the loss of energy due to turbulence, which occurs just downstream in the case of the orifice is largely avoided in the Venturi meter by means of the *gradual* divergence beyond the throat. Hence the overriding advantage of this meter compared with the pipe orifice is that it can be manufactured so as to incur almost negligible loss of energy. Compare Figs. 12.7 and 12.9.

For the arrangement of a horizontal meter shown in Fig. 12.9 $Z_1 = Z_2$. Neglecting losses and applying Bernoulli's equation gives

$$\frac{p_1}{\rho} + \frac{v_1^2}{2} = \frac{p_2}{\rho} + \frac{v_2^2}{2}$$

Dividing through by g and rearranging

$$\frac{p_1}{\rho g} - \frac{p_2}{\rho g} = \frac{v_2^2}{2g} - \frac{v_1^2}{2g}$$

But $\frac{p_1}{\rho g} = h_1 =$ the pressure head at section (1)

and $\frac{p_2}{\rho g} = h_2 =$ pressure head at section (2) (Chapter 11)

Also $h_1 - h_2 = H$, the loss of pressure head between the upstream and throat tapping points.

$$\therefore \quad H = \frac{v_2^2}{2g} - \frac{v_1^2}{2g} \qquad (12.7)$$

Now from the equation of continuity

$$A_2 v_2 = A_1 v_1$$

where A_1 and A_2 are the cross-sectional areas at the respective tapping points.

Thus, substituting $v_2 = \frac{A_1}{A_2} v_1$ in eq. (12.7) gives

$$H = \left(\frac{A_1}{A_2}\right)^2 \frac{v_1^2}{2g} - \frac{v_1^2}{2g}$$

$$= \frac{v_1^2}{2g} \left[\left(\frac{A_1}{A_2}\right)^2 - 1\right]$$

If $\frac{A_1}{A_2} = \mathrm{n} =$ a constant for the meter,

then $H = \frac{v_1^2}{2g}(\mathrm{n}^2 - 1)$

$$\therefore \quad v_1 = \sqrt{\frac{2gH}{(\mathrm{n}^2 - 1)}}$$

Therefore theoretical discharge $= A_1 v_1 = A_1 \sqrt{\frac{2gH}{(\mathrm{n}^2 - 1)}}$ (12.8)

Since some small loss of energy is unavoidable the actual volumetric rate of flow will be less than that given by eq. (12.8). Therefore introducing a discharge coefficient C_d, the actual rate of flow is given by

$$Q = C_d A_1 \sqrt{\frac{2gH}{(n^2 - 1)}} \qquad (12.9)$$

Now for any one meter $\sqrt{2g}$, A_1 and $(n^2 - 1)$ are constant and since C_d is also reasonably constant it follows that eq. (12.9) may be expressed in the form

$$Q = k\sqrt{H}$$

where k is called the meter coefficient.

The discharge coefficient C_d has a value of the order of 0·98.

Example 12.5

A Venturi meter measures the flow of water in an 80 mm diameter horizontal pipe. The difference of head between the entrance and the throat of the meter is measured by a **U**-tube, containing mercury, the space above the mercury in each limb being filled with water. Determine the diameter of the throat of the meter in order that the difference of the levels of the mercury shall be 250 mm when the quantity of water flowing in the pipe is 36 m^3/h.

Assume the discharge coefficient is 0·97.

Relative density of mercury is 13·6.

Introduction

In the Venturi eq. (12.9), H is the pressure head difference in metres of the liquid flowing, in this case water. This head difference must therefore be expressed as a head difference which is equivalent to 250 mm of mercury with water in both limbs of the **U**-tube above the mercury, as shown in Fig. 12.10.

Consider the conditions at the common surface XX in the **U**-tube shown enlarged in Fig. 12.11.

For the left-hand limb:

$$\text{pressure at XX} = p_1 + \rho g h + \rho g h'$$

For the right-hand limb:

$$\text{pressure at XX} = p_2 + \rho_m g h + \rho g h'$$

where ρ = density of water and ρ_m = density of mercury

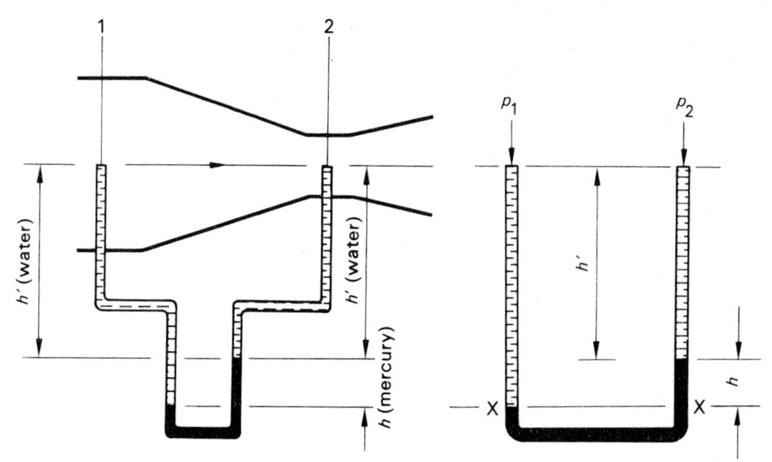

Fig. 12.10 Example 12.5

Fig. 12.11 Example 12.5

Then for equilibrium, the pressures in the two limbs, at XX are equal.

$$\therefore \quad p_1 + \rho gh + \rho gh' = p_2 + \rho_m gh + \rho gh'$$

and
$$p_1 - p_2 = gh\,(\rho_m - \rho)$$
$$= \rho gh\left(\frac{\rho_m}{\rho} - 1\right)$$

or
$$\frac{p_1}{\rho g} - \frac{p_2}{\rho g} = h\left(\frac{\rho_m}{\rho} - 1\right)^*$$

Note that this expression is independent of the head h′ as both limbs contain water above the mercury.

Since $\dfrac{p_1}{\rho g} - \dfrac{p_2}{\rho g} = h_1 - h_2 = H$

where H is the difference in pressure head between the upstream and throat tapping points,

then $H = h\left(\dfrac{\rho_m}{\rho} - 1\right)$

Substituting this value of H in the Venturi discharge eq. (12.9) gives:

$$Q = C_d A_1 \sqrt{\frac{2gh\left(\frac{\rho_m}{\rho} - 1\right)}{n^2 - 1}}$$

* This expression could have been deduced directly from eq. (11.2), since for a horizontal meter the difference in height between the tapping points h_0 is zero. See also Fig. 11.5.

Solution

The only unknown in the above equation is n^2 which is the ratio of areas. This will give the required throat area.

In this case $\frac{\rho_m}{\rho} = 13{\cdot}6 \quad \therefore \frac{\rho_m}{\rho} - 1 = 12{\cdot}6$

$$A_1 = \frac{\pi}{4}(0{\cdot}08)^2 = 0{\cdot}005 \text{ m}^2$$

$$Q = 36 \text{ m}^3/\text{h} = 0{\cdot}01 \text{ m}^3/\text{s}; \quad h = 0{\cdot}25 \text{ m}$$

Substituting these values:

$$0{\cdot}01 = 0{\cdot}97 \times 0{\cdot}005\sqrt{\left(\frac{2 \times 9{\cdot}81 \times 0{\cdot}25 \times 12{\cdot}6}{n^2 - 1}\right)}$$

$$n^2 - 1 = 14{\cdot}15$$

$$n = 3{\cdot}89$$

Now $\quad n = \frac{A_1}{A_2} = \frac{d_1^2}{d_2^2}$

$\therefore \quad 3{\cdot}89 = \frac{80^2}{d_2^2}$

$$d_2 = 40{\cdot}6 \text{ mm}$$

The diameter of the throat for a manometer reading of 250 mm is 40·6 mm.

Example 12.6

A Venturi meter with a throat diameter of 450 mm is fitted into a 900 mm diameter pipe. It is placed with the axis of the meter inclined at 30° to the horizontal. The head difference across the meter is measured by a mercury manometer. Calculate the discharge of oil of relative density 0·8 through the pipe when the manometer reading is 180 mm of mercury. Assume that the meter coefficient is 0·97.

Introduction

Referring to Fig. 12.12 it is first necessary to determine the effect of the inclination of the meter on the discharge.

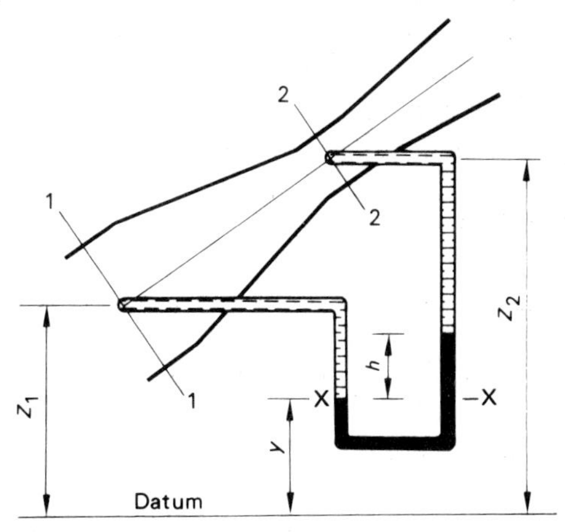

Fig. 12.12 Example 12.6

From Bernoulli's equation,

$$\frac{p_1}{\rho g}+\frac{v_1^2}{2g}+Z_1=\frac{p_2}{\rho g}+\frac{v_2^2}{2g}+Z_2$$

$$\frac{p_1}{\rho g}-\frac{p_2}{\rho g}+Z_1-Z_2=\frac{v_2^2}{2g}-\frac{v_1^2}{2g}$$

But from eq. (12.7) $\quad \dfrac{v_2^2-v_1^2}{2g}=H$

$$\therefore \quad \frac{p_1}{\rho g}-\frac{p_2}{\rho g}+Z_1-Z_2=H \qquad (12.10)$$

Now for the **U**-tube in Fig. 12.12, with the pipe liquid in both limbs above the mercury, the pressures at the level XX are the same in both limbs.

Thus $\quad p_1+\rho g(Z_1-y)=p_2+\rho g(Z_2-y-h)+\rho_m gh$

and $\quad \dfrac{p_1}{\rho g}-\dfrac{p_2}{\rho g}+Z_1-Z_2=h\left(\dfrac{\rho_m}{\rho}-1\right)^*$

Comparing this with eq. (12.10) gives:

$$H=h\left(\frac{\rho_m}{\rho}-1\right)$$

which is the same relationship as that found for a horizontal meter.

* This relationship could have been obtained direct from equation 11.2 in which Z_2-Z_1 corresponds to h_0.

Thus $Q = C_d A_1 \sqrt{\dfrac{2gh\left(\dfrac{\rho_m}{\rho} - 1\right)}{n^2 - 1}}$ as before.

This relationship is independent of Z_1 and Z_2 and indicates that the manometer reading h, for a given value of Q does not depend on the inclination of the meter.

Solution

$$A_1 = \frac{\pi}{4}(0{\cdot}9)^2 = 0{\cdot}637\ \text{m}^2,\ A_2 = \frac{\pi}{4}(0{\cdot}45)^2 = 0{\cdot}159\ \text{m}^2$$

$$n = \frac{A_1}{A_2} = 4,\quad \frac{\rho_m}{\rho} - 1 = \frac{13{\cdot}6}{0{\cdot}8} - 1 = 16$$

$$\therefore \quad Q = 0{\cdot}97 \times 0{\cdot}637\sqrt{\left(\frac{2 \times 9{\cdot}81 \times 0{\cdot}18 \times 16}{16 - 1}\right)}$$

$$Q = 1{\cdot}2\ \text{m}^3/\text{s}.$$

For a meter reading of 180 mm of mercury the discharge of oil is $1{\cdot}2\ \text{m}^3/\text{s}$.

Problems

The following values should be used:
Density of water 1000 kg/m^3. Relative density of mercury 13.6.

1. Water is pumped 'uphill' in a pipe inclined at 30° to the horizontal. At a point A the internal diameter of the pipe is 150 mm and at a higher point B, 2 m from A, the internal diameter is 75 mm – the change being uniform. If the pressure at A is 400 kN/m^2 find the pressure at B when the volumetric rate of flow is 30 l/s.

2. A pipe line AB is 100 m long and is laid on ground rising with a slope of 1 in 15. The diameter of the pipe at the lower end A is 0·15 m. Water flows along the pipe at a rate of 240 m^3 per hour and discharges to atmosphere at B where the pipe diameter is 0·1 m.
Neglecting losses, determine the static gauge pressure at A.

3. State Bernoulli's theorem for the steady flow of a liquid and explain the meaning of the terms and the units of the quantities given.
Water flows with steady motion through a pipe inclined downwards. At the upper end A the pipe is 100 mm diameter and the pressure is 65 kN/m^2. At a point B which is 3 m below A the diameter of the pipe is

60 mm and the pressure is 40 kN/m^2. Assuming that there are no losses due to friction between A and B calculate the volumetric rate of flow.

4. Water is flowing up a pipe inclined at 45° to the horizontal. At a point A the internal diameter of the pipe is 160 mm and at a higher point B, 4 m from A, the internal diameter is 80 mm, the change in diameter being uniform. If the pressure difference between A and B is 42 kN/m^2 calculate the velocities at A and B and the rate of flow assuming there are no losses.

5. A vertical pipe AB carrying fuel of density 800 kg/m^3 tapers slowly through a distance of 2·5 m. At A the static gauge pressure is 600 kN/m^2 and the velocity 10 m/s. If the static gauge pressure at B is 400 kN/m^2 determine the velocity at this point.

6. Water is pumped vertically upwards through a pipe AB by means of a pump situated at A.

The pipe diameter at A is 180 mm and the diameter decreases uniformly to 90 mm at point B which is 10 m above A. At B the velocity of the water is 6 m/s and the gauge pressure is 0·85 bar. Neglecting losses determine:

(*a*) the gauge pressure at point A;
(*b*) the power required to drive the pump given that the pump efficiency is 75%.

7. Water flows downward through a pipe 1 km long which has a slope of 1 in 100. The diameter of the pipe changes uniformly from 800 mm at the upper end to 400 mm at the lower end.

If the pressure in the pipe is 290 kN/m^2 and 250 kN/m^2 at the upper and lower ends respectively and there is an estimated energy loss of 90 J/kg due to friction determine the volumetric rate of flow.

8. The velocities of flow down a vertical pipe 2 m long are 5 m/s at the upper end and 2 m/s at the lower end. The pressure head at the upper end is 3 m of the liquid. Given that the loss of head in the pipe may be expressed in the form $0{\cdot}35\,(v_1 - v_2)^2/2g$, where v_1 and v_2 are the velocities in the pipe at the upper and lower ends respectively, calculate the pressure head at the lower end of the pipe.

9. A pipe line inclined upwards at 30° to the horizontal consists of two pipes, 300 mm diameter and 250 mm diameter connected by a valve. The flow passes through the valve from the larger pipe. A pressure gauge is attached to the pipes on each side of the valve at a distance apart of

4 m measured along the centre of the pipe line. When water flows through the pipe line at the rate of 0·25 m^3/s the gauge attached to the larger pipe reads 400 kN/m^2 and that attached to the smaller pipe 220 kN/m^2. Determine the loss of head across the valve in metres of water.

10. In an experiment to determine the coefficients of velocity, discharge and contraction for a 10 mm diameter sharp-edged orifice located in the vertical side of a tank, a constant head of water of 690 mm was maintained above the centre line of the orifice. The jet issued horizontally from the orifice and in a horizontal distance of 450 mm from the vena-contracta it dropped a distance of 80 mm. Given that the measured discharge from the orifice was 55·2 kg in 5 minutes calculate the values of the orifice coefficients.

11. A jet of water issues from a sharp-edged orifice 15 mm diameter in the vertical side of a tank. The horizontal and vertical co-ordinates of a point in the jet relative to the centre of the orifice are 1 m and 200 mm respectively. Assuming that the coefficient of velocity for the orifice is 0·97 estimate the value of the head of water above the centre-line of the orifice.

Taking the coefficient of contraction as 0·63 estimate the rate of flow per hour from the orifice.

12. A horizontal Venturi meter has an entry diameter of 240 mm and a throat diameter of 80 mm. The coefficient for the meter is 0·96. The difference in pressure between the entrance and throat is equivalent to 50 mm of mercury. Determine the quantity of water flowing through the meter.

13. A Venturi meter having a meter coefficient of 0·96 is fitted into a horizontal water main of 100 mm diameter. If the reading of a U-tube manometer connected across the inlet and throat tapping points is 153 mm of mercury when the flow through the meter is 12 litre/s determine the throat diameter of the meter.

14. A calibration test on a Venturi meter having inlet and throat diameters of 200 mm and 100 mm respectively provided the following information:

Rate of flow of water = 3120 kg/min.

Difference in mercury levels in the limbs of the mercury-water U-tube manometer = 180 mm.

Determine the discharge coefficient for the meter.

15. A pressure difference of 20 kN/m^2 is recorded between the tapping points of a Venturi meter having inlet and throat diameters of 150 mm and 75 mm respectively.

If the meter discharge coefficient is 0·97 determine the rate of flow of water through the meter in cubic metres per hour.

16. A mercury **U**-tube manometer is connected to the inlet and throat tapping points of a Venturi meter which is used to measure the flow of oil of relative density 0·8.

Derive an expression for the pressure difference between the tapping points in terms of the difference in mercury levels in the manometer and the densities of the oil and mercury.

Given that the inlet and throat diameters of the meter are 120 mm and 60 mm respectively, estimate the rate of oil flow in cubic metres per hour when the difference of mercury levels in the manometer is 80 mm. Take 0·97 as the meter discharge coefficient.

17. The cooling water from a small condenser passes along a horizontal pipe 90 mm diameter to a measuring tank from which it is discharged through a sharp-edged orifice of 40 mm diameter. For the condition of maximum flow the head in the tank over the centre of the orifice is 1·5 m. If the coefficient of discharge for the orifice is 0·62 determine the corresponding velocity of flow in the pipe. A Venturi meter is to be used instead of the measuring tank and one with a throat diameter of 30 mm is selected for fitting into the pipe line. Estimate the pressure drop between the entry and the throat of this meter for the same rate of flow.

Answers

Chapter 1 (page 29)

1. $R_A = 12{\cdot}4$ kN; $R_C = 57{\cdot}6$ kN
2. 2·5 kN m anticlockwise
3. 107·6 N
4. $R_B = 183$ kN at 38° 36′ to the horizontal; $R_C = 28{\cdot}4$ kN
5. $R_X = 81{\cdot}44$ kN at 41° 48′ to XY
6. $T_A = 172{\cdot}9$ N; $T_B = 92{\cdot}9$ N; $T_C = 65{\cdot}7$ N
7. $R_A = 844$ N at 4·4° to AB; 972 N
8. Forces in kN: $R_A = 37{\cdot}5$; $R_B = 42{\cdot}5$; AB = –62·5; BC = CD = –45·84; DE = –70·83, EF = FG = 56·66; GH = HA = 50; BH = DF = 0; BG = –16·66; DG = –25; CG = 25
9. Forces in kN: $R_1 = 65$; $R_4 = 75$; 2-5 = 40·41; 3-5 = 28·86
10. Forces in kN: $R_A = 61$; $R_B = 31$; (1) = –70·44; (2) = –65·82; (3) = –38; (4) = –62; (5) = (6) = 53·7; (7) = 35·22; (8) = –4·62; (9) = 16; (10) = –24
11. Forces in kN: (1) = –70; (2) = –45·8; (3) = –28·6; (4) = 10·3; (5) = 46·9; (6) = 66; (7) = 27·6; (8) = –12·5; (9) = 50·8
12. Forces in kN: $R_A = 51{\cdot}96$ Horizontally; $R_E = 60$ in direction EC; AB = 51·96; BC = 17·32; CD = DE = BE = –34·64; BD = 34·64
13. (1) = 127·28 kN; (2) = 90 kN; (3) = 56·57 kN; (4) = –130 kN
14. Ties: BJ, IJ, FH, IH, CI, EI; No force in AJ, DI, HG; CI = 48 kN
15. BF = –71·25 kN; FG = 155·9 kN; GD = 42·22 kN
16. Forces in kN: $R_A = 17{\cdot}9$; $R_B = 41{\cdot}8$ at 37·6° to horizontal; (1) = –12·5; (2) = –22·5; (3) = 38
17. Forces in kN: $R_Y = 10{\cdot}67$ horizontal; $R_Z = 15{\cdot}32$ at 46° to horizontal; BE = 6·67; CY = –1.33; DG = –3; AG = –13·33; AE = –8·33; EF = –10; FG = 12.37

18. Forces in kN; AB = BC = −8; CD = 11·31; BD = 4; AD = 16·97; DE = 28·28; R_A = 23·32 at 31° to AB

19. Forces in kN: AB = 140; BC = 210; CD = 60; AD = 70; CA = 100. Components at C are 130 horizontally to left and 120 vertically upwards.

Chapter 2 (page 61)

1. R_A = 72·66 kN; R_B = 97·33; M_{max} = 268·66 kN m at 20 kN load

2. R_B = 125 kN; R_D = 85 kN; M_{max} = 90·62 kN m at 4·75 m from A; contraflexure occurs at 1·74 m from A

3. R_A = 80 kN; R_D = 10 kN; M_{max} = 300 kN m at C

4. R_B = R_E = 90 kN; M_{max} = 110 kN m at C; Contraflexure at 2·8 m from A

5. Reactions: 4·27 kN and 7·23 kN; M_{max} = 4·54 kN m under 6 kN load. Contraflexure at 2·79 m from left hand end

6. Reactions: 85 kN and 100 kN; M_{max} = 35·41 kN m at 2·83 m from left hand end

7. Reactions: 125 kN and 200 kN; Moments: −120 kN m at left support, −277·5 kN m at right support, −36·7 kN m at 3·33 from left support

8. Reactions: 85 kN and 100 kN; Moments: −15 kN m at left support; −35 kN m at right hand support; maximum is 35·4 kN m at 2·83 m from left hand end

9. Reactions = 19 kN; M_{max} = 30 kN m at centre span

10. R_A = 65·9 kN; R_E = 99.1 kN: moments in kN m; M_B = 85·5; M_C = 58·2; M_D = 9·1; M_E = −60; M_{max} = 86·9 at 2·636 m from A

11. 18 kN at 3·5 m from R; M_{max} = 38·75 kN m under 18 kN load

12. R_B = 26 kN; R_C = 30 kN; M_B = −10 kN m; M_C = −20 kN m; M_{max} = 47·6 kN m at 9·8 m from A

13. R_A = 88 kN; R_C = 105 kN; M_B = 78 kN m; M_C = −48 kN m; M_{max} = 116 kN m at 2·95 m from A; Contraflexure at 6·35 m from A

14. R_B = 120 kN; R_C = 150 kN; Contraflexure positions are 1·36 m from A and D

15. R_B = 9·5 kN; R_D = 10·5 kN; M_B = −7 kN m; M_C = 5 kN m; M_D = −8 kN m; M_{max} = 11·75 kN m at 8·5 m from A

16. M_{max} = 60 kN m at centre; Contraflexure at 1·17 m from each end; Supports should be 1·656 m from each end (Solution is obtained by making the moment at the support numerically equal to the moment at the centre span)

17. Reactions: 126 kN and 224 kN; M_{max} = 132·3 kN m at 2·1 m from left hand end

18. W = 130 kN/m; Contraflexure at 1·382 m from each end

Chapter 3 (page 103)

1. (*a*) 0·56 mm (*b*) 4·8
2. (*a*) 426 MN/m^2 (*b*) 663 MN/m^2 (*c*) 37·4% (*d*) 197 GN/m^2
3. (*a*) 485 MN/m^2; 495 MN/m^2 (*b*) 74·4 GN/m^2
4. 84·1 MN/m^2; 50·5 MN/m^2; 32·8 mm
5. 51·9 kN
6. (*a*) 11·49 MN/m^2 (*b*) 0·0261 mm; 59·2 MN/m^2
7. (*a*) 914 mm^2 (*b*) 0·208 mm
8. (*a*) 79·3 MN/m^2; 43·65 MN/m^2 (*b*) 0·595 mm (*c*) 16·67 J
9. 0·011 mm; σ_{steel}: σ_{brass} = 0·6875:1
10. 341 MN/m^2
11. (*a*) Steel = 17·1 MN/m^2; Brass = 16·91 MN/m^2 (*b*) Steel = 29·4 MN/m^2; Brass = 34·61 MN/m^2
12. Gunmetal = 47·75 MN/m^2; Steel = 38·2 MN/m^2; 144·2 MN/m^2
13. (*a*) 694 kN (*b*) 2 mm
14. 22·8 kN; 11 mm
15. 70 MN/m^2; 7
16. 2005 rev/min
17. 661 rev/min
18. 3·48 mm
19. 756 kN/m^2
20. (*a*) 9·9 J (*b*) 49·5 kN (*c*) 0·4 mm
21. 79·6 J
22. (*a*) 35·7 MN/m^2; 71·4 MN/m^2; 0·535 mm (*b*) Cast iron = 8·6 J; Steel = 11·46 J (*c*) 12·87 kN
23. AC = 73·3 kN; AD = 30·2 kN; CD = 90·4 kN; BD = 83·3 kN; 236 J.

Chapter 4 (page 134)

1. 5·25 m
2. 2·67 m
3. 100 MN/m^2; 60 N m
4. 113 MN/m^2; 139 m
5. 9 mm; 18 mm; 14·58 N m
6. 755 N/m
7. 9.66 kN; maximum load for solid shaft is 3·53 kN
8. 26·25 kN
9. (*a*) 49·18 MN/m^2 (*b*) 523 m

10. (*a*) 21·37 kN m at 90 kN load (*b*) 137 MN/m^2 (*c*) 102 m
11. 2·7 kN/m; 85·7 MN/m^2
12. 1202 N
13. (*a*) 100 mm (*b*) 5·33 x 10^6 mm^4 (*c*) 7·02 MN/m^2; 14·04 MN/m^2
14. σ_t = 9·32 MN/m^2; σ_c = 14·29 MN/m^2
15. M_{max} = 5·78 kN m at 2·312 m from support nearest to concentrated load; width of beam = 86·5 mm
16. 63·3 MN/m^2; 352·25 m
17. 134 MN/m^2; 123·5 MN/m^2; 34·7%
18. Flange at bottom; 14·5 kN

Chapter 5 (page 156)

1. 38 GN/m^2
2. 17·45 MN/m^2; 25·72 kW
3. 5·873 MW; 1·49°
4. 200 mm; 36·4 MN/m^2
5. 15·2 MN/m^2
6. 1·5 MW
7. 208 mm; 124·8 mm
8. 50 mm; 32·8 mm
9. 117·8 MN/m^2; 78 GN/m^2; 110 mm
10. 73·25 MN/m^2
11. (*a*) outside diameter = 175 mm (*b*) 1·23°
12. 11·76 MW; 62·8 MN/m^2
13. 26·75 MN/m^2; 0·478°
14. 0·303 m; 1·103:1
15. 1·25 mm; mass-hollow:solid = 0·302:1; twist-hollow:solid = 0·564:1
16. 13·88 kW; 0·149°/m
17. 78·3 mm
18. (*a*) 1082 N m (*b*) 44·1 MN/m^2; 60 MN/m^2 (*c*) 2·2°
19. AB = 100 mm; BC = 118 mm
20. 1025 N m; 31·5 MN/m^2

Chapter 7 (page 192)

1. v_A = 33·65 km/h; v_B = 56·34 km/h; v_C = 72·22 km/h
2. 3 m/s^2; 90 m
3. 86·4 km/h; 14·74 km; 0·533 m/s^2

4. (*a*) 1573 m (*b*) 163·2 s
5. 639 m/s; 11° 19′
6. (*a*) 16·3 km (*b*) 610 m/s at 19° to the hillside
7. (*a*) 1 m/s; –9 m/s (*b*) 2 m, –18 m (*c*) 20·5 m
8. (*a*) 0·9 s or 4·43 s (*b*) –7 m/s^2
9. (*a*) 2·4 m; 2·8 m/s^2 (*b*) 3·38 s (*c*) 3 s
10. (*a*) 19·25 m/s; 50·4 m (*b*) 183·3 m
11. (*a*) 19·8 m (*b*) 5·36 s (*c*) 0·96 s
12. 18·32 rad/s^2; 1728 m/s^2
13. 7.33 rad/s; 100·5 s
14. (*a*) 0·582 rad/s^2 (*b*) 133·4 s (*c*) 297·3 (*d*) 0·291 m/s^2 (*e*) 219 m/s^2

Chapter 8 (page 231)

1. (*a*) 0·2 m/s^2 (*b*) 1 in 65·5 (*c*) 1 in 39·3
2. (*a*) 6° 15′; 1·066 m/s^2 (*b*) (i) 6 m/s (ii) 144 m
3. 41·95 kN; 596 kN
4. 35·65 km/h
5. 0·73 Mg; 79·5 km/h
6. 10° 40′; 0·189
7. 0·804; 0·846 m
8. 45·1 km/h; 96·1 km/h
9. 0·63; 48 km/h
10. 66 km/h
11. 135 rev/min; 0·039 m
12. 10° 52′; 29·6 N
13. 1·1 m/s in direction of initial velocity of A; 101·4 kJ; 10·1 m
14. 0·266 m/s, 0·71 m
15. (*a*) 430 kJ (*b*) 3·33 m
16. (*a*) 5·48 m/s (*b*) 140 kN
17. (*a*) 18·18 km/h (*b*) 33·65 kJ
18. (*a*) v_A = –0·7 m/s; v_B = 3·8 m/s (*b*) v_A = 3·56 m/s at 82° to line of centres; v_B = 2·685 m/s along line of centres
19. 52·5 m/s
20. 696 kN
21. (*a*) 5·85 m/s (*b*) 13·2 m; 69·7 kJ
22. (*a*) 41·4 MJ (*b*) 1·48 km
23. 208 m; 666 m; 3120 W; 3785 W
24. 51·8 km/h; 0·192 m/s^2

Chapter 9 (page 253)

1. 15 737 N m
2. 1·054 rad/s^2; 298 J; 15·82
3. 21 mm; 0·107 N m
4. (*a*) 5941 N m (*b*) 71·29 kW
5. (*a*) 20·8 rad/s^2 (*b*) 58·57 N
6. (*a*) 1552 J (*b*) 9·05 rev/s (*c*) 1·193 rad/s^2; 47·62 rev
7. 0·688 m/s
8. 1·014 m/s^2; 15·15 kN
9. 0·365 m (*a*) 0·783 rad/s^2 (*b*) 5640 rad; 40·6 kJ
10. 126 kJ; 70 N
11. (*a*) 6·95 rev/s (*b*) 245 J (*c*) 48 rad/s^2; 11·52 N m
12. 34 rev/min
13. 30 220 N m
14. 679 W; 68·78 kJ
15. 88·49 rev/min; 441·7 J
16. (*a*) 6·6 rev/min (*b*) 35·6
17. 531 kJ; 1·66 kN; 75·6 m

Chapter 10 (page 276)

1. 74·5 mm
2. (*a*) 0·76 m; (*b*) 4·87 s; (*c*) 13·3 N
3. (*a*) 132 N, 1·94 m/s; (*b*) 242 N; (*c*) 2·31 m/s
4. (*a*) 10 s, 340 mm; (*b*) 0·134 m/s^2
5. (*a*) 358 mm; (*b*) 190 mm; (*c*) $\frac{1}{12}$ s
6. (*a*) 1·13 m/s; (*b*) 0·94 m/s; (*c*) 212·7 m/s^2; (*d*) 106·3 m/s^2; (*e*) 127·62 N
7. 0·385 s, 0·49 m/s
8. 1·55 kg, 8·54 m/s^2
9. 1096 N/m, 0·337 m/s
10. 0·156 s, 0·97 m/s
11. 0·99 m
12. (*a*) 2 s; (*b*) 0·856 m/s^2; (*c*) zero; (*d*) 1·31 m
13. 3·384 kg m^2
14. (*a*) 0·45 m; (*b*) 2·5 kg m^2
15. (*a*) 2·38 s; (*b*) 0·423 m; (*c*) 2·15 s

Chapter 11 (page 297)

1. (*a*) 3178 N/m^2; (*b*) 9064 N/m^2
2. 76.34 mm
3. 63·765 kN/m^2
4. 19·816 kN/m^2
5. (*a*) 25·5 kN; (*b*) 16·6 kN; (*c*) 8·3 kN
6. 331·09 kN, 1·89 m from bottom of gate
7. 29·43 kN/m^2, 482 Nm
8. 11·56 kN, $\frac{1}{24}$ m below centre of plate
9. 33·65 kN
10. 14·99 kN, 0·83 m from base
11. Hinge: 11·32 kN, Fastening: 12·51 kN
12. 1·217 m
13. 39·24 kN, 2·125 m
14. 17·343 kN, 1·0625 m

Chapter 12 (page 320)

1. 368·5 kN/m^2
2. 94·47 kN/m^2
3. 0·031 m^3/s
4. 1·04 m/s, 4·16 m/s, 0·0212 m^3/s
5. 23·47 m/s
6. (*a*) 2 bar; (*b*) 10·24 kW (Energy at A = 201·125 J/kg; mass flow rate = 38·2 kg/s. Theoretical power is product of energy/kg and mass flow rate)
7. 1·28 m^3/s
8. 5·91 m of liquid
9. 15·65 m
10. $C_v = 0·96$, $C_d = 0·64$, $C_c = 0·67$
11. 1·33 m, 1·99 m^3/h
12. 0·056 m^3/s (Note—difference in pressure given as *equivalent* to 50 mm of mercury)
13. 50 mm
14. 0·963
15. 100·8 m^3/h
16. Pressure difference = $\rho_0 gh(\rho_m - \rho_0)$, 51·08 m^3/h
17. 0·663 m/s, 1·8 m of water or 17·7 kN/m^2

Index

Acceleration, 174 et seq.
  angular, 179
  centripetal, 189
  uniform, 177
Acceleration - time graph, 176
Amplitude, 259
Angular,
  impulse, 250
  momentum, 250
  motion, 236 et seq.
  velocity, 179
Auxiliary circle, 258

Bending,
  of beams, 110 et seq.
  stresses, 111
Bending moment, 37 et seq.
  definition of, 42
  notation for, 42
Bernoulli's equation, 304
Brinell hardness, 168

Centre of,
  oscillation, 274
  percussion, 237, 274
  pressure, 290
  suspension, 273
Centrifugal force, 204
Centripetal,
  acceleration, 189
  force, 204
Charpy impact test, 173
Coefficient of,
  contraction, 310
  discharge, 310
  velocity, 309
Complementary shear stress, 95
Compound bars, 77
Compound pendulum, 273
Compression test, 165
Conical pendulum, 215
Conservation of,
  energy, 217
  momentum, 219
Continuity,
  equation of, 301
Contraction,
  coefficient of, 310
Contraflexure, 53

Density, 280
Discharge,
  coefficient of, 310
  rate of, 301
Displacement, 174
Distance - time graph, 175

Energy, 217
  conservation of, 217
  kinetic, 217
  potential, 217
  strain, 91
Equations of,
  continuity, 301
  motion, 174
Equilibrium, 1
  structural, 11
Extensometers, 164

Flow,
  measurement of, 308
  rate of, 301
  steady non-uniform, 301
  steady uniform, 300
Force,
  polygon, 1
  vector diagram, 15
Frequency of oscillation, 259
Funicular polygon, 2

Hardness test, 168
  Brinell, 168

Hardness test–*continued*
- Rockwell, 169
- Shore Scleroscope, 170
- Vickers, 169

Head,
- kinetic, 305
- potential, 305
- pressure, 305

Hooke's law, 68

Impact, 219
- oblique, 222
- of a fluid jet, 225

Impact tests, 172
- Charpy, 173
- Izod, 173

Impulse, 218
- angular, 250

Intensity of pressure, 281
Izod impact test, 173

Joints,
- resolution at, 13
- riveted, 96

Kinetic,
- energy, 217
- energy of a liquid, 303
- energy of rotation, 242
- equation of motion, 197
- head, 305

Laws of,
- motion, 196
- universal gravitation, 198

Link polygon, 2
Liquids in motion, 300 et seq.

Manometer, 284
Measurement of pressure, 283
Method of sections, 16
Motion,
- angular, 236 et seq.
- equations of angular, 179
- equations of linear, 179
- kinetic equation of, 197
- laws of, 196
- of liquids, 300
- periodic, 258
- simple harmonic, 258
- under varying force, 228

Modulus of,
- elasticity, 69
- rigidity, 69, 95

Moment of resistance, 115, 142
Momentum, 196
- angular, 250
- conservation of linear, 219

Neutral axis, 113

Orifice, 308
Oscillation,
- centre of, 274
- frequency of 259

Parallel axis theorem, 118
Pendulum,
- compound, 273
- simple, 271

Percussion,
- centre of, 274

Periodic,
- motion, 258
- time, 259

Perpendicular axis theorem, 120
Piezometer, 284
Polar second moment of area, 144
Potential energy, 217
- of a liquid, 303

Potential head, 305
Polygon,
- force, 1
- funicular, 2
- link, 2

Power,
- transmission of, 144

Pressure,
- absolute, 283
- atmospheric, 283
- centre of, 290
- energy of a liquid, 303
- gauge, 283
- head, 284
- intensity of, 281
- on an immersed surface, 289
- variation with depth, 282

Radius of gyration, 237
Rate of flow, 301
Relative density, 281
Rockwell hardness test, 169

Second moment of area, 116
- polar, 144

Shaft,
- compound, 153
- hollow, 147
- solid, 147

Shear force, 37 et seq.
- definition of, 41
- notation for, 39

Shear,
- strain, 94
- stress, 93

Shore scleroscope hardness, 170
Simple harmonic motion, 258 et seq.
Simple pendulum, 271

Spring,
  deflection of, 263
  stiffness of, 262
Steady flow, 300
Steady non-uniform flow, 301
Steady uniform flow, 301
Strain,
  direct, 68
  energy, 91
  shear, 94
Stress,
  complementary shear, 95
  direct, 66
  due to bending, 111
  due to temperature, 81
  due to torsion, 140
  hoop, 87
  in thin cylindrical shells, 86
  in thin rotating rings, 89
  longitudinal, 87
  proof, 74
  shear, 93
Strut, 12
Suspension,
  centre of, 273

Tensile,
  strength, 70, 71
  test, 69, 161
  testing machines, 161-3
Thrust, 281
  on immersed surface, 289
Tie, 11
Torsion,
  of circular shaft, 140 et seq.
  stress, 140
  test, 166
Translation,
  linear, 200
  in a curved path, 203
Transmission of power, 144

Vehicle,
  on a curved horizontal track, 205
  on an inclined curved track, 210
Vena-contracta, 309
Velocity, 174 et seq.
  coefficient of, 309
  -time graph, 175
Venturi meter, 314
Vickers hardness test, 169

$$\sigma = \frac{M_{max}\, y}{I_B}$$

From the max Bending moment diagram

I = (Second moment of area)